AUTOBIOGRAPHY OF
SIR JOHN RENNIE, F.R.S.

LECTOR HOUSE PUBLIC DOMAIN WORKS

AUTOBIOGRAPHY OF SIR JOHN RENNIE, F.R.S.

JOHN RENNIE

ISBN: 978-93-90294-35-0

Published: 1875

LECTOR HOUSE LLP
E-MAIL: lectorpublishing@gmail.com

AUTOBIOGRAPHY OF SIR JOHN RENNIE, F.R.S., PAST PRESIDENT OF THE INSTITUTION OF CIVIL ENGINEERS.

COMPRISING THE HISTORY OF HIS PROFESSIONAL LIFE, TOGETHER WITH REMINISCENCES DATING FROM THE COMMENCEMENT OF THE CENTURY TO THE PRESENT TIME.

BY

JOHN RENNIE

1875.

PREFACE.

The following Autobiography was written by Sir John Rennie in 1867, shortly after he had retired from the active duties of his profession. As will be perceived in the sequel, it was composed wholly from memory. Sir John was subsequently unable to revise it as he would have desired, and it has since been found impossible to do so. Nevertheless it is believed that but few substantial errors will be found; while the kindliness with which the autobiographer invariably speaks of every person with whom he came in contact, is a guarantee that there can be nothing to offend the most sensitive person, or which might tend to injure the just claims and reputations of others. It is now presented to the public in its original state, having undergone merely some necessary correction, in the hope that the memoirs of the man who was perhaps unrivalled in his branch of the profession—and which comprise valuable hints as to the neglected art of hydraulics, as well as advice to engineers commencing their career, the result of the experience of a lifetime of no ordinary duration—together with the reminiscences of one who had seen much both of men and things,

"Qui mores hominum multorum vidit et urbes,"

may not be unacceptable either to the profession or the world at large.

London, *September, 1875.*

CONTENTS

AUTOBIOGRAPHY OF
SIR JOHN RENNIE.

CHAPTER I.

My birth and early education—I enter my Father's office—Commencement of Waterloo and Southwark Bridges—Anecdotes of Mr. Ferguson, of Pitfour—The Stockton and Darlington Railway and Surveys between Port Patrick and Donaghadee—Account of the mode of erecting the arches of Southwark Bridge—Journey to the Continent and Field of Waterloo—Account of the building of Waterloo Bridge—It is opened in State by the Prince Regent, 1817.

I was born at 27, Stamford Street, Blackfriars, London, on the 30th of August, 1794. Having been taught my letters at home, I was sent to the care of Dr. Greenlaw, who kept a boys' school at Isleworth. It was a large house, formerly belonging to the Bishop of London. To the house were attached excellent gardens and playground. The situation, moreover, was open and healthy, and the total number of boys was about fifty, ranging from eight to sixteen years of age. They were well fed and taken care of by the Doctor's excellent wife, and his sister-in-law, Miss Hodgkins. The Doctor's eldest daughter, Miss Greenlaw, taught the youngest boys their letters; whilst the Doctor and his assistants devoted themselves to the education of the others, which education consisted chiefly of classics, writing, arithmetic, French, and occasionally geography and the elements of astronomy. During the time that I was there the most remarkable scholar was the celebrated poet, Percy Bysshe Shelley, who was then about twelve or thirteen (as far as I can remember), and even at that early age exhibited considerable poetical talent, accompanied by a violent and extremely excitable temper, which manifested itself in all kinds of eccentricities. His figure was of the middle size, although slight, but well made. His head was well proportioned, and covered with a profusion of brown locks; his features regular, but rather small; his eyes hazel, restless, and brilliant; his complexion was fair and transparent; and his countenance rather effeminate, but exceedingly animated. The least circumstance that thwarted him produced the most violent paroxysms of rage; and when irritated by other boys, which they, knowing his infirmity, frequently did by way of teasing him, he would take up anything, or even any little boy near him, to throw at his tormentors. His imagination was always roving upon something romantic and extraordinary, such as spirits, fairies, fighting, volcanoes, &c., and he not unfrequently astonished his schoolfellows by blowing up the boundary palings of the playground with gunpowder, also the lid of his desk in the middle of schooltime, to the great surprise of Dr. Greenlaw himself and the whole school. In fact, at times he was considered to be almost upon the borders of insanity; yet with all this, when treated with kindness, he was very amiable, noble, high-spirited, and generous; he used to write verse, English and Latin, with considerable facility, and attained a high position in the school before he left for Eton, where, I understand, he was equally, if not more, extraordinary and eccentric.

Cotemporary with Shelley there was another peculiar character, named Tredcroft, from the same county, viz. Sussex; he also had considerable poetical talent, but unfortunately lost his health, and ultimately, I understand, died completely imbecile at an early age. I remained at this school until the year 1807, by which time I had acquired a tolerable knowledge of the Greek and Latin classics, and arithmetic as far as vulgar fractions and decimals. I was then sent to the celebrated Dr. Burney's, at Greenwich, where there were about 100 boys, varying from ten to eighteen.

Dr. Charles Burney was considered one of the best Greek and Latin scholars of the day, and was the intimate friend of Porson and numerous other literary celebrities. His school was therefore very highly esteemed for classics, but for little or nothing else; for although a certain quantity of arithmetic and the elements of algebra and geometry were taught, yet these were quite secondary to the classics.

I therefore made little further progress in anything but classics, in which I became a tolerable proficient, and had Homer, Thucydides, Euripides, Sophocles, Virgil, Horace, &c., at my fingers' ends, whilst I could scarcely demonstrate the Pons asinorum of Euclid; in fact, in those days a knowledge of Greek and Latin was considered as including everything else, and anything like a science or physics was considered of secondary consequence. I made the acquaintance of two men, who afterwards much distinguished themselves by their scientific acquirements, namely, the late Herbert Mayo, the well-known surgeon and physiologist; also the late Sir George Everest, the scientific Director of the Triangular Survey of India; and Dr. Milman, late Dean of St. Paul's. Dr. Burney's school was by no means so well managed as that of Dr. Greenlaw in everything which regarded the comfort of the boys, neither were they so well fed or looked after, and it was a great relief to me when I left the school in 1809.

It then became a question with my father whether I should go to Oxford or Cambridge, or whether I should finish my education at home, under the superintendence of proper masters. About this period, and ever since the year 1802, there was nothing but war heard or talked of all over the world. The whole country was as it were turned into a camp; every man capable of bearing arms became a volunteer, and at school even we were regularly drilled to the use of arms; and I was so excited by the extraordinary victories of Nelson and the early career of Wellington that I determined to enter the army, but to this my father was decidedly opposed, as he wished to bring me up to his own profession. I was therefore reluctantly obliged to give up all idea of the military profession and follow that of a civil engineer; and my father wisely determined that I should go through all the gradations, both practical and theoretical, which could not be done if I went to the University, as the practical part, which he considered most important, must be abandoned; for, he said, after a young man has been three or four years at the University of Oxford or Cambridge, he cannot, without much difficulty, turn himself to the practical part of civil engineering. All idea, therefore, of my going to Cambridge or Oxford was given up. My father at that period had one department of his business exclusively devoted to practical mechanics, that is, to the making of machinery of all kinds; this department, although it formed by no means the principal part of his profession, nevertheless enabled him to make experiments which were of great value in the other departments of his business, and was by no means unprofitable, as the importance of machinery and mechanical contrivances was then to a certain extent appreciated, and was daily becoming more so. My father always said that theoretical and practical mechanics were the true foundations of all civil engineering; and he therefore insisted that as I had to a certain extent learned the theoretical, so I must now learn the practical part. I was therefore sent into the mechanical department, and commenced work planing and sawing boards, making patterns, and other similar works. After this I was put to turning both wood and metal; and although I did not attain complete practical efficiency in these departments, which would have required several years, nevertheless I learned sufficient to enable me to become a tolerable judge of workmanship.

I was then put into the drawing office, where I learned to copy geometrical plans, by which, in a short time, in combination with what I had acquired in the workshop, I gained a general knowledge of design and construction.

My time was employed in this manner about eight hours daily, but my evenings were devoted to the acquisition of geometry, algebra, and trigonometry, plane and spherical; also astronomy under the late Astronomer Royal, Mr. Pond, and his father-in-law, Dr. Bradley, and in learning French, Italian, and German. Having acquired to a certain extent a proficiency in the mathematical sciences, I was placed under the direction of the late Mr. Francis Giles, a land surveyor of considerable experience and ability, who was generally employed by my father to make his various hydraulic surveys for canals and harbours under his immediate direction, which Mr. Giles executed with his usual fidelity and ability. Under Mr. Giles I learned the use of the chain, level, and theodolite, and was enabled to apply my theoretical knowledge in trigonometry, plane and spherical, to practice. About this period, viz. the year 1813, having obtained a tolerable knowledge of the rudiments of my profession, both theoretical and practical, my father determined to place upon my shoulders a certain degree of responsibility, and put me under the direction of that late worthy and excellent man, Mr. James Hollingsworth, whom my father had appointed to be resident engineer of the Waterloo Bridge, which was then building. I felt the responsibility of this office a good deal, and entered upon it with every determination and desire to meet my father's approbation; and during the inclement winter of 1813-14, when the frost lasted about two months, and the Thames above London Bridge was frozen over for several weeks, I was obliged to attend the piling of the foundations of the first and second piers on the Surrey side of the river night and day for three days each week, which severely tried my constitution.

At this period Vauxhall Bridge was also in course of construction, and I was directed by my father to attend to this also, under Mr. Jones, the resident engineer; but they had scarcely finished the Middlesex abutment up to the springing of the first arch, and were preparing the caisson for founding the first pier, when the Company found that they had not sufficient funds to carry into effect Mr. Rennie's design, which was very beautiful. The bridge was to be made entirely of the fine blue sandstone from Dundee, and was to consist of seven arches, segments of circles, the centre arch being 110 feet span, with a rise or versed sine of 19 feet, and depth of keystone 4 feet 6 inches; piers 18 feet 6 inches thick at the springing of the arch, the two arches next the centre being 105 feet span each, with a rise or versed sine of 17 feet, keystone 4 feet 5 inches, and springing stones 9 feet long, and the two piers 17 feet 6 inches thick each. The two next arches were 100 feet span, with a rise or versed sine of 15 feet, keystones 4 feet 4 inches, and springers 9 feet, and piers 17 feet thick each; the two sub or shore arches 90 feet span each, with a rise or versed sine of 13 feet, keystones 4 feet, and springers 8 feet, abutments 21 feet thick at the springing, having a total width of waterway of 700 feet. The arches were surmounted by a Roman Doric cornice and plain block and plinth parapet, and the projecting points of the piers were surmounted by solid square pilasters, with a niche in the centre. The roadway was 34 feet wide between the parapets, and was formed by a very flat segment of a circle rising 1 in 53. The piers were intended to be founded by caissons resting upon a platform supported by bearing and surrounded by sheeting piles. This was upon the whole a very elegant, light, and chaste design. Finding that the Company had not sufficient funds to carry into effect the stone design, Mr. Rennie proposed another wholly of iron, consisting of eleven arches, with a total waterway of 732 feet, supported upon cast-iron columns filled with masonry and resting upon a platform supported upon piles and surrounded by sheeting piles. The centre arch was to be 86 feet span and 8 feet rise, and the others diminishing regularly to each end so as to enable the roadway to be formed into a graceful curve rising 1 foot in 60. This also was an extremely light, elegant, and economical design. The total cost of this elegant design was estimated at 100,000*l.*, and would have been executed first, but at that time even this amount was not forthcoming. The works then stopped, and some time elapsed before the Company was resumed, and ultimately constructed the present bridge.

In the year 1814-15 my father was appointed engineer-in-chief of the Southwark Bridge Company, and as this was proposed to be constructed in the narrowest part of the river between Blackfriars and the Old London Bridge, considerable opposition was made to the Act of Parliament for its construction by the Corporation of London and the Conservators of the river, on account of the obstruction which they said the bridge would offer to the navigation; this however was finally overcome, but it was decided by Parliament that the bridge should be constructed with as large arches as possible. Accordingly Mr. Rennie submitted a design consisting of three cast-iron arches, the centre being 240 feet span, with a versed sine of 24 feet, and two side arches of 210 feet each, with a versed sine or rise of 18 feet 10 inches each, with piers of 24 feet wide each at the springing, thus giving a clear lineal waterway of 660 feet, which was a great deal more than that of the Old London Bridge at that time existing. This design was approved of and ordered to be carried into effect. By this time, with the experience of the Waterloo and Vauxhall bridges and my other studies, I had gained considerable knowledge in bridge building, and my father was anxious to give me as much encouragement as possible; although, therefore, he appointed a worthy and practical man, Mr. Meston, as nominally the resident engineer, yet he confided to me the arduous task of making out the working drawings under his own direction, and of carrying them into effect. I therefore felt highly gratified with this great mark of confidence, and devoted my whole energies to the work night and day. The ironwork was carried into effect by Messrs. Walker, of Rotherham, under the able management of their experienced and able superintendent, Mr. Yeats, and the masonry and piling under the well-known contractors, Messrs. Jolliffe and Banks; and Mr. Meston, the resident engineer, faithfully discharged his duties.

As these arches were the largest of the kind ever constructed, considerable doubts as to their stability occurred to many, and the subject was discussed amongst scientific men with considerable energy; and amongst others, the celebrated Dr. Young undertook to investigate Mr. Rennie's calculations, and came to the conclusion that the bridge was well designed, and would be a perfectly safe and stable structure, and equal to the support of any weight or amount of traffic which could be brought over it. But in order to fulfil these conditions, it was absolutely necessary that every detail of materials and workmanship should be worked out with the greatest skill and accuracy.

As the arches were of such great span with so small rise, the pressure upon the piers and abutments was chiefly lateral; it therefore became necessary to construct them in such a manner that they should offer the most effectual resistance to this pressure. In consequence, the foundations of the abutments were made on an incline, and the masonry from thence upwards to the springing of the arches was made to consist of a series of courses radiating upwards until they reached the angle of the springing courses; so that, in point of fact, the abutments formed, as it were, a continuation of the side arches to their base; and in order to connect the courses of masonry more solidly together, the courses were connected with each other from the top to the bottom by several series of vertical bond stones, thus forming one solid immovable mass. These abutments were supported on a platform composed of piles, double sleepers, and planking, the piles being 20 feet long, 12 inches in diameter in the middle, and driven solidly into the ground at right angles to the inclination of the foundation. As the pressure upon the piers was nearly equal on both sides, it was necessary that the foundations should be laid level. These also rest upon a wooden platform of double sills and planking, lying upon piles of the same dimensions as those of the abutments, driven vertically into the ground below, and the courses of masonry, which were laid horizontally, were connected together in the vertical direction by a series of bond stones in a similar manner to those of the abutments. The abutments and piers were founded many feet below low-water mark of spring tides, so as to be below the reach of any possible scour of the river. Those parts of the piers from immediately below the springing of the arches to a point above the top of the main solid ribs of the arches were composed of large blocks of stone set nearly vertically, breaking bond laterally and vertically with each other, and in the centre of this part of the piers there was a set of keystones 12 feet long and 2 feet thick, tapering on each side, forming so many stone wedges. These were very finely worked on all sides. These wedge stones broke bond laterally with the blocks in front of them, and were firmly driven into their places for a depth of 2 feet by means of heavy wooden rams. The masonry of the pilasters and salient angles of the piers is of the same character as those of the interior of the pier before described, and worked into them in the same manner, so as to form one solid bond from one point of the pier to the other. The whole of the exterior of this part of the piers, as well as of the abutments, is cased with granite from Scotland or Cornwall; and it was necessary that the blocks forming this casing should be of the largest kind, which hitherto was quite unusual, particularly for the facing of the abutments from whence the arches were to spring, which required blocks from 15 to 20 tons. These were of such unusual magnitude, and nothing of the kind had hitherto been used in London, or even elsewhere in England, that the contractors made considerable objection to obtaining them, and even went so far as to say that it could not be done. I was perfectly convinced that it could be done, and that it was merely a question of a little extra expense, and strongly recommended my father to insist upon it, as it was absolutely necessary for the security of the bridge; and he did so, and directed me to proceed to Aberdeen for the purpose of obtaining them. I accordingly started for Aberdeen; and when there, carefully examined all the quarries in the neighbourhood, which I found had only been opened up on a small scale, and were merely adapted for getting paving stones, the commerce of which with London was then upon a considerable scale; but the idea of obtaining blocks of the size required for the Southwark Bridge was considered to be entirely out of the question; or, even if they could be obtained, the price would be such that the contractors would not consent to pay. In fact, so many difficulties and objections were made that I found nothing could be done in that quarter. I therefore determined to proceed to Peterhead, 30 miles farther northward, where the red granite abounds in large masses near the coast, and where I was told that I should probably succeed; but still, they said, even there it would be very difficult to get them. Upon arriving at Peterhead I immediately set to work exploring the adjacent country for several miles round, and soon found that blocks of the size required could readily be obtained, and even larger ones if necessary. I accordingly selected, by way of experiment, a mass of solid rock about four miles to the south of Peterhead, lying within a quarter of a mile of the sea coast, and about 200 feet above the main turnpike road to Peterhead, which ran along the sea shore. This block, weighing about 25 tons, was accordingly marked out, and was soon detached from the main mass of rock by means of wedges, and was 10 feet long and 5 feet square. The workmen who executed this task were rewarded with ample wages and a good supply of whisky, and were extremely proud of their achievement. Then came the important question, how they were to convey it to Peterhead. To get it to the turnpike road was soon accomplished by means of a wooden inclined tramway formed of stout planks moved upon wooden rollers. Good wages and whisky settled this, and the workmen considered it a further great triumph; but still the greater difficulty remained, how to get this vast block (as it was then considered) four miles to Peterhead. I then went back to Peterhead, and after numerous inquiries, and as many failures and objections on all sides, at last found two large single bogies, each consisting of a pair of strong wheels 8 feet in diameter, connected by a strong axle shaft and a double pair of shafts in front. These two pairs of wheels I joined together at the axle shafts by two strong beams, cased with wrought iron, and strengthened the wheels and axles in other respects as far as necessary. I then took this carriage to the block of red granite already mentioned lying in the road, and slung the stone, by means of strong chains, to the two longitudinal bearers of the carriage. Some twelve or fourteen horses were then attached to the carriage, and off we departed in great triumph for Peterhead. The toll-keeper, never having seen such a mass of stone before, did not know what to charge. However, having at length satisfied his demand, we proceeded onwards, and we had scarcely advanced a mile when we came to a soft piece of road, which yielded under the great weight of the stone, and the wheels stuck fast, buried about 9 inches in the ground. This accident created general dismay amongst the attendant workmen, and they began to consider the task hopeless. However, nothing daunted by this mischance, I soon rallied their courage, and with plenty of screw-jacks, wedges, and levers judiciously applied, we raised the wheels out of the ground, and placed strong wooden beams under them, forming a rough kind of railway, over which we dragged and pushed the carriage with its stone in safety, until we had passed the unsound part of the road. This operation detained us about a day. Everybody worked with the greatest ardour and goodwill, which was aided not a little by a plentiful supply of ale and whisky, and the men were determined, for the honour of Scotland, that they would not be beaten. Having overcome this serious obstacle, we started again on our journey, and reached Peterhead about four hours afterwards, making the total length of the journey—four miles—a day and a half. The whole town of Peterhead, having never seen such a sight before, and having considered our task impracticable, turned out to see us, and welcomed us with the most enthusiastic acclamations.

The next thing to be done, having succeeded so far, was to get a vessel that would take this monstrous block of stone, as it was termed, to London; and although there were a considerable number of vessels in the harbour, I could not at first prevail upon any of the captains to take the charge. All sorts of objections were made, and amongst the rest, it was impossible to get the stone on board, and if they did, it would make a hole in the bottom, and the vessel would founder with all on board. At last, after a great deal of difficulty, I found a brig of about 200 tons burthen, the captain of which, after a good deal of persuasion, consented to take the block of stone to London, provided that I would put it on board at my own risk and expense, and indemnify him against all risk or loss on the voyage, which I accordingly agreed to do.

Then came the last important question, how was the block to be got on board? There was no crane in the port capable of lifting above 2 or 3 tons.

I immediately set to work to supply this deficiency by means of two sets of strong sheer-poles, capable of bearing 10 to 15 tons each. The vessel engaged was accordingly brought alongside the quay where the 4-ton crane was fixed, so that it should nearly plumb the centre of the hatchway of the vessel, which it was necessary to enlarge and strengthen considerably before it could receive the stone. I then secured the sheer-poles well at the top, and placed one set on each side of the crane, a short distance from the extremity of each end of the hatchway. The legs of the sheer-poles were firmly fixed in the bed of the harbour, striding over the vessel, so that they were perfectly independent of the vessel, and the top of each pole was directly over the centre of the hatchway. To the top of the sheer-poles I applied a pair of strong treble sheave-blocks, capable of receiving a thick rope; each block was worked by a double purchase crab or windlass manned by eight men each, besides four to work the crane, so that the block would be suspended at three points, the sheers taking the greatest weight. After a good deal of trouble I got the whole of this apparatus as complete as circumstances would permit, which were not the most favourable. From the quay a strong timber gangway was constructed over the hatchway, the outer end being supported, clear of the vessel, by piles driven into the bed of the harbour on each side, in order that the ship might be kept perfectly steady until the stone was placed within the hold, because otherwise the stone resting upon any part of the deck might have upset it. Everything being ready, the stone was brought alongside the vessel and the tackle of the crane and of the two pairs of sheer-poles was made fast to three sets of strong chains fastened round the stone, which was transferred upon rollers over the centre of the hatchway of the vessel, the purchases of the crane and sheer-poles being kept sufficiently tight so as to prevent any undue pressure upon the platform. The stone was then raised clear of the platform, when I heard a crack; in fact, one of the sheer-poles had bent and partially yielded; it was then blocked, and, the sheers having been first spliced with strong rope, the stone was again hoisted and swung clear of the platform, which was removed, and the stone was lowered into the hold of the vessel and properly secured without any further delay or accident. The whole of these operations were witnessed with intense interest by many of the inhabitants of Peterhead, and when so successfully completed the quays resounded with cheers. The gallant workmen who laboured so arduously and with such goodwill, and to whose exertions the success may be mainly attributed, were plenteously regaled, together with their friends, with all the good things which Peterhead afforded, in which the worthy inhabitants joined, and the remainder of the day was passed in mutual goodwill and festivity.

My readers will, I trust, excuse the detailed manner in which I have described these operations, which at the present day would be considered trifling to a degree, but, at the time above mentioned, more than half a century since, operations of this kind had not been attempted, and were entirely novel, and were considered extraordinary; they must, therefore, be viewed as the pioneer to the far greater operations of the kind which have followed. For now such stones are considered mere trifles, and blocks of almost any reasonable size can be quarried, polished, and transported to their destination, however distant, at comparatively much less cost and with greater facility. The cutting and polishing of granite at that time was accomplished at great expense, as it was done entirely by manual labour; now it is performed by means of machinery at greatly reduced cost, and polished granite of every kind is introduced into buildings, which was formerly considered impracticable; and thus the architect is provided with additional means of ornamenting his structures. It is true that many centuries before the Egyptians had shown the way; but then the whole power and resources of the nation had been devoted to this object, and incredible sums of money and great labour had been expended, regardless of the misery and oppression of the people. But in Great Britain it has been considered a true maxim of political economy, that every article should be produced at the least possible cost, and no work should be undertaken unless it would yield a fair profit for the capital expended; and whenever there is a reasonable prospect of obtaining this satisfactory result, any amount of capital which may be required is always forthcoming. Witness the vast sums which have been expended on railways alone, besides steam-vessels, manufactures, machinery, and other similar undertakings.

Whilst at Aberdeen and Peterhead, my father gave me an introduction to his old and intimate friend, the well-known James Ferguson, of Pitfour, the member for the county, and the intimate friend of Mr. Pitt. Mr. Ferguson possessed a large fortune; he was an old bachelor of the most amiable and charitable disposition, beloved by everybody and universally popular throughout the county. As illustrative of the manners of those days, I will simply mention that when I presented my father's letter of introduction he received me most kindly, and invited me to spend a few days under his most hospitable roof, which I accepted, and on the first day there was assembled a large party of the most influential gentlemen of the county; as was usual wines of all kinds flowed in abundance, and universal hilarity prevailed. The consequence was that not long after dinner several of the guests fell off their chairs and took their nap under the table, from which after a short time they recovered and resumed their seats, and again set to work at their potations, which continued until long past midnight; by this time another considerable batch of guests were under the table, leaving their glasses full. I was so much amused at this unaccustomed scene, that, by way of frolic, I took the full glasses of some of the guests on the floor and poured them down their throats, which had no other effect than to make them sleep sounder. In this manner the evening passed merrily away, and it was late in the morning before the whole of the company found their way to their beds. The amiable host allowed everyone to do as he liked, and when he had had enough, which was not very soon, he retired to bed, leaving his guests to take care of themselves; in fact, Pitfour was *"Liberty Hall,"* and was open to all comers, the only limit being the amount of sleeping accommodation.

To give some idea of his hospitality, Mr. Ferguson seldom had less than thirty-six pipes of fine port wine in his cellars, besides claret, burgundy, sherry, champagne, brandy, and whisky, in proportion. He was so fond of Mr. Pitt, that it is believed that if Mr. Pitt had survived him, he intended to have made him heir to his estates, which were said to have been worth above 20,000*l.* a year.

When attending his duties in the House of Commons, he lodged in apartments in St. James's Street, and after the parliamentary labours of the week were over, Mr. Pitt, the late Harry Dundas, First Lord Melville, and Mr. Ferguson used to retire to a country house at Wimbledon, and spend the Saturday, Sunday, and Monday in the greatest conviviality, until it was time to return to their parliamentary labours. Mr. Ferguson rarely, if ever, spoke in the House of Commons, but when he did, it was always to the purpose: his speeches, although exceedingly short, were replete with much common sense, accompanied by a terseness of wit, humour, and drollery, which convulsed his hearers with laughter, so that he was a general favourite. He used always to say that he had heard "mony a gude speech, but that he never changed his vote, he aye voted with Mr. Pitt."

Having shown the good people of Peterhead how to get and ship the large blocks of red granite for the Southwark Bridge, and feeling that there would be no further difficulty about the matter, I returned to London, after an absence of two weeks, and resumed my duties at the Southwark Bridge. In the spring of 1814 my father, being desirous that I should be initiated into the practice of marine and trigonometrical surveying upon a large scale, sent me, under the direction of the late Mr. Francis Giles, who had then been appointed by Mr. Rennie to make an extensive survey of the different places where it was practicable to construct proper artificial harbours on the south-west coast of Scotland, such as Port Nessock, Ardwell, and Port Patrick bays, on the Scotch side of the channel, and Ballantrae, Donaghadee, Ballyhone, and Bangor bays, on the opposite coast of the Irish Channel, in order to decide which were the best places on either side of the channel to make permanent good artificial harbours for packets, for the purpose of establishing the best and most direct communication between the south-west coast of Scotland and the opposite coast of Ireland. In addition to making the surveys of the different ports above mentioned, it was absolutely necessary to make a correct chart of the channel within the above limits, comprehending a coast-line of about 30 miles on each side, including the Copeland Islands. It was also necessary to determine the exact distances between the different ports, together with the soundings, currents, rises of tide, prevailing winds, and all the other attendant hydraulic phenomena. This was a very extensive survey, and required great skill, judgment, and experience, and Mr. Giles was fully competent to undertake it.

As the shore, particularly on the Scotch side of the channel, was very precipitous, rugged, and mountainous, it was impossible to measure with anything like accuracy a base line from which a series of triangular observations could be made so as to connect the two coasts together; Mr. Griles was therefore obliged to resort to the Bay of Luce, situated about 10 miles to the east of Port Patrick; it had this disadvantage, however, that in consequence of the intervening mountains neither the Scotch nor Irish coast could be seen. But there was no alternative, for no other convenient place could be found to measure a base line. The head of the Bay of Luce, however, at low water consists of an extensive district of flat sand 6 miles long, admirably adapted for the purpose of measuring a correct base line. This plan was accordingly adopted, and a base was measured 6 miles in length, first by the common chain, then by another chain consisting of steel links each 5 feet long, and lastly by well-seasoned deal rods each 10 feet long; these measurements were repeated with great accuracy several times, and a mean was then struck by which the variation was reduced to a fraction of an inch. It should also be observed that the measurements were taken when the atmosphere was about the same temperature, so that the final measurement was reduced to as near accuracy as practicable. Having established the base, strong vertical staffs with flags at their summits were then accurately fixed upon the summits of the neighbouring mountains which overlooked the Irish Channel, and from which on a clear day the opposite coast of Ireland could be distinctly seen. From each end of the base the above-mentioned angles were taken with one of Troughton's best 7-inch theodolites, between the different mountain stations, and the distance between the extreme points of these stations was accurately calculated, so that it gave a base line of about 30 miles along the Scotch coast, from whence correct sights were taken on flagstaffs fixed on the high land above Bangor, Donaghadee, and Ballantrae, including the Copeland Islands on the opposite coast of Ireland. These observations being taken on different levels, were subsequently reduced to the same plane, so that the exact distance was obtained between the different stations on each side of the channel, and a correct chart was made. In addition to the general survey, detailed hydraulic charts were made of the places on each coast which were best adapted for making harbours of the size required, viz. Port Nessock, Ardwell, and Port Patrick, on the Scotch side, and Bangor, Ballyhone, Donaghadee, and Ballantrae, on the Irish side of the channel; and as there was a great variety of interests concerned, each proprietor being desirous of having the permanent ports established on his own property, it was finally decided by the Government, at the recommendation of my father, that the whole of the surveys should be submitted to the arbitration of the Trinity Board, to select one port on each side of the channel which was best adapted for establishing a communication by packets between the two countries for letters and passengers in the most expeditious, convenient, and effective manner. The Trinity Board, after having given the subject their most mature consideration, ultimately decided on Port Patrick on the Scotch, and Donaghadee on the Irish side. This decision was approved of by the Government, and Mr. Rennie was desired by them to prepare designs for harbours in both of the above places; he did so, and the Government ordered them to be carried into effect under his direction: for a more particular description the reader is referred to my work on British and Foreign Harbours. During the progress of these surveys I learned a great deal in this important department of civil engineering, and personally surveyed the bay of Port Nessock and the Copeland Islands, which gave me an excellent lesson, as, on account of the rugged nature of the coasts, they were attended with considerable difficulties.

As already observed, in the following year, 1815, I was placed under Mr. Giles' direction during the whole period of the survey of the river Tyne, which had been entrusted to him by my father also. In the same year I was under Mr. Giles during the surveys for the Barnard Castle, and Stockton and Darlington railways and canal; and I subsequently made a hydraulic survey of the port of Blythe for my father, and for which he afterwards made a design for its improvement. I had been previously employed under Mr. Giles in the surveys for the eastern extension of the Kennet and Avon Canal. During a part of this time our head quarters were at Windsor; and one Sunday afternoon I recollect very well attending the promenade on the terrace at Windsor Castle when His Majesty King George the Third, accompanied by the Queen and Royal Family, made their appearance with their attendants and joined the promenade, and were received in the most loyal and affectionate yet respectful manner. I also about the same time assisted Mr. Giles in the survey of the Thames in the vicinity of Woolwich Dockyard, the accumulation of the mud in front of which was so great at that time that it threatened to render that dockyard useless. The evil to a certain extent has since been remedied by removing a number of the projections which interfered with the currents of the ebb and flood tide, according to the plan laid down by the late Mr. Rennie, although, as he clearly pointed out, it would be impossible to improve the river to such an extent in front of Woolwich and Deptford dockyards as to render them fit for the construction and accommodation of large vessels of war; and therefore he recommended that they should be abandoned and sold, and that a proper establishment should be made at Northfleet capable of accommodating at all times of tide any number of the largest vessels of war at that time in the navy, or that might be built hereafter. This dockyard was intended to be so arranged that all classes of vessels of war could be built there; and it would contain establishments for manufacturing cordage, sails, anchors, guns, smith work of every kind, together with depôts of provisions, and stores of all sorts, all of which were to be arranged in such a manner that each operation of building, repairing, storing, and fitting out for sea should be completed in the order required; so that a vessel, after having been built, might be transferred at once into a dry dock to be coppered, then take in her masts, ballast, rigging, sails, stores, provisions, guns, boats, and seamen, and sail at once complete to her station from the centre dock-gates; in the same manner, when returning from her station, she might be repaired and fitted out again ready for sea. The saving effected by such an establishment would have been immense, and the service would have been performed in a far more expeditious and effective manner. Mr. Pitt was then Prime Minister, and his master mind at once acknowledged the advantages of such an establishment; it was accordingly approved of by the Government, the land was bought, and it was ordered to be carried into effect, but the unfortunate death of that great minister and the change of Government effectually put a stop to all further proceedings in this direction. Since that time the old dockyards of Woolwich and Deptford, Sheerness and Chatham, have been improved and enlarged, at an expenditure as great as would have completed Mr. Rennie's establishment at Northfleet, without half the efficiency or accommodation. The old useless dockyards of Woolwich and Deptford, which the late Lord Melville, when First Lord of the Admiralty, condemned and ordered to be sold, have been retained, and the increase of unnecessary expense has been enormous, whilst the evils complained of have not been remedied till quite lately.

Having by this time a considerable knowledge of surveying, practical and theoretical, in all its departments either on land or water, so that I could undertake either, my father deemed that I had learned sufficient, and directed me to return to my practical duties in the construction of the Southwark Bridge. As this was considered a work of great importance and difficulty I felt highly honoured by my father's confidence, and devoted my energies to it with the greatest anxiety and with a determination to do everything in my power to make it successful. The difficulty of obtaining the large blocks of granite and other stone had been successfully overcome, and they arrived with great punctuality, and the masonry of the abutments and piers was successfully carried into effect as previously described. Then came the important question of erecting the cast-iron arches; each arch consisted of eight main ribs, and each rib consisted of thirteen pieces, the lower or main part of which formed the chord or arch upon which the whole of the superstructure was to be supported. These thirteen pieces were solid, 2½ inches thick in the mass and 3 inches thick at the bottom, and 2½ thick at the top, and formed so many radiating blocks, like arch-stones. At the end of every block there was a transverse frame extending from one side to the other through the whole width of the bridge; against each side of these frames the main ribs abutted and were nicely fitted to them, and in order to prevent them from moving laterally there were projecting dovetailed cheeks cast on the frame, and between these cheeks and the ends of the main ribs solid cast-iron wedges the whole depth of the main rib were fitted, then drawn home against the ends of the rib; by this means the ribs were kept firmly within their places, and as an additional precaution strong diagonal braces, having a strong feathering rib on each side, were inserted diagonally between the ribs from one end of the arch to the other, and secured to the ribs by projecting dovetailed cheeks on them, and wedges and bolts, so that these cast-iron arches were constructed in the same manner as a stone arch, being almost as it were a solid mass depending upon the equilibrium of the different pieces for its stability. The depth of the main rib of the centre arch at the crown is 6 feet and 8 feet at the piers, whilst the depth of the ribs of the side arches at the crown is 6 feet and 8 feet at the abutments. As these ribs with their attendant transverse frames and diagonal braces formed the main part of the arches upon which the whole of the superstructure depended, it was necessary that they should be extremely well put together and properly united to the piers and abutments. Contrary to the usual mode of constructing stone arches, they were commenced at the centre instead of the sides, a strong wooden centring supported upon tiers of piles having been previously constructed between the piers and abutments to support them whilst being put in their places. Each piece of each rib was carefully placed upon the centre, resting upon nicely-adjusted strong wooden double wedges, and connected together as they proceeded by the transverse frames and diagonal braces before mentioned. By this means the whole of the arches were constructed at the same time from the centre to the skewbacks or bearing parts of the piers and abutments; but in order to connect them properly with them it was necessary to devise a particular arrangement. For this purpose a transverse frame, similar to those already described for connecting the rib pieces together, was accurately imbedded and fitted to the skewbacks or bearing places on piers and abutments, resting on a bed of sheet lead, and the joints were filled in with melted lead also; this formed a solid and to a certain extent elastic bearing, upon which the main ribs were ultimately to rest.

At the ends of the arched rib-plates next to the piers and abutments there was another transverse frame plate of the same kind as those previously described, and fixed there in a similar manner; this brought the ends of the arches within 6 inches of the abutting or bearing plates fixed in the skewbacks or springing places of the piers and abutments.

Between the frame plates fitted on the skewbacks or masonry of the piers and abutments, and those fitted on the ends of the rib plates of each arch, solid cast-iron wedges, 9 feet long and 6 inches thick at the back, and 2 inches thick at the bottom, 9 inches wide, three being behind each rib, were accurately fitted by chipping and filing, so that it would slide down to within 12 inches of the bottom; when these wedges were all accurately adjusted at the same temperature to the same depth, they were simultaneously driven home by wooden rams to their full depth, so as to reach about an inch below the bottom of each rib; by this means the whole of the three arches were gradually brought to their bearing without being raised wholly from their centres. Matters were then allowed to remain in this state for a few days in order to give time for every part to come to its bearing and to ascertain whether there was any defect in any part.

After the minutest search in every part no defect could be discovered; the wedges between the centres and the under sides of the ribs were then gradually slackened until the whole of the arches came to their full bearing, and were removed entirely, leaving the arches perfectly free of support. During the whole of these operations, from first to last, which occupied about a week, not the slightest accident or fracture occurred; the total subsidence of the main arch barely exceeded 2½ inches, whilst the subsidence of the two side arches barely exceeded 2 inches, which had been allowed for in the construction.

In order to ascertain the effects of expansion and contraction of the arches by the variation of the temperature of the atmosphere, I constructed steel, brass, and wooden gauges, accurately divided into decimal parts of an inch, and erected them upon different parts of the centres, where the effects were most likely to be apparent, and I kept the register for several weeks, during the height of summer, autumn, winter, and spring. I found that the variation in the rise and fall of the crown of the arches, the abutments being fixed, was 1/10th of an inch for every 10° of temperature, so that, taking the extremes of temperature at London to be 10° below freezing point of Fahrenheit in winter, and 80° in summer; the utmost rise and fall of the arches may be taken at 7/10ths, or at most one inch; but as any variation in the temperature, unless continued for some time, has no sensible effect upon such a large mass of iron, so, in our variable climate, the rise and fall of the crowns of the arches may be taken upon the average somewhat below the amount above given.

After the arches had been brought to their bearing and had been relieved from the centres, the superstructural framework was carried up and firmly connected and bracketed together by diagonal ties and wedges; in doing this the ends of the superstructural frames were too tightly wedged to the masonry of the piers, without my knowledge, so that they would not allow the main ribs of the arches to play freely, and some of the masonry courses above the main ribs were slightly splintered and deranged; the wedges were then slackened, and some of them removed entirely, and thus the evil was immediately remedied; the whole structure has ever since remained in a perfect state.

The bridges and approaches were finally completed and opened to the public traffic in March, 1819, the ceremony being performed by Sir John Jackson, the chairman, the directors, and a few friends.

In the month of August, 1818, having worked very hard, I may say almost night and day, for some time, I was nearly worn out, and was permitted to have a short holiday. I therefore determined to go to Belgium and visit the celebrated field of Waterloo, which closed the long and eventful revolutionary war, and attracted the admiration and interest of the whole civilized world. I accordingly started for Dover, in company with my old friend, the late Mr. Joseph Gwilt, architect, and crossed over to Calais.

On landing we repaired to Dessin's hotel, at that time one of the best in Europe, and rendered famous by Sterne, whose rooms are still shown as one of the most interesting curiosities (to Englishmen at least) of the place. Here we passed the remainder of the day very agreeably. Everything was new to us—the people, their language, their manners, their mode of living. We had been so long considering the French as our deadly foes that we could hardly believe ourselves to be at peace with them, and to be actually in France and so civilly treated by them. Then the living was so good and cheap, compared to that of England; champagne, which with us was considered the greatest luxury, and only within reach of the highest and most wealthy, was here obtainable for four francs a bottle, whilst in England it was twenty-four, and almost everything in the same proportion. In the town of Calais there is nothing to see. The harbour is but indifferent, and almost dry at low water, and is chiefly maintained open by sluicing and dredging. On such a flat, sandy shore, with the prevailing winds and currents always driving on and accumulating the sands, it is very difficult to make and maintain a good harbour; still I think that a great deal more might be effected by pursuing a different course, and having a proper system of open piers and breakwaters on the outside. After breakfast on the following day we started in a travelling calèche, with two horses, for St. Omer, about 26 miles distant. On our way we stopped to examine the celebrated quadrangular bridge, called the Pont Sans Pareille, across the junction of the two canals of Picardy. This is certainly a meritorious, well-executed work, but the idea is by no means new, as the Gothic triangular bridge of Crowland, in Lincolnshire, across the junction of the Welland, was executed many centuries previous to the Pont Sans Pareille, and is still extant in a perfect state. Near this bridge is the celebrated maiden fortress of Ardres, which is said to have never been taken. Here also, according to Froissart, was the Champs d'Or, or Field of the Cloth of Gold. Ardres is a poor, miserable little fortress, surrounded with earthworks and ditch, on a flat plain. In ancient times it might have been formidable, with their means of attack, but now it would be utterly defenceless. From Ardres we passed through St. Omer and Lille, and after visiting Tournay and Ghent we proceeded to the capital of Belgium, where we took up our quarters at the Hôtel de Flandre, in the highest and best part of the city. We were particularly struck with the magnificent Hôtel de Ville and its lofty spire; also the remarkable place or square in front of it, which, looking to its picturesque mediæval buildings and the remarkable historical events which have taken place there, renders it one of the most interesting in Europe. The fine old Gothic cathedral of St. Gudule, the museum, fine canals, &c., particularly attracted our attention.

On the next morning we started for the scene of the celebrated battle of Waterloo, which had occurred about two months previously. In this place, like every Englishman, I took the greatest possible interest, and pictured to myself the whole of that terrific and stirring scene as being enacted before me. Notwithstanding the lapse of time since which that battle had taken place considerable traces of it were still visible, particularly in the blood-stained walls and ruined, desolate, and half-consumed buildings of the keys of the position, Hougumont and La Haye Sainte, and the remnants of shakos, arms, and military clothing which strewed the field on all sides, and the fresh-made graves, where many thousand gallant fellows lay entombed. The whole field and neighbouring villages were crowded with guides to explain the different particulars of that memorable struggle, and to sell the numerous articles which they had raked up from the field of battle; we bought some of these as mementos, and wandered for hours over every part of this field of desolation, until we fancied that we had mastered every detail of the conflict, and were almost fit to take the command of an army ourselves. We then returned to Brussels, highly gratified and instructed by the excursion.

We left Brussels much pleased with that pretty little industrious capital, and proceeded to Malines, where the fine old cathedral and town rewarded us well during our short visit. From thence we journeyed on to Antwerp, where we stopped at the Grand Labourer, a celebrated old-fashioned hotel. This famous old city, the great emporium of the Belgian trade, interested me much, with its magnificent cathedral and other churches, its fine old Hôtel de Ville, and spacious quays and docks; the Scheldt is here a fine river. There was a good museum of Dutch and Flemish pictures, but the *chef-d'œuvres* of Rubens and Vandyke had not yet returned from Paris. I was much struck with the extent and strength of the fortifications; also with the costume of the natives, particularly that of the women, which still resembled a good deal that of their former masters, the Spaniards. At the time of our visit everything was in a depressed state. Its trade had not yet recovered from the effects of the great war, and its then silent streets contrasted greatly with their former activity. The Roman Catholic religious ceremonies were conducted with great magnificence, and struck us simple Protestants, who had never witnessed anything of the kind before, with considerable astonishment.

We left Antwerp for Ghent, and took our departure for Bruges in one of the trackschuyts or barges, by means of which the great bulk of the goods and passenger traffic of the kingdom was carried on. These canals are magnificent specimens of the kind, and, being upon a much greater scale than our own, particularly struck me with admiration. The canal was bounded on each side with spacious banks, and was of great width, with a towing path and carriage way for general traffic. These banks were bounded by rows of trees, which serve for shade; at the same time their clippings and timber yield a considerable amount of profit.

We reached Bruges about noon, and had just time to examine the fine cathedral and townhall, as well as the interesting town, its quaint old buildings, quays, mercantile warehouses, all in the architecture of the Middle Ages, during which Bruges attained its greatest prosperity. We left this interesting town in the afternoon by another trackschuyt for Ostend, about 12 miles distant, by a canal of the same dimensions as the one above described. Most of these canals being connected with each other, Brussels, Ghent, Malines, Louvain, and the other large towns, possessed a complete network of water communication with the ports of Antwerp and Ostend, and trade is carried on with the greatest facility.

We reached Ostend in the evening, and learned that a packet was about to sail for Margate, of which we determined to avail ourselves. We had, however, sufficient time to examine this indifferent port, the second in the kingdom, which, in addition to a badly-contrived entrance by two guide piers, has two small docks. It is a mere tidal harbour, with an awkward bar at the entrance, and numerous shoals on the outside. The town possesses nothing remarkable; it was then garrisoned by English troops, and there was constant communication with England.

From this period I devoted my time almost exclusively to the Waterloo and Southwark bridges, but particularly to the latter, which was almost entirely under my direction, subject to the orders of my father. Besides the above works, I was a good deal employed in the drawing office in making drawings and estimates and calculations for a variety of new works upon which my father was engaged. I also occasionally visited the rolling and other machinery of the Royal Mint on Tower Hill, which my father at that time was constructing for the Government, and during the evenings I was employed in learning mathematics under Dr. Bradley and Dr. Firminger, and the Italian, French, and German languages, so that my time from morning to evening was fully employed. I must not omit to say that at this time I attended the lectures of the celebrated Sir Humphry Davy, the Professor of Chemistry at the Royal Institution, who at that day astonished the world by his wonderful discoveries.

The first, second, and third arches of the Waterloo Bridge being completed, Mr. Rennie determined to slacken the centres of the first arch, which was on the Surrey shore, where the bridge commenced. This was when the arches were entirely relieved from the centres, and the total subsiding of this arch was 2½ inches, which is nearly half an inch less than had been allowed; the centres were then removed from the first arch to the fourth arch, only three centres being employed. Each centre consisted of eight ribs, upon the truss principle, resting upon a compound system of wedges, supported upon struts placed upon the offsets of the piers and abutments; all the ribs were well connected together by transverse and diagonal ties, as well as the planking upon which the arch-stones rested. The trestles or bearers of the centre ribs, together with the wedges, having been first fixed in their places upon the offsets of the pier and abutment where the centre was to be fixed, four ribs of the centre were transferred, and fixed upon them in the following manner.

The ribs of the centre having been constructed upon a platform upon the shore near the bridge, a large barge or floating stage, capable of carrying four complete ribs, which weighed 40 tons each, was built to receive them. This floating stage was extremely strong, and transversely across the centre of it there were four strong stages at the same distance from each other as the ribs of the centre were intended to be when fixed in their position, to support the arches whilst building. These stages were supported by double transverse beams, resting upon powerful screws

15 inches diameter, in boxes resting upon the bottom of the vessel. Above each of these stages, yet securely attached to them, was a framework, to which the ribs of the centres were lashed whilst being transported to their places. When the centre ribs were finished and all was ready, the floating stage, at high water, was brought alongside the platform, upon which the ribs of the centre had been constructed, and were lying ready to be transferred to their places. Each rib was then raised by means of powerful sheer-poles, to which were double-purchase crabs, treble blocks, with all the necessary ropes, chains, and other tackle, by which means each rib of the centre was readily raised from the platform where it was built and transferred to its proper stage in the floating barge, and there secured in an upright position, when the ribs had been fixed in their places. The barge was then floated into the opening where the arch was to be constructed, which was generally done about half an hour before high water, so as to allow ample time to adjust and fix the ribs over the corresponding pair of wedges and trestles upon which it was ultimately to rest, which was done as the tide fell, and adjusted to the greatest nicety by the screws before mentioned; when the rib had been fixed in its place, the barge returned to bring another, which was served in the same manner as the first, and thus the fixing of one centre occupied only six days. This system answered most effectually, and was subsequently adopted by Mr. Robert Stephenson for fixing the great tubes for the Menai and Conway bridges.

The Waterloo Bridge, as well known, consists of nine equal semi-elliptical arches, 120 feet span each, with a rise of 34 feet 6 inches, the keystones at the crown being 4 feet 6 inches deep and 10 feet at the spring, and 18 inches thick at the soffit; inverted arches on each pier between the main arches 4 feet 6 inches deep. The piers were 20 feet wide, each having projecting buttresses, supported by two three-quarter Doric columnar pilasters, over each pier, the whole being surmounted by a Doric block cornice and balustrade parapet, level from end to end, the same as the roadway. The roadway above the piers was supported by six brick walls, 2 ft. 3 in. thick, covered with corbel stones. The shores being low on both sides of the river, the approaches are constructed so as to form an inclined plane rising 1 in 30 on the Surrey side, and nearly level on the north, or Middlesex side, with the Strand, upon a series of brick arches 16 feet wide each. These arches serve for storehouses. The roadway was formed by a layer of well-puddled clay 15 inches thick, then a layer of lime and of fine gravel 3 inches thick, then a layer of equally broken granite, in pieces 2 inches in diameter, 1 foot thick. Through the centre of the masonry of each pier a hole 18 inches in diameter was cut, entering the river on one side of the pier at low water, and from the top of this hole inside the pier cast-iron branch pipes of the same diameter were carried to side drains on each side of the roadway, so that all rain and surface water was effectually carried off into the river, thus preventing leakage.

The piers and abutments were founded in the solid bed of the river, which is strong gravel; they rest upon a wooden platform, supported upon piles 12 inches in diameter driven 20 feet into the bed of the river. The whole of the arches and exterior face of the bridge are built of Cornish granite, from the vicinity of Penryn, and the balustrade is made of fine grey Aberdeen granite.

The contract for the Cornish granite was taken by a very worthy man of the name of Gray, and the price was such as on so large a quantity ought to have enabled him to realize a very handsome profit; but he had no system or machinery adequate for the purpose, and instead of opening quarries properly upon an enlarged scale in the solid rock, by which he would have saved a great deal, he chiefly confined his operations to the loose outlying blocks, which reduced his profits considerably, and in the end it is very doubtful whether he did more than cover his expenses. As the dressing of granite for masonry was entirely new at that time, nothing having been built of this material in London, it was extremely difficult to find masons who would undertake it, even at such enormous prices as 1s. 9d. to 2s. per cube foot, so that the contractors, Messrs. Jolliffe and Banks, could not afford to pay it. Workmen were therefore obtained from Aberdeen, and the price was ultimately reduced from 2s. to 1s.; notwithstanding, however, the prime cost of the stone, the freight, dressing, mortar, and setting complete in the bridge cost about 7s. 2d. to 7s. 3d., so that the total cost was near 7s. 6d. It should be observed, however, that at that time there was a duty of threepence per cubic foot (or ton?) on stone, which has since been taken off. The interior stone consisted of hard sandstone from Derbyshire and Yorkshire.

The bridge and approaches were completed and opened with great ceremony by George IV., then Prince Regent, on the 15th of June, 1817, in commemoration of the battle of Waterloo, after which it was especially named. Twenty-five pieces of artillery were placed on the centre of the bridge, which fired a salute as His Royal Highness, the directors of the Company, and a brilliant suite walked over in procession, when he christened it Waterloo Bridge, and declared it open to the public. His Royal Highness came by water in his state barge, accompanied by the Admiralty and other barges, in which were the ministers and suite; he landed at the stairs on the south-east side of the bridge, and walked over it from south to north; he embarked again on the north-east side, and returned to Whitehall and Carlton House. The sight was very brilliant, the weather magnificent, and everybody seemed to be satisfied.

The total cost of the bridge was 565,000l., which was 10,000l. more than estimated by Mr. Rennie; the approaches, besides the land and buildings, cost a further sum of 112,000l.; so that the total cost of the bridge and approaches was 677,000l., and the land and buildings and contingencies 373,000l., making a total of 1,050,000l. This is certainly a very large sum for a bridge and its approaches; but when its extent is considered, the bridge alone being a quarter of a mile long, and the approaches nearly three-quarters of a mile more, also the great cost of materials and labour of every kind, the stone-cutting costing from 4s. to 6s. a cubic foot in the rough state, timber from 7l. to 14l. per load, and labour in the same proportion (which is more than double the present price), we cannot be surprised at the total cost.

I still continued my duties at the Southwark Bridge, which was completed in March, 1819, and was opened without any ceremony by Sir John Jackson, the chairman of the Company, and the other directors.

CHAPTER II.

Travels in Switzerland, Italy, Greece, Asia Minor, Constantinople, and
Egypt—Return to England—Death of Mr. Rennie.

I had now received a tolerably good education, both theoretical and practical, as a civil engineer; but before entering fully into practice on my own account, my father thought it advisable that I should travel for a time to study what had been done in ancient and modern times both in architecture and engineering. I accordingly left England on the 7th of June, 1819, in company with my cousin, Colonel, now General, Sir J. Aitchison, and the late Lord Hotham, his friend. We had a thirty-three hours' passage from Brighton to Dieppe, during which time, having exhausted the captain's store of bread and cheese, not very abundant, we were glad to fall back on a dozen mackerel, which Lord Hotham's servant was fortunate enough to catch. We passed through France without much incident; but when the view from the summit of the Jura suddenly burst upon us, the magnificent scene made a most lasting impression upon my memory. The valley of the Rhone, the Lake of Geneva, backed by Mont Blanc and its splendid range of mountains, rose before us as if by magic; we were totally unprepared for it, could scarcely believe our senses, and stood enraptured for nearly half an hour. We then descended to Geneva, where we passed several days very agreeably, examining what was then the picturesque old town, with its clumsy, old-fashioned waterworks; for the improvement of these my father, through the well-known Dr. Marcet, had just made a design for the municipality, which was much approved of, and which I understand has since been partially carried into effect. I here made the acquaintance of General Dufour, so well known for his scientific acquirements, and after a few days went on an expedition to Chamounix, where, as no one then thought of ascending Mont Blanc, we climbed the Montanvert and the Mer de Glace, where I made some experiments with Leslie's hygrometer. Having returned to Geneva, we started again on the 7th July for a more extended tour in the mountains, going by the lakes of Morat and Bienne, the scene of the great battle of 1476. I examined these two lakes, which were evidently rapidly filling up, but by lowering the outfalls a great part of the whole might be recovered, whereas at present the borders are to a great extent covered with reeds, and the marshes exhale unwholesome effluvia. After passing through Freibourg, where the bridge, then newly opened, was considered one of the wonders of Switzerland, and which, as a most remarkable work of its kind, I examined attentively, we passed on to Berne, and going through the mountains, returned to Geneva. On our way, being at Meyringen, and short of ready cash, we proposed either to return direct to Geneva, or to change one of Herries' circular notes; but on offering one of these notes to the landlord, he at once said there was no occasion for it, as we were Englishmen, and that was enough. Having produced a large bag of five-franc pieces, he told us to help ourselves, and was with difficulty persuaded to take one of Herries' notes in exchange. I merely mention this to show how high the name of Englishmen then stood on the Continent.

Leaving Geneva for Italy, we proceeded by the route of the Simplon, the construction of which I had promised to observe very attentively for my father. The first portion only presents, as objects of interest, the excavations through the rock of St. Gingough, near the upper end of the lake. From Martigny we started up the valley of the Rhone, where, though the ground is generally level, the road yet encounters considerable difficulties from the river, which here assumes the character of a torrent, and when swollen by floods sweeps almost everything before it; wandering from side to side it deposits the *débris* of the one side on the banks of the other, forming, we may say, alternately rapids and almost still pools, which renders it extremely difficult to confine its course within any reasonable limits; so that the art of the engineer is taxed to the uttermost. I thought that in many places the works were not designed with that solidity which is so necessary under such circumstances, and that sufficient precautions were not taken to arrest the progress of the *débris*. I considered that by providing depositories for it at certain favourable stations, the violence of the floods might have been considerably controlled, a much greater extent of land on both sides of the river rendered available for cultivation, and the extensive marshes, which, operated upon by a powerful sun, now produce most injurious exhalations, might have been deprived of their baneful influence.

Having passed Sion, we left Brieg early on the 10th of August, and as soon as we began the ascent I descended from the carriage, and with line and rule I measured every bridge until we reached Boveno, on the Lake Maggiore. I sent off from Milan, as I had promised, a detailed account of this celebrated road to my father, giving a drawing and account of every work, which I afterwards had the gratification of knowing afforded him great pleasure. The whole Pass must have at first sight been appalling to the engineers who traced the line of road; and although many other works of the kind of greater magnitude have been since executed, nevertheless, all things considered, it is worthy the approbation of mankind, and does great credit both to those who designed and those who executed it.

We reached Milan[1] on the 12th of August, and I was much struck with the fine

---

[1] Let me here relate an anecdote of the almost incredible instinct of the dog. Passing by the palace of the Austrian viceroy, I observed a dog sitting with an air of profound melancholy before one of the sentry-boxes. Colonel Brown, our representative, who was then with me, said that this dog formerly belonged to a soldier of the body-guard of Eugène Beauharnois, the viceroy, and accompanied his master to Moscow. The man never returned, but upwards of two years afterwards the dog did, and resumed his station before his former master's sentry-box. After a time the dog came to be talked about, and at length the viceroy, an Austrian archduke, had him brought into the palace and tried to domesticate him, but he always returned to the sentry-box, where he lay motionless, and at times moaning. Seeing this, the archduke ordered him daily rations, and he was placed in the sentry's orders for protection, and in this state I saw him; but a short time after the dog died,

canal which unites it with Pavia, and which can only be compared with the canals of Belgium. It has been said that the first pound lock was invented and executed by the famous Leonardi da Vinci, but subsequent inquiries have induced me to believe that our own country is entitled to the honour in the Exeter ship canal. I was also greatly pleased with the system of irrigation employed generally throughout Lombardy. This system was originally introduced by the Italians themselves, and during the Austrian rule was carried to the greatest perfection, Lombardy being by nature peculiarly well adapted to it. The vast and fertile valley of the Po is for the most part destitute of rain during the summer, when it is most wanted; but it fortunately happens that at this season water is most abundant from the melting of the snows on the Alps, which descend into the adjacent lakes and rivers, and would be otherwise wasted and thrown away if not employed for fertilizing the land. The water, therefore, at this season is conducted by an elaborate system of artificial canals, and distributed over the adjacent lands according to their respective levels, at a certain price, varying with the quantity distributed. Thus the constant supply of water, the high temperature, and the fertility of the soil combined, produce the most abundant crops, and the plains of Lombardy are rendered the most productive and valuable in Europe; whereas in winter, when the temperature is lowest, the snow is congealed on the Alps, and comparatively little water comes down when it is least wanted.

From Milan we reached Verona, where the bridge, consisting principally of brick, with binding courses of marble, can boast of one of the largest brick arches in the world, an excellent example of what may be done with this material when properly handled. After passing Vicenza and Padua, we reached Venice early in September, 1819.

The extraordinary and at the same time most beautiful and novel appearance of the city, with numerous towers and spires, about which I had read and heard so much, and had so long wished to see, now stood before me, and its loveliness more than realized my most sanguine expectations. When I considered its origin, a few fishermen's huts built upon the mud banks of the lagoon by men flying from the invasion of Attila, then the rise of the great republic whose wealth, conquests, and influence were destined hereafter to play such an important part in the world, and lastly the fallen and degraded state in which it then presented itself before me, I was lost in astonishment; I was for a while transported as it were in a dream, and could scarcely believe where I was.

The Grand Canal first attracted our notice,—perhaps there is no thoroughfare in the world lined with so many magnificent palaces,—and along which we passed until we came to the Rialto, a drawing of which I made and sent to my father. But if I was delighted, and I may almost say astonished, at the Grand Canal, I was still more so with the Place of St. Marc and its surrounding buildings, so varied in their architectural styles, yet each so picturesque and elegant in itself, and combined together forming at once the most interesting and beautiful scene of the kind in the world. It is one of those sights, at least speaking for myself, that never satiates—the more I looked the more I admired it. As to the details of these

apparently inconsolable.

different buildings which we saw, they are so much better described in the numerous guide books that it is unnecessary to repeat them here.

I visited every part of the lagoons, including the various islands, all of which are more or less deserving of notice, particularly the island of Murano, the seat of the celebrated glass manufactories; and also the Moravian establishment. But what really most interested me were the lagoons, and the means which must have been resorted to for keeping them open, notwithstanding the numerous causes which were and are constantly in operation to fill them up with the alluvial matter brought down from the mountains and plains by the various rivers and streams which discharge their waters into this portion of the Adriatic, also from the alluvium brought in from the adjacent shores, by the tide, which rises from 2 to 4 feet, and at times, during heavy gales from the southward, as much as 6 feet, overflowing the quays of St. Marc's Place.

There was a long-continued discussion amongst the numerous distinguished mathematicians, engineers, and others who have written upon this subject, as to the best way of preventing the filling up of the harbour. Some contended that the only method of effecting this was to admit all the rivers into the lagoons freely; for although they might deposit a certain amount of alluvial matter, nevertheless the great quantity of water discharged would alone be sufficient to carry away this deposit. But they forgot that when the rivers met the sea the current would necessarily be checked and rendered powerless to carry forward any matter which might be held in suspension, and that consequently the detritus would be deposited and form banks and shoals which the waters could not remove; thus in time the lagoons would be filled up, grass marshes would be formed, the city of Venice would be united to the mainland, and the harbour would be destroyed. On the other hand, it was argued, for the reasons above mentioned, that the only way to preserve the lagoons and the port of Venice was to exclude the rivers when densely charged with alluvial matter, and only to admit their waters at certain times, when they were comparatively clear; thus all the advantage would be obtained from the scour of these rivers, without the disadvantage arising from their deposits. Ultimately the arguments of the latter prevailed, the rivers were excluded from the lagoons by making a capacious canal all around them with sluices at their mouths, by means of which the waters were discharged into the lagoons when they were tolerably clear of alluvial matter; the surplus waters were discharged into the adjacent sea clear of the lagoons, and any alluvial matter which was brought in from the sea was removed by dredging from the main channels of the lagoons, so that they were in a fit state to admit the tidal waters and thus to keep the lagoons open.

But there was another important agent to be provided against, namely, the alluvial matter brought in by the winds, waves, and currents from the scouring of the adjacent shores of the Adriatic; this is done to a certain extent by dredging. Originally these banks contributed materially to the formation of the outer banks, which protect the lagoon on the sea side. If these banks were broken through or completely swept away, which the storms of the Adriatic frequently threatened to do, the lagoon, and with it the port, would be seriously injured or totally destroyed. This was remedied by defending this outer barrier bank of the lagoon by facing it with stone, and where the sea was most violent by constructing a solid breakwater of stone, and protecting it further by stone filters carried out a sufficient distance into the sea in order to divert the current, and to enable the alluvial matter to be deposited between them so as to form an additional protection to the main breakwater; this was accordingly done, and thus an extraordinary work at great expense has been constructed between Lido and Malamocco, the principal entrances from the sea to the lagoon, for a length of four miles, where the effects of the sea are greatest.

Malamocco is the principal entrance for large vessels, and the channel from thence to Venice has been deepened, chiefly by dredging, to the extent of 24 feet at low water. Lido, which is the next chief entrance and the nearest to Venice, being about 1½ mile distant from it, serves for the general class of merchant vessels. The other entrances of Foggia, Tre Porte, and the Piave, are seldom used except for fishing vessels, and it is not necessary to do more to these than to keep them in their present state, that is, to prevent deterioration, as it is an object of importance to allow the great mass of water by which the lagoon is chiefly preserved to pass in and out of the main entrances, Malamocco and Lido. Upon the whole it appears to me that this latter plan is the wisest that could be adopted, and the result has proved that it has been so far successful, although attended with considerable expense. It is in fact a choice of two evils, and the least has been chosen. It is, I believe, admitted that the port of Venice is now capable of receiving as large, if not larger, vessels than she ever received before; for it should be recollected that in the most flourishing times of Venice there were no ships drawing 23 or 24 feet, and vessels of this size can now enter and depart at all times. Hence Venice has been converted into a port fit for modern requirements; but it must always be borne in mind that so strong are the natural obstacles against its maintenance that nothing can preserve it in its present condition but the most constant vigilance and care. Fortunately the method of dredging by steam has been introduced, and this may be done to any extent, at a comparatively moderate cost, but it can only be compensated for when there is a sufficient amount of trade to pay for it. Still, in whatever way we may consider the question, it must be admitted that the port of Venice has been preserved in a most extraordinary manner during so many centuries, notwithstanding the natural obstacles against it; and now that it has been connected with the kingdom of Italy, there is no further drawback to its full development, and it only requires self-reliance and energy to render it what it once was, one, if not the most important, of the commercial cities of Italy.

At Bologna we had the pleasure of making the acquaintance of the celebrated linguist Mezzofanti, whose modest and simple manner, accompanied by his extraordinary acquirements, quite enchanted us. The singular fact is that at the time of our meeting him he had never been out of the province of Bologna, and yet he had acquired the knowledge of twenty-four languages, and, as far as our limited acquaintance went, he spoke the English, German, French, Greek and Latin tongues perfectly; and those conversant with the Oriental languages informed us that he spoke them equally well. He appeared to be completely absorbed in languages, and was scarcely acquainted with any other branches of knowledge; still his wonderful mastery of this branch of study was a great acquirement, and must have required vast powers of memory as well as indefatigable study.

From thence we passed through Ravenna to Ancona, the position of which is good, occupying as it does a salient point of the coast. The water is deep and there is a commanding height for a citadel. The Roman emperors resolved to take advantage of the situation, and built a town here, the place being well adapted for a seaport. They accordingly made a mole on the south side to protect the harbour against the most dangerous winds, namely, those blowing from the south; they commenced the work from the shore by throwing down large blocks of rough stone, which were obtained from the rocks in the vicinity, in order to form a base for the superstructure; these blocks of stone were deposited promiscuously in the sea and left to the action of the waves, which in a comparatively short time consolidated them until they formed a mass that at length became immovable, always adding more stone as required until the whole became solid; this foundation was carried up to the level of low-water mark. They then commenced the superstructure of masonry, of squared stone and brickwork cemented together by pozzolana or hydraulic mortar, which was best adapted to make the work permanent. The inside or quay wall was formed by first placing a close framework of timber in front and at the angle or slope at which the quay wall was to be formed. They then threw down a mass of pozzolana, lime, and small rubble stone, mixed together, between the wooden frames and the rubble stone which had been previously deposited to form the base of the mole. This in a comparatively short time become solid, as the mixture of pozzolana, lime, and stone possesses the quality of setting under water. When it had become sufficiently solid, the timber frame was removed and transferred to another section to form a continuation of the quay wall, and thus the whole line of inner quay wall which formed the roadway was made by backing the quay wall to its full height with rubble, and a parapet of masonry was erected on the outside; this superstructure, although rough, stood very well with occasional repairs. In this manner the mole was carried out to a great extent. A fine triumphal arch was erected at the end to commemorate the completion of the work.

It is singular that the same emperor should have constructed a like work in a similar situation and in the same manner at Civita Vecchia, on the opposite coast of Italy, after every attempt had been made to establish a port by his predecessors at Ostia, which was built at extraordinary expense, and has been filled up by the alluvial matter brought down by the Tiber and from the sea by the littoral currents, so that it is now three miles from the sea.

The principle of construction used in these works resembles a good deal the mode adopted by the Phœnicians at Tyre, and subsequently by the Carthaginians at Carthage, and by the moderns. The Romans also employed the hollow mole, that is, a mole constructed with arches, by means of which the current charged with alluvial matter was enabled to pass through the mole, and thus any deposit within or without the mole was to a great extent prevented. The harbour of Civita Vecchia remains serviceable at the present day for vessels drawing 20 feet of water, although, as may be naturally expected, a certain deposit has taken place during the lapse of so many centuries which requires to be dredged out occasionally. At

Ancona only one pier was built on the south side, and consequently an eddy and stagnation took place on the north side, as the littoral current runs from south to north, and therefore there is a tendency to deposit on the north side. The French when masters of Italy commenced another mole on the north side, thus enclosing a considerable space of sea so as to form a close harbour, which no doubt is of great service; still, from the nature of the local circumstances, a certain deposit may always be expected. This, however, can be removed by dredging, and the harbour may always be preserved in a state of efficiency if the extent of trade will warrant the expenditure necessary for the purpose.

On reaching Rome, it is difficult to express the emotions I felt on a first sight of the Holy City, surrounded by the desolate Campagna, the Tiber rolling in front, the Castle of St. Angelo, and the numerous towers of churches rising out of the mass of houses, crowned by the gorgeous dome of St. Peter's elevated in proud majesty above the whole, backed by the magnificent views of the ancient city, once the mistress of the world.

This most interesting, I may say, most thrilling, sight lay before me in all its solemn majesty. When I considered that it rose from a few insignificant shepherds' hovels to imperial splendour, then became the prey of the barbarian hordes of the north, and lastly, the throne of the Catholic Christian world, I was awed by the wonderful decrees of Providence, and at the instability of all human grandeur. I lay restless all night; I could scarcely realize the fact that I was actually in this wonderful, all-absorbing city, to visit which from my childhood had been one of the greatest objects of my ambition. It would be vain to attempt to describe in detail its numerous splendid buildings, both ancient and modern, its museums, and its countless treasures of priceless art. I will merely remark that there is no city in the world more worthy of a visit than Rome, or where greater gratification and instruction can be derived; for example, the aqueducts and the Cloaca Maxima show how thoroughly the Romans understood the importance of sewerage and good water for the preservation of human life, an importance that is only now beginning to be recognized in England, while there still remains very much to be accomplished.

My friend Colonel Aitchison was obliged to hasten his departure. I then took lodgings in the Piazza de Spagna, and devoted myself to the study of the Italian language, architecture, and drawing, and in my leisure moments entered into society; and fortunately at that time there were assembled there some of the most distinguished characters in science, literature, and art, besides diplomatists and leaders of fashion, from the various capitals of Europe. Amongst the first may be mentioned the celebrated chemist and philosopher, Sir Humphry Davy, and his talented wife, under whose hospitable roof I passed many happy days, and at the same time received much valuable instruction. I made acquaintance with the Marchese Martinette, the scientific engineer, who at that time devised the extensive hydraulic works for the improvement of the rivers and drainage of the marshland districts of Bologna and Ferrara, from whom I derived much information. I also met the well-known antiquary, Sir W. Gell, who to his interesting memoir of Greece afterwards added much towards explaining the antiquities of Rome, and in his agreeable society I spent many a pleasant day. Dodwell too had just returned from Greece; and I made the acquaintance of that prince of sculptors, Canova, to whom my father had given me a letter of introduction. I frequently went to his studio, where he always received me with the greatest kindness. He was then at work on the model of his famous dying Madeleine, which struck me as a master-piece of elegance and beauty, combined with the resignation and piety which so pre-eminently distinguishes it as a beautiful specimen of art. Nobody could be more kind, amiable, and modest in his manner than that distinguished sculptor. Flattered by emperors, kings, and the great and cultivated of every land, he never for a moment forgot himself or appeared to be elated, or to be put out of his or-dinary simple, unobtrusive manner; whilst to his brother artists he was equally kind and familiar; to his inferiors he was always gentle and considerate, giving his humbler fellow-workers every encouragement and advice, cheering them on their way, and not unfrequently assisting them with his purse when their necessities required it. He was devoted to his art, of which he was so eminent a professor, and with it combined all those amiable and charitable feelings which rendered him an universal favourite and a benefactor to mankind.

I made acquaintance also with the great Dane, Thorwaldsen, who frequently admitted me to his studio. He was totally different from Canova. His square, mas-sive head, covered with a redundance of flowing locks; his finely-developed coun-tenance, beaming with talent and firmness of purpose; his colossal and well-pro-portioned figure, and erect and commanding gait, all combined to raise in the mind of the spectator a degree of respect and admiration not usually to be met with. Yet with all this apparent sternness there was combined a happy mixture of gentleness, genial sociability, and good-nature, which, after a little acquaintance, soon made you feel at home with him; and the more you knew him the more you liked him. He was then at work upon the colossal figures of the Twelve Apostles, for the church at Copenhagen, a commission given to him by the King. Several of these were finished, and magnificent specimens of art they were. I have since seen them in their places, and have looked at them with increased admiration. His Triumph of Alexander had just then been completed, and a finer specimen of bas-relief it is impossible to see. It is singular that with all this nobleness of char-acter, and being withal a perfect master in his art, this great man—for certainly he was so in his sphere—should have looked upon Canova as his rival, and disliked him with a dislike almost amounting to hatred, whereas the gentle Canova had no such feeling towards Thorwaldsen.

Their styles, moreover, were so totally different from each other, that there could be no reason for jealousy between them. Canova excelled in the female form, where nothing but elegance, gentleness, and grace are required, although he by no means failed in the male figure; whereas Thorwaldsen excelled in that of the male, where force, manliness, and dignity are mainly requisite. Both were at the head of their profession in their respective styles, and both have left behind them numer-ous masterpieces of art, which have never been surpassed in ancient or modern times.

In addition to these celebrities, there were Lawrence, Chantrey, and Turner, all engaged in their respective walks of art. These I had known in England, and I received much kindness and instruction from them, and enjoyed the greatest pleasure in their delightful society.

Numerous great houses were at that time open for the reception of all the celebrities of Roman and foreign society, to which I had access. Amongst many others may be mentioned those of the beautiful Princess Borghese, the Duchess of Devonshire, and Lord William Bentinck. At the house of the Duchess of Devonshire I was introduced to the polished courtier and priest, Cardinal Gonsalvi, at that time Minister of Foreign Affairs to the Pope, whose graceful manners and elegant language, combined with a vast fund of information, enchanted everybody. I also met Thomas Moore, who enraptured the whole audience by singing his fresh melodies, with a degree of pathos, taste, and feeling which was peculiarly his own. Neither must I forget Lady Morgan, whose lively and constant prattle afforded much amusement, not altogether destitute of information.

At the hospitable house of Lord and Lady William Bentinck I saw the *élite* of English society, where I was frequently a welcome guest. Amongst others, appeared the beautiful Miss Canning, afterwards the Marchioness Clanricarde, and the silent, amiable, and unfortunate Lord William Russell, who afterwards fell a victim to his servant, the assassin Courvoisier.

To enumerate all the agreeable and talented persons of whom I had the honour of making the acquaintance at Rome would be foreign to my purpose, and out of place here. Suffice it to say, that the winter of 1819, which I passed at Rome, was one of the most agreeable and instructive of my life, and will ever be remembered by me with the most lively feelings of satisfaction. Before leaving I had the honour of being presented to the benevolent Pius VII., and was much captivated with his kind and unostentatious reception. After witnessing the Church ceremonies of Christmas, I left Rome just before the Carnival. I passed through Castel Gandolfo, and afterwards came upon the Pontine Marshes.

The Pontine Marshes consist of a low tract of land, extending from the elevated ridge of Castel Gandolfo to a spur of the Apennines, which approach the Mediterranean near Velletri, being a length of about 20 miles, whilst on the east they are bordered by the Apennines, from whence numerous streams and torrents descend into the marshes; and on the west, or sea side, they are bounded by a line of sandbanks, or dunes, thrown up by the waves and storms from the west; so that they form, as it were, a basin into which all the waters flow, without a natural outfall through which they can escape. To attempt to make a direct outfall to the sea was perfectly practicable, but there was no rise of tide, so that though the outfall would be considerably improved, there would yet remain the difficulty of keeping it open, in consequence of the constant influx of the sand. In similar, and indeed in all cases of this class, there are two points to be considered, namely, the water which comes from the highlands, and that which falls directly upon the lowlands. Now the former, coming from a higher level, and necessarily having greater velocity, will naturally descend to the lowest level, which is the marshes, and will force its way to the outfall before the lowland water; so that until the highland water is discharged, the lowland water, having less inclination, cannot escape, and as the outfall was extremely deficient, neither the highland nor lowland water could be discharged, and both combined accumulated upon the lowlands and flooded them. It therefore was evident that so long as this state of things remained it was impossible for the lowlands to be properly drained, as the outfall was deficient for the discharge of both waters when combined.

The only feasible mode of attaining the desired object was to separate them; then, at least, the lowlands could not be incommoded by the highland water, and would only have to discharge its own drainage. Thus, if a catchwater drain had been made along the base of the highlands, all that water might be discharged into the sea at any level required, for there is ample fall or inclination; and if another and separate outfall had been made for the discharge of the water falling upon the lowlands, although with much less fall, it might have been discharged, at least, to a great degree, if well embanked and of a proper size, because it would have been unobstructed by the highland water. But even supposing that any portion of the lowlands had been below the level of the outfall, this defect could easily have been remedied by pumping machines, worked either by wind, animal, or steam power, connected with main and subsidiary drains of proper capacity. This plan, which has been so effectively exemplified in the drainage of the great level of the Fens and elsewhere, was not adopted, but the whole of the waters, highland as well as lowland, were thrown into one drain and outfall, and thus before the highland water could be discharged, the lowland water stagnated upon the land. Until Mr. Rennie's system is adopted, there never will be a perfect drainage; and the more imperfect the outfall, the greater is the necessity for adopting his system.

At Terracina we were terrified with the account of the brigands whom we should have to encounter in crossing the spur of the Apennines before we entered the kingdom of Naples. I here examined the old Roman port, consisting of two artificial piers of stone carried out from the shore, with the entrance pointing west. Both piers are long, and are well constructed of solid masonry, with a parapet and roadway; there being large mooring stones, with holes through their extrem-ities, fixed solidly into the inner or quay walls of each pier. Upon the whole this is a substantial work, although, as the piers are curved, they are badly adapted for breaking the waves on the outside, and for preserving tranquillity within the harbour. The whole space within and without the piers was filled up with sand to within two feet of the quay walls. It is a common error, even at the present day, to make curved piers; the consequence is, that when the waves strike them they accu-mulate as they move forward, until at length they break with increased force and carry all before them. The same takes place with the waves entering the harbour, which produce such a degree of agitation within that vessels cannot lie in safety; whereas by making the piers in several straight arms the waves strike them on the outside, and they are broken, and neutralize each other at the angles.

I reached Naples through a road closely patrolled by soldiers, with blockhous-es for ten or twelve men at very short intervals. Being bound for Greece, my time was very short, yet I managed to ascend Vesuvius and visit Herculaneum with Lord Guildford, as well as the Museo Borbonico as often as possible.

I examined the modern harbour of Naples, and I doubt if anything could be more badly designed. There are two small piers, the entrance between which is difficult; it is exposed to the prevalent and dangerous winds, and is scarcely safe inside. This is the more extraordinary, because the numerous remains of the ancient harbours of the Romans, dispersed throughout the bay, might have served as models for a port adapted for all the requirements of modern trade as well as for war vessels. The harbour must be improved before Naples can be provided with that accommodation which her increasing trade imperatively requires.

I started from Naples with several others on the 31st January, 1820, and after an alarm from a threatened attack by brigands on the summit of the pass of Baveno, from which we were rescued by the timely arrival of the Receiver-General of the province, on his official journey, accompanied by a numerous escort of dragoons. We reached Lecce, the capital of the province, on the evening of the 8th February, having been eleven days on the journey, a distance of about 300 miles, rather fatigued, but much pleased with this novel and interesting country, so seldom visited by travellers. I alighted at the best hotel in the place, which was but very indifferent, and the next day called upon General Church, the Governor of the province, and was most kindly received by him, he insisting upon my making his house my home. Here I again met with Lord Guildford, Lord William Russell, and Chevalier Bronsted, with whom I had previously made acquaintance at Rome. Lord Guildford was on his way to the Ionian Islands on official business, and Sir Thomas Maitland, the Governor, had ordered a vessel to be sent to Brindisi to convey his Lordship, who very civilly offered me and our two mutual friends a passage. As the vessel was not expected for several days, we in the meantime became the guests of General Church, and were most hospitably entertained by him. General Church was an extraordinary man. He was below the middle size, about the age of five-and-forty, extremely well built, spare, sinewy, and active, with a well-proportioned head, sharp piercing eyes, rather aquiline nose, and a closely-compressed mouth, denoting great firmness and resolution. He commanded a regiment of Albanians and Greeks, as an auxiliary corps in the British employment, during the great war, and in that position assisted the operations of the British cruisers on the coast of Italy; and hence he became subsequently attached to the army of Lord William Bentinck, after his conquest of Sicily. Church was a proficient in the Greek, Italian, and French languages, and, having considerable military talent, and being a great disciplinarian, soon brought the rough and savage elements of which his corps was composed into tolerable order, and rendered them of considerable service in the wild warfare in which they were engaged. At the conclusion of the war he retired on half pay to Naples, where, being well known to the Government, he was made Governor of the province of Otranto, at that time overrun with brigands. Church was appointed to the command with unlimited control, and by his vigorous and energetic conduct soon spread terror and dismay amongst them; he was here, there, and everywhere; when they least expected, he came upon them suddenly, dispersed them, and destroyed the leaders without mercy. He had many narrow escapes himself from sharing the same fate. Once it is related that he and his aide-de-camp, Captain Kusini, entered unknowingly a small town, of which one of the most able and daring brigand chiefs, with a powerful band of followers, was in possession. Church, when he found this, determined to make the best of it, being perfectly aware of his danger. He entered the chief inn and sent for the landlord, who recognized him at once, and asked him if he knew that the brigand chief and his followers were actually in the town. "Yes," replied Church, with imperturbable coolness, "I am come expressly to meet him; tell him that I want to see him immediately." The brigand chief accordingly came, astonished to see the General, whom he least expected; he began to be alarmed, thinking he was surrounded. The General, addressing the brigand by name, informed him that there was no chance of his escape, but that if he and his followers would surrender, he would pardon them and get them employed in the King's army. The brigand chief and his men declined this generous offer. Church then told him that he was sorry, for their own sakes, as in future they could expect no mercy. The brigand and his followers then withdrew, determined on their side to show no quarter to the troops if ever they should get them within their power; but for the moment, being ignorant of Church's position, they were afraid to attack him, little supposing that he was at that time theirs. This extraordinary interview having terminated, Church felt that he had played the game far enough, and the sooner he and his aide-de-camp made their escape the better. The landlord, fortunately for them, kept the secret. The General and his aide-de-camp escaped by the back of the house, climbed over some fences, reached their horses, and then galloped off, never pulling bridle until they had reached their own followers, who, when they heard what had passed, were astonished at their escape, which was due alone to the coolness and courage of the General. Scarcely had Church and his aide-de-camp departed when the brigand and his followers learned how completely they had been deceived, and at once set off in pursuit; but they were soon obliged to retreat, being themselves pursued by a superior force, from which they narrowly escaped capture; the band was afterwards destroyed. This is but one out of the numerous instances of Church's extraordinary adventures. In a short time he extirpated brigandism, the province regained its tranquillity, and the people pursued their several employments in peace without fear of molestation, blessing the General who had relieved them from their oppressors.

Being particularly desirous of seeing Brindisi, from my recollections of Horace, I obtained an escort of two dragoons from General Church, for, said he, "You may meet some unwelcome visitors on the way; but if they see the uniform of my dragoons they will not trouble you with their acquaintance." Brindisi, which I need not now describe, struck me as being an excellent port on the whole, and now that it is made the terminus of the railway from Naples, which connects it with the entire railway system of Europe, it will no doubt become a place of considerable importance.

I embarked with Lord and Lady Guildford and others in a Government vessel, and reached Corfu on the 27th February. The day was fine, and we were much struck with the beauty of the surrounding sea and mountains, together with the view of the magnificent inlet forming the harbour of Corfu, comprising as it did the ancient and picturesque town on the bold outline of St. Salvador and the rich undulating fields. As this was the first time I had seen anything of Greece, I particularly remarked the handsome appearance of the population and their picturesque costume.

There was nothing worthy of notice in the town except the old fortifications, so that I determined to make my stay as short as possible and take the first conveyance for Zante, and from thence to the mainland of Greece. Corfu, having been under so many different governments, partook in some measure of the character of all—Turkish, Venetian, Russian, French, English, but the ruling feature was Greek and Albanian. Our Government, urged on by that amiable, excellent, and enthusiastic person, the late Lord Guildford, had determined to establish an university at Zante, to revive classical learning, and Lord Guildford was appointed Lord Rector. He was an excellent scholar and linguist, and a most good-natured person; he was anything but a man of the world, and little acquainted with the real character of the modern Greeks. Sir Thomas Maitland, the Lord Chief Commissioner and Governor of the islands, was a totally different character, and knew the Greeks well. He was a stern, uncompromising soldier, with great talent, courage, and firmness, joined to long experience in war, politics, and governing mankind in every part of the world; he was not to be deceived by plausible appearances; to use his own phrase, he would stand no humbug, and would make his commands obeyed, although he was a strict administrator of justice. He treated Lord Guildford's plan with great ridicule. "They were clever and learned enough, but they had already a great deal to unlearn; the first thing was to make them honest and obedient to the law." When Sir Thomas became Lord Chief Commissioner of the Ionian Islands, the population was in a most disorganized state; pillage, murder, and piracy, were very common, and the malefactors were triumphant, and defied the law. It therefore required a strong hand to keep them in order, and Sir Thomas was just the man to do it—which he did. Under his strict but just rule they soon became more manageable, as they found he would not be trifled with.

On the day after my arrival I left my card at the palace, and received an invitation to dine with him on the 2nd March. I had taken no letter of introduction to him, although he knew my father; but I had previously heard that he disliked nothing so much as letters of introduction, and seldom paid them any respect, even when coming from the highest quarters. I accordingly made my appearance at the palace punctually at the time appointed, and soon after a brilliant assemblage was collected, in full costume, all waiting for the Governor. He shortly after appeared in full uniform, covered with orders. He was a rather fine-looking man, about sixty-five, of the middle size, with a strong, square, well-built, well-proportioned, figure. His countenance and speech betrayed his Scotch origin; and his large bluish-grey eyes shaded by shaggy eyebrows, his well developed nose, and compressed mouth, evinced a decided strength of will. His speech was terse and blunt, but with a strange mixture of kindness and severity, and altogether he was evidently made to command and be respected. I kept in the background, but he came up to me, and in his dry Scotch manner said, "I suppose you are Mr. Rennie; and pray, sir, may I ask what brought you to this country?" "To study the antiquities of ancient Greece." He at once replied, with a certain degree of vehemence, "A pack of nonsense; gang awa' back to your worthy father in England, he will teach you more in two days than ye will learn here in all your life." Then, shaking me very cordially by the hand, he said, "But I am very glad to see you, and if I can do anything for you I shall be very happy, for your gude father's sake." Dinner was then announced, and a capital dinner we had, Sir Thomas making himself very agreeable, and cracking his Scotch jokes right and left with a great deal of glee. I sat nearly at the bottom of the table, and immediately opposite to me was Colonel Napier, afterwards the celebrated Sir Charles Napier of Scinde. I was much struck with his countenance; his keen piercing eyes, his prominent aquiline nose, and his restless, quivering lips, marked him out as a man of great character. I said to myself, if opportunity offers, that man will much distinguish himself hereafter, and time has shown that I was not mistaken. Sir Thomas is reported to have said, "Napier is a great Radical, and a friend to liberty and equality; he has considerable talents, and I will give him an opportunity of showing himself, and I will wager that before six months are over there will be a petition to the House of Commons against him for tyranny and oppression of the people." Sir Thomas was a true prophet; he made Colonel Napier Governor of the island of Cephalonia; a most troublesome set he had to deal with, and he was obliged to use harsh measures to bring them under control; true enough a petition was got up by the inhabitants of the island complaining of him for his severity and cruelty, and requesting the Government to remove him from his command. When Sir Thomas heard of this he laughed heartily, and said, to his staff, "Did not I tell you so? the Radicals, however much they may preach about liberty and equality, are always the greatest tyrants." Sir Thomas was held in high respect at head-quarters, and whenever he went to England and asked for instructions at the Colonial Office, was told to write them for himself, as the Government had entire confidence in him.

I here made the acquaintance of Captain Smyth, of the Navy, and his amiable and accomplished wife. Captain Smyth was then employed in making a survey of the Ionian Islands for the British Government, which was afterwards published; and certainly for elaborateness of detail and completeness of execution it has seldom been surpassed. Every plate is ornamented with a view of the architectural remains of the most remarkable buildings in the district, and also views of the coast, which were drawn by Mrs. Smyth with great taste, beauty, and fidelity, and form an important feature in this great work. Captain (afterwards Admiral) Smyth was a man of considerable scientific acquirements; and after having been most actively employed in different parts of the world (always with distinction), he retired to Bedford, where he erected an observatory, and published his celebrated astronomical observations, which, in addition to his public service, entitle him to a high place among literary, scientific, and professional men. He was of a most amiable and jovial disposition, ever ready for fun and amusement whenever they did not interfere with his duties; he was, moreover, always ready to serve a friend, and was universally beloved and respected by his numerous acquaintance. For myself, I regarded him as a valuable and sincere friend, whose loss I afterwards most deeply regretted.

Count Lunzi, one of the Greek nobles, a most agreeable and talented young man, and a large proprietor on the island, who had travelled with us, invited me to his country house, and I set off with my friend the Chevalier Bronsted to pay him a visit on the following day, the 10th February. We started on horseback on a fine day, and after riding through a rich, level plain for about ten miles, we reached the volcanic pitch-wells, and on our return found Count Lunzi awaiting our arrival at his Villa Sarachina. He received us most cordially, and conducted us into the house, and we were introduced to his family, by whom we were most hospitably entertained. We then took a kind farewell of our host at 5 P.M., mounted our horses, and proceeded homewards. At first we rode along leisurely, and gradually quickened our pace; at last we got into a full gallop. My horse, which was little better than a pony, although a very handsome, strong, well-made animal, by this time became so excited that he fairly got the better of me, and proceeded at a furious rate, so that I lost all control over him. Away he went helter-skelter over land and water, driving all before him. In vain I attempted to stop him. At last, finding it useless, I let him have his way; and arriving at the bay of Zante, he made direct for the sea. I allowed him to proceed until he began to swim; then, finding he was in no disposition to return, I dismounted, and partly swam, partly waded ashore, where I awaited his return; he soon got tired of swimming, and came to land. I then caught him, and mounted, and returned to Zante. My friend Bronsted, who was completely distanced, joined me as I remounted, and we rode back together. As we entered the town I saw a number of soldiers looking on and laughing; I did not know why. But it appeared that this pony was noted throughout the place for his tricks, and they wondered how, after venturing upon his back (which I certainly should not have done if I had known his vicious character), I had come back safe and sound. However, I had punished him pretty well, and he became quieter afterwards. We had a hearty laugh at the adventure; and, being thoroughly drenched to the skin, I changed my clothes, and joined Sir Patrick Ross's dinner-party, when again everybody laughed at me, and congratulated me upon my happy escape.

Before quitting this subject I cannot help saying a few words about my friend Bronsted, the Dane. He was a most excellent person, and a first-rate scholar and antiquary, well known for his researches. He had acquired some most valuable bronzes, being portions of helmets found at Cortona, of which he published an elaborate account. The figures and other ornaments are most elaborate in design and execution; in fact, they are masterpieces of art, and were afterwards bought by the British Museum (for a considerable sum), where, I believe, they still remain.

I made several other excursions to various parts of the island; amongst others, to the top of the mountain which forms the south-west promontory of the bay. The summit rises far above the Mediterranean, and the view from it is magnificent, commanding the whole of the beautiful island, which lies at its base like a rich garden, and some other islands and adjacent coasts.

The inhabitants are a good-looking, active, and industrious race, but, like their neighbours, inclined to be very turbulent, and require to be kept in order by the strong arm of the law.

While here, just at the equinox, we had, as usual, some very severe storms; and one night I was roused from my sleep with a violent shaking of the bed, which lasted several seconds. It was at the same time raining tremendously, and blowing very hard, accompanied by violent thunder and lightning. I jumped out of bed, not knowing what it was; the house trembled, and I thought it was coming down. The other inmates were also alarmed. After waiting anxiously for some time the storm abated. I went to bed again, and slept soundly until morning, and then inquired of the landlord the cause, when he informed me that it was the shock of an earthquake, which they frequently felt in the island; and on the following day we learned that the same earthquake had been most severely felt at the neighbouring island of Santa Maura, that many houses had been thrown down, and a number of the unfortunate inhabitants destroyed. I had frequently been told by persons who had lived in volcanic countries of the extraordinary sensation produced by earthquakes on the human frame, but until I had experienced it I never could appreciate its effects; certainly they are most remarkable; the whole nervous system is convulsed, and one fancies that the last hour has arrived, so that it takes some time before the system recovers its usual tone. I certainly should not wish to experience another shock, and was extremely happy to have escaped with nothing more than a shaking and a severe fright.

Having now spent fourteen days very agreeably, being most kindly entertained by the Governor, Sir Patrick Ross, the officers of the 75th, and others, and having seen enough of Zante, I was anxious to proceed on my journey, only waiting for a vessel to take me to Greece, when fortunately I heard that an English mercantile brig, commanded by Captain Burgess, a rough old Scotchman, was about to sail for Patras, so I at once took my leave of Sir Patrick Ross, his family, and all my other kind friends, packed up my baggage, embarked at 3 P.M. on the 23rd March, and bade adieu to Zante. After a pleasant passage, but rough accommodation, we arrived at the entrance to the Gulf of Corinth, and anchored in the roadstead about a mile from the shore. I paid my passage, landed, proceeded up to the town, called upon Mr. Green, the Consul, and soon found tolerably comfortable quarters in a Greek house, but the beds were sadly infested with bugs, which annoyed me much. Before leaving Zante, I ought to say that I here engaged a Greek, who had been well recommended to me. His name was Demetrius Papandriopolo. He was about thirty-five, of the middle size, sharp, active, sober, intelligent, and honest, and served me faithfully through the greater part of my journey, for above twelve months, until I returned to Malta on my way back to England. I found him a most valuable servant, and he rendered me the most essential services.

Patras was then a trading town of some importance, in fact, the principal one in the Morea, and it contained a considerable number of inhabitants, almost wholly Greeks, without including the Turkish garrison. It is situated on a hill about a quarter of a mile from the gulf of the same name, where there is a small rubble jetty for boats to land their goods, &c. The roadstead outside is safe and well protected. The town is surrounded by a rich, fertile plain, well cultivated with olives, vines, and currants. The entrance to the Gulf of Corinth is protected by two old castles, the distance between them being about a mile and a half; and three miles east is the modern town of Lepanto (it was formerly the ancient Naupactus), celebrated for the victory over the Turks by John of Austria; it contains 2000 souls; the surrounding views are very fine. The town was under the command of a Turkish Governor, and is situated upon the side of a hill rising to the east, encompassed by a rude wall and ditch, and crowned at the summit by a citadel, in which the Governor and the garrison resided. Besides the citadel there were no remarkable buildings of any kind: the streets were narrow and dirty. There were a few mosques interspersed here and there, whose graceful minarets, rising above the mass, gave the place a most picturesque appearance. The town contained numerous bazaars, where all the trade of the place was carried on, and the neighbourhood of the main street was filled with cafés, crowded with Greeks, Turks, Ionians, French, English, and various other nations, all smoking their long pipes, drinking coffee, sherbet, and various other liquors, apparently in great comfort. The Turk, as lord supreme, moved leisurely about with the most solemn dignity, having the greatest contempt for his neighbours, and every now and then, as a Christian passed by, he greeted him with a lofty scowl, as if he considered him unworthy of notice, uttering the simple word Giaour! and passed on without further ceremony. I frequently wandered amongst the bazaars, then to me a perfectly novel sight, and was much amused with the gravity and solemn dignity with which the Turks sat cross-legged behind their counters; if you asked for any of their wares, they quietly, with a monosyllable, ordered their assistant to show it to you, as if they considered it a favour. This was so different to what I had been accustomed to in the civil, well-bred shopkeepers in Europe, that I could not help laughing outright, which I soon found would not do, as it gave great offence.

One morning, whilst walking along the shore in front of the roadstead, whom should I stumble upon but my old friend Captain Smyth, who had come to Patras on a surveying expedition in one of the small auxiliary boats attached to his larger vessel, which lay at anchor in the roadstead. This auxiliary boat, although small, was fitted up with every convenience for the purpose, and adapted to enter shoal water, in order to complete the details of the survey. In this he was accompanied by one or two assistants, and a few men to work it, so that he could be absent from his ship for several days without inconvenience. Captain Smyth was delighted to see me, and asked me to dine with him on board his boat, and I agreed to, with great pleasure. I accordingly went there at the time appointed, and amongst the guests met Captain Hunter and his wife, and our worthy Consul (Mr. Green). We had a most cordial welcome and an excellent dinner, and afterwards passed a delightful evening. Towards sunset we adjourned to the beach in order to take ices, which we got from the town, and enjoy ourselves by smoking our cigars. Whilst we were thus happily seated, laughing and talking together, we were surrounded by numerous Greeks from the town, amongst whom were several Turkish soldiers, armed to the teeth, and carrying their long pipes, which they never abandoned for a moment. The jolly old Captain said, "Now I will astonish these fellows," and immediately dispatched one of his officers for his seven-barrelled pistol, which was brought, and duly loaded. He then told his interpreter to inform the Turks that this was a pistol which he could discharge as many times as he liked without reloading. The Turks held up their hands in astonishment, exclaimed "In Shallah!" and with a scornful look at us, said such a thing was not possible. Smyth, however, at once rose from his seat, deliberately discharging his pistol one, two, three, four, five, and six times, to the utter amazement of the Turks, who could scarcely believe what they saw. He then, with the utmost coolness, asked whether they were convinced, to which they all replied, with the exclamation of "In Shallah!" perfectly so. Smyth then offered them coffee and ices, of which they readily partook, and, after a hearty salutation and shaking of hands, they returned to the town, saying what wonderful people these Ingleses are. This pistol resembled the modern Colt revolver, although differing in some particulars. Before leaving I visited the celebrated cypress tree, three miles to westward. Its base is about 40 feet in circumference, and it is 35 to 40 feet high, the upper part having been blown off during a gale. I also visited St. Andrea, where were the remains of the well and some fragments of sculpture.

I took leave of my excellent and talented friend, the Captain, and determined to start the next day, the 27th of March. Upon returning to my quarters at Patras, I was informed that there was a French gentleman of the name of Prevot, an artist, who was about proceeding to Athens to make a panoramic view of that city, and was desirous of knowing whether I would allow him to accompany me. Finding him a very agreeable person, and having been introduced by Mr. Green, and being glad of a companion upon my lonely and novel journey, I willingly consented to his joining me.

Having got all my baggage on board, attended by my servant Demetrius, we started early on the following morning, the 27th, at seven. The day was unfortunately very rough, with violent gusts of wind, heavy rain, accompanied by thunder and lightning; notwithstanding which we started on our journey, and when we got into the gulf there was a heavy swell (which threatened at times to overwhelm the boat and all in it), with baffling winds, which materially retarded our progress. We struggled on all day: at last our boatmen said that it was impossible to proceed farther; we therefore determined to run ashore, and finish our journey the next day. We accordingly did so, and landed in a small bay, the only safe one on the south side, near Vostizza, at about four o'clock in the afternoon, two hours before sunset. Then came the question, where were we to sleep? Vostizza was about one or two miles distant, and there were no means of transporting our baggage there; and to leave it in the boat was to expose it to being plundered, as the boatmen said they would not be responsible for it, in consequence of the numerous bad characters lurking about, pirates, robbers, &c. As there was no possibility of getting our things to Vostizza, and there were the remains of a convent (which had been ruined by an earthquake) close by on the shore, on a cliff about 50 feet above the gulf, I proposed at once that we should place our things there, and sleep, to which my companion, who was very tired and sick of the voyage, readily consented. We got the boatmen to land the things, and place them in the convent; but when we got there, to our great surprise and mortification, found that it was uninhabited and half in ruins. There was no alternative, however, as there was no other building near us; so we made the best of it, and prepared to pass the night there. We went upstairs, explored the ruined building all over, and at length we discovered one large room tolerably perfect, and capable of keeping out the rain. This we chose for our night's quarters, placed the baggage in order, and then commenced to prepare our supper. Upon examining our provisions, such as we had, we found them considerably damaged by the wet and unfit for use. I then decided that Demetrius and myself should go to the bazaar of Vostizza, and purchase what we could get, whilst we left M. Prevot in charge in the convent; and I gave the boatmen some money on account to provide themselves for the night where they could: the boat, everything having been taken out, was firmly moored to the shore.

Demetrius and myself went to Vostizza, and returned to the convent just after dark, laden with a bag well furnished with wine and provisions, the best we could get, and immediately set about preparing our mattresses for beds, and lighting a good fire to cook the supper. Demetrius was a capital cook, and about nine o'clock had got ready an excellent repast. Having had scarcely anything all day, and being miserably wet and cold, after having washed, and dressed in dry things, with a roaring wood fire before us, we enjoyed our supper excessively, forgot all discomforts, and about eleven lay down upon our mattresses for the night. Before doing this, warned by what our boatmen told us of the insecurity of the place, we fastened the door of our room as well as we could with an English padlock, and placed our heavy baggage and the table against the door, examined the priming of our pistols, which were well loaded, and lay down to sleep. My mattress was placed immediately opposite to the door, Demetrius was in the other corner, and M. Prevot near the fireplace, with an understanding between us that if any noise occurred during the night we were not to leave our berths, for fear of mistaking each other for an enemy. We slept tolerably sound until about two o'clock in the morning, when I awoke and thought I heard the movement of footsteps on the landing outside our door. We had kept a light burning all night. I listened again attentively, and thought I heard voices outside, then something like a push against the door. I awakened my companions, and told them to be ready with pistols, but not to fire until we were attacked, and then if possible to keep our positions. I had scarcely done this when a violent attack was made against the door, which in a moment was burst in, the table, portmanteaus, &c., being hurled into the middle of the room; this was immediately followed by a rush of several armed men. I instantly fired into the midst of them, Demetrius seconded me by another shot, and I repeated mine. We then heard a shriek as if somebody was wounded, and the whole of our assailants precipitately beat a retreat, scampering down the stairs as rapidly as possible. I reloaded, and we jumped up, rushed to the stair-head, and were going to follow them, but upon second thoughts we considered it best to remain where we were. We then secured the door again and tried to sleep, but in vain; we therefore lay watching until daylight, when our boatmen called us, and we prepared to depart, rejoicing heartily at our lucky escape from being robbed and murdered, as we assuredly should have been if we had not been armed and defended ourselves so stoutly. Before leaving, we examined the landing and stairs, and discovered some traces of blood, so that some one must have been wounded. Our boatmen told us that we had a very narrow escape; they said that they would not have slept there, and indeed told us before that we had better not, as there were numerous pirates and rascals in the vicinity, who waylaid all travellers, and seldom allowed them to escape even with their lives.

Early in the morning, which was very fine after the storm, we set sail with a fair wind, and bid adieu to Vostizza, with a remembrance of its inhospitable shore not likely to be soon forgotten. We had a delightful voyage, and enjoyed the beautiful mountain scenery on both sides of the gulf with much satisfaction, and reached the custom-house of Corinth, three miles from the town, before the close of the day. We then discharged our boatmen, and finding no horses, were obliged to stop at this nasty, wretched place, and passed a most miserable night. Next morning, the 29th, having got horses, we started at eight and reached the caravanserai at Corinth at nine, and having breakfasted and dressed, called upon the Bey, who received us very civilly. We wandered about all day examining the place and its antiquities. I arose early in the morning, awakened by the Muezzin as he was calling the faithful to prayers from the galleries of the minarets attached to the mosques. I had not heard this at Patras, having slept too late, and therefore it appeared to me for the first time most novel and singular. I listened to it with peculiar interest, as the contrast was so different to our own, where this office is performed by bells. On the same day, after breakfast, I examined the town, a poor miserable place domineered over by the Turks, with no trade, and the wonder was how the people lived. There was nothing worthy of examination except the beautiful remains of the Temple of Venus, of the simple, severe, yet commanding Doric order. A few columns only were left, surmounted by the cornice and entablature; these have been so often measured and engraved, that it is unnecessary to say more about them here, except that they are eminently beautiful, and an excellent specimen of that bold and expressive style of architecture. I was anxious to mount the Acropolis, crowned by an old fort and garrisoned by Turks, but this was not permitted: I regretted it much, for, seated as it is on a commanding eminence, eighteen hundred feet above the level of the adjoining gulf, the view over the sea and the fine mountainous surrounding country must be magnificent; but the Turks in those days were so proud, insolent, and domineering, that the sight of a Christian was an abhorrence to them, and one was glad to get out of their way, although even then they had considerable respect for the English, more indeed than any other nation; for we had saved them from the French and had protected them against the encroachment of the Russians, who had already begun to be very troublesome. Having nothing further to see, we were anxious to get away as early as possible.

There were two routes to Athens, to which we were destined, one entirely by land over the isthmus, the other by sea, after crossing the six miles of isthmus which separates the Corinthian Gulf from the Ægean Sea. We determined therefore to take the latter, and on the 30th March, having engaged the necessary number of horses, and loaded them after a considerable wrangling and noise with their keepers, started at 1 P.M. for Recrees Ceneres, on the borders of the Ægean Sea. At 3 P.M., in crossing the isthmus, we came upon the remains of the wall said to have been built by the Venetians for preventing the incursions of the Turks into the Morea. Here we dismounted, and engaged one of the boats of the country and embarked. However, as there was little or no wind, we were out all night, and did not reach the Piræus before four o'clock in the afternoon of the 31st. This is a fine natural port, but the town was then a miserable place, surrounded by a few wretched buildings, at the head of which was a Turkish custom-house, where the authorities at first made considerable opposition to our landing. However, I soon silenced them with a respectable backshish, or present, when they became as civil as possible, and not only allowed us to land with our baggage, but did not subject us to any examination; we therefore at once inquired for horses to take us to Athens, but finding none to be had, were obliged to remain that night in a wretched plastered room at the custom-house. Next day, 1st April, having got horses, we started at six, with no end of wrangling with muleteers, which the Turks soon silenced by laying about right and left, to our great amusement. I at once interfered, and rewarded the Turks with another backshish, when they heartily saluted me, and we started without further molestation, and reached Athens, six miles distant, over a miserable road, within an hour and a half, at half-past seven o'clock. There were then no hotels, so we took up our quarters at the house of a respectable Greek, Toagrafos, with but poor accommodation. I by this time had become so accustomed to bad fare and lodging that I did not mind it, but my friend, the French artist, was not so easily satisfied; we therefore parted, and he endeavoured to find better apartments elsewhere. I never met him again, and consequently do not know how he succeeded with his panorama, or whether he ever completed it, but I heard that about a year afterwards a panorama of Greece had been exhibited at Paris with considerable success. I hoped that it was that of my friend, for he appeared to be an excellent person, and was certainly a very pleasant companion, with all the characteristic ardour and enthusiasm of his countrymen. He had never been out of France before, and therefore felt the *désagréments* of the journey a great deal more than I did; nevertheless, by the time he reached Athens he had become tolerably well accustomed to it, and, with the usual French *sang froid*, made up his mind to it, and enjoyed the journey as much as myself.

We arrived at Athens on Friday, the 1st April, in the midst of spring; the weather was delightful, the sun shining brightly, the sky cloudless, the vegetation bursting into full luxuriance, the plain and mountains covered with flowers and shrubs of the most brilliant and varied hues. Nature seemed to be in all her glory, and man to sympathize with her. I was in most exuberant spirits, and was pleased with everything around me. The beauty of the country, and the very idea that I was in Athens, which from my boyhood I was most anxious to see, and which I scarcely ever expected to behold, aroused me to such a pitch of enthusiasm that I could scarcely believe my eyes; which was not extraordinary, considering the brilliant and memorable scene which burst upon my sight, and by which I was surrounded. I soon sallied forth again, called upon the English and Austrian consuls and delivered my letters, and hurried about all day, taking a hasty view of the Parthenon, the Temple of Theseus, Jupiter Olympus, and the Temple of the Winds, which I admired excessively; then, thoroughly tired and exhausted, I returned to my lodgings, took a hasty supper and went to bed, dreaming of the glories of ancient Greece. After breakfast I went out, and at once made my way to the Acropolis, where the glorious Parthenon stood out prominently. I was riveted to the spot with admiration, and could go no farther. At last I came to myself, and found the scorching rays of the sun rather more than agreeable. I then returned to my quarters, dined, reposed during the heat of the day, and towards evening I again sallied forth, and enjoyed a magnificent sunset, such as I had been totally unaccustomed to in our frigid climate of England. I again went to the Acropolis the following day, and there enjoyed the splendid scene. The Acropolis is a detached rock, standing high above the plain or valley on which Athens is situated; upon this rock is the citadel, where the Parthenon, the Erechtheum, and all the other principal buildings are placed, and in addition to this it is a citadel or fortress, where, in case of attack, and the city below being taken, the Acropolis would form a refuge for the garrison and the inhabitants. From the summit you command a view of the country all around: on the north Mount Olympus and the Pindus range of mountains; on the south the Mediterranean, the island of Syra, and several others; on the east the plain of Athens, flanked by Mount Hymettus, and Mount Hybla on the north; on the west the Piræus, the Ægean Sea, and the mountains of the Peloponnesus or Morea; in fact, whichever way you look, the view is delightful, and you are satisfied.

I at once removed into a new lodging in a Greek house, which appeared tolerably clean and comfortable but as usual I found it full of fleas, which tormented me day and night. However, when I went to bed I hit upon a plan to relieve me at least of a good many of them, which fortunately proved successful. I placed a number of large plates filled with oil, having a wick in them; I then lighted the wicks, and the fleas, attracted by this, leaped into the plates, which in the morning were black with them, and by this means I got rid of the greatest part, and at length enjoyed some comfort.

I now determined to study the antiquities more at leisure, and set regularly about it. Every morning, immediately after breakfast, I proceeded to the Acropolis, and employed myself in drawing and measuring the Parthenon, Erechtheum, and other buildings until dinner time; and after, in the evening, I took walks and rides round the adjacent country. Having finished the Acropolis, I then proceeded to the Temples of Theseus and Jupiter Olympus, the Temple of the Winds, and the other antiquities, which occupied me about a month. All these splendid specimens of architecture have been so accurately described in the elaborate works of Stuart, Cockerell, Dodwell, Gell, and others, that it is unnecessary to describe them here. With regard to the Piræus, and the other ports of Athens, I examined them minutely, and for plans and description of them I refer to my work on 'British and Foreign Harbours.' I also saw a little of Athenian society. Amongst others, I made the acquaintance of the Maid of Athens (so celebrated by Byron), who was very pretty, ladylike, and agreeable. I was also most kindly received by Mr. Gropeus, the Austrian Consul, Mr. Logotheti, the English Consul, and by the well-known artist, Mr. Luzieri, who had made some fine views of Athens for Lord Elgin.

The Turks treated the Greeks with great contempt and very little ceremony. By way of illustration, I was present at the arrival of a new Governor or Vaivode. His Excellency came mounted on a beautiful white Arabian horse, surrounded by about a hundred well-mounted cavalry, all in the magnificent Turkish costume of the day; next there followed the hareem, in crimson-coloured carriages, each slung between two horses, and escorted on foot by numerous eunuchs and blacks, all splendidly attired; then the baggage, which was carried on the backs of horses, led and guarded by Greeks, mixed with Turkish soldiers on foot. The whole town turned out to witness the sight and welcome their new Governor, who scarcely deigned to look at them; whilst the attendants going before cleared the way with sticks, which they by no means used sparingly, dealing their blows right and left on everyone within reach. A number of Turkish lads amongst the crowd amused themselves by firing off their guns and pistols, which were loaded with ball and shot. They took no care which way they fired; sometimes in the air, sometimes on the ground, and sometimes straight forward amongst the crowd; of course everybody got out of the way, myself amongst the number. I saw one Greek who had his leg broken by a shot from a Turkish lad near me, at which he, the Turk, and the whole of his companions, shouted with joy, and seemed to think it great fun: at this I could not restrain my indignation, and being armed with a good stout stick, I rushed amongst them, striking right and left: some Greeks tried to stop me, but I stuck to them, and had the satisfaction of seeing that no more shots were fired except in the air. Nothing impressed the Turks with so much respect as courage and vigour; the more you humbled yourself, the more they bullied you, being anything but fond of fighting. I found from experience that this was the only way to treat them; for though frequently amongst them, and under the most trying circumstances, I was rarely molested. Afterwards I witnessed the entrance of the Pacha of Negropont, which was in the same style, but upon a more extended scale.

I occasionally visited the convents, inhabited by a few worthy monks, who kindly receive strangers and lodge and board them for a moderate sum. The situation is excellent, being in the higher part of the town, close to the beautiful remains of the Temple of the Winds, with a splendid view. Athens at the time of my visit was a poor miserable place, without an inn worthy of the name. It was surrounded by a crenellated wall, which even the artillery of that day would have demolished with a few rounds of shot. The Ilyssus and Cephisus are trifling streams, which, passing on the south side of the town, discharge their waters into the sea a few miles below. During summer there is scarcely any water, as it is conducted away by different channels for irrigation, so that the beds of these two rivulets are dry.

Having explored Athens sufficiently for my purpose, I determined to make a tour of the Morea, and accordingly hired a small decked vessel of about 30 tons to cross the Ægean Sea, leaving Athens for the Piræus on the 25th of March; and after waiting some time on board in the harbour, at seven on the 25th started, reached Ægina at eight, landed after breakfast, and proceeded on foot to the Temple of Jupiter, on a hill about a mile from the shore. We spent here nearly three hours measuring and examining the remains of this beautiful specimen of Greek architecture, and sketching the surrounding picturesque scenery; then started for the town and port of Ægina, whither I had sent the vessel to meet us, and was told that the distance was only three or four miles, which turned out to be nearly twelve, over a rough path, so that we did not reach Ægina until nearly 6 P.M., very tired and hungry, having had no lunch. The Temple of Jupiter is a fine specimen of the Doric, and is well described in Mr. C. R. Cockerell's excellent work, to which the reader is referred. The small town of Ægina is situated about the middle of the west side of the island, facing the Morea, from which it is separated by a channel about eight or ten miles wide. The town and island of Ægina were governed by a Vaivode, who had a few Turkish soldiers, and was inhabited chiefly by fishermen and small traders. The surrounding country is hilly, with fertile valleys producing olives, vines, and grain, and beans of several kinds; and there are a few sheep and goats.

The next morning, the 26th March, I rose early and examined the port, which consisted of two artificial harbours, formed by two piers, each carried out from the shore, enclosing a considerable space of water, with a depth of about 10 or 12 feet. This was made by the ancients, and Mr. Cockerell gave me a plan of it, which is inserted in my work on 'British and Foreign Harbours.' It was then blowing very hard north-west and contrary; however, soon after noon the wind abated, when we embarked, and arrived at the little port or bay of Pilascro, the ancient Epidauros, on the opposite shore of the Morea, about five o'clock in the afternoon. Here we disembarked, and walked round the place, and could only discover what appeared to be the remains of a bath, and some mutilated figures near it; and a little farther some similar remains, but nothing remarkable; so that it would require considerable time to make anything out of them. In Sir W. Gell's 'Itinerary of the Morea' what is there is well described.

We slept on board. On the 27th we rose at daylight, and found that the baggage horses, Greek servants, and an excellent Arabian horse, which I bought at Athens, and which I sent round by land, had arrived at Epidauros. Having disembarked our baggage and discharged the vessel, we loaded the baggage on the horses, after which we started for Napoli di Romania, a small town situated on the Gulf of Argos, not many miles distant from Epidauros, through an undulating, picturesque, and rich country. On account of the guide taking the wrong road, we were obliged to stop at the village of Tero, where there are a considerable number of Roman and Greek remains.

On Friday, 28th March, we started at half-past six, and at noon reached the picturesque fortress of Napoli di Romania, on the east side of the Gulf of Argos, situated on a lofty hill commanding the entrance to the gulf. The miserable town lay beneath the citadel, which, as well as the town itself, was strongly fortified according to the Venetian system, and was a place of considerable strength, garrisoned by a set of bigoted Turkish troops, imbued with a thorough hatred of Christians, so that I was strongly advised not to stop there. There was, however, no alternative, as there was no other place near. I therefore rode into the town, and took up my quarters at the English Vice-Consul's, who was a Greek. Immediately after my arrival I went out, and tried to get into the citadel, in order to have a view of the town and surrounding country, which was very fine, but the Turks refused me admittance in the most insolent manner, and told me to get about my business as quickly as possible. I found that it was of no use to attempt to proceed farther, and returned at once to the town, and after having examined it, and found nothing worthy of notice, returned to my lodgings. Here the Vice-Consul came to me in a great fright, and said that the Turks had been much annoyed at my arrival, and that I must depart directly; strongly urging me to do so, as he said that he could not protect me, and if I stayed he would not be responsible for the consequences. I replied that as it was night it was too late to depart then, but that I would do so the next morning at daybreak. To this he consented. I accordingly supped, and went to bed. On the following morning my servant and myself were in the saddle before daybreak, and were at the gates of the fortress waiting for them to be opened. In Turkish fortresses at that day it was the custom to make every foreigner and Greek dismount and walk over the drawbridge. I thought that this was degrading, and determined not to do it. As soon, therefore, as the gates were opened, and the drawbridge let down, I started off at a rapid pace, and told Demetrius to follow me. In passing the Turkish guard, consisting of three or four badly-armed men, they called out to me to stop, which I disregarded, and galloped off as fast as I could. I had scarcely proceeded a couple of hundred yards when, turning round, I saw Demetrius in the hands of the Turks, who were belabouring him with sticks, and trying to pull him off his horse. He kicked and fought with all his might. I immediately rode back with a loaded pistol in each hand, and dashed amongst the Turks, who soon released my man, and spurring our horses we galloped off as hard as possible, never pulling rein for a full mile, and then, finding we were not followed, stopped and joined our men and baggage, whom I had left behind in a small Greek house before I entered Nauplia the preceding evening. Thus I got happily out of this affair, which might have been rather serious; but I always found the best way was to show a bold front to the Turks, and not allow them to insult you. After a little halt we started again, and stopped to examine the ruins of the ancient city of Tiryns, said to have been built by the Cyclopeans. These are very curious and interesting, and are supposed to be the oldest specimens of architecture in Greece. They are constructed of rough, only partially hewn blocks of stone, rudely put together without cement. We have here an attempt to construct a Gothic arch; but the stones are not radiated, the beds being flat, each stone projecting over another until they meet in the centre. These ruins are extensive, covering a surface of several acres. They are well described and represented in Gell's 'Itinerary.'

From Tiryns I proceeded to Mycene, about 12 miles farther northward. These ruins are also very extensive, and once formed the capital of Agamemnon, with a small surrounding territory. The chief building worthy of remark is the Hall of Atreus. This, of the kind, is a fine work, constructed of solid masonry, of large blocks of stone well squared, and put together without cement. The hall is circular, and 47 feet 6 inches in diameter in the inside, and 42 feet 6 inches high, the top covered by a single stone, thus forming a Gothic arch inside of squared blocks of stone, the beds being horizontal, and the stones projecting over each other in the inside until they meet in the centre, the angles of the projecting stones in the inside being cut off and dressed smoothly on the face, so that, when viewing it from within, the whole has the appearance of a nicely-constructed dome, lighted from the top by a circular opening. The side walls are 18 feet 6 inches at bottom, diminishing to 18 inches at top. It is a remarkable work, and may be considered as one of the earliest specimens of the arch, and a considerable improvement upon that of Tiryns. The entrance to the hall is by a massive doorway, the sides being formed of single stones well dressed, and the top of another single block equally well prepared; this is surmounted by another large block, upon which is sculptured in bold relief two lions rampant. The approach to the doorway is between two solid walls of masonry, forming a passage open to the air. This being near the summit of a hill, there is a fine view from it over the plain, or rather valley, beneath, backed by the ruins of the ancient city of Argos and its Acropolis, on the Gulf of Nauplia, and on the west by a fine range of mountains, the highest of which towers prominently above the rest, standing far above the level of the Mediterranean.

After having measured the Hall of Atreus, and having made a sketch of the surrounding country, during which my people had managed to cook an excellent dinner in the open air, I dined, and started about 4 P.M., and reached Argos at 5 P.M. The ruins of the ancient city contain little worthy of remark, and the modern one is a miserable village, seated amongst the ruins. The rich country was not half cultivated, merely dotted here and there with a few vineyards and olive woods. Argos is now about a mile from the head of the Gulf of Nauplia, but it is very probable that in former times the sea washed the walls of the ancient city; the intervening space has been filled up by alluvial deposit, and it is now little better than a marsh.

I slept at the caravanserai, and next morning started for the Temple of Jupiter at Nemea, passing through a rich valley with vines, grain, and olives, and the village of Agioz Georgious, near which is a convent on a hill. These convents are not the seats of learning, for the monks are lazy and ignorant; but they form comfortable resting-places for travellers who pay well for their accommodation. The ruins of the Temple of Jupiter, well described by Gell, consist only of three columns of an indifferent style of Doric.

Having returned to Argos to sleep, on Sunday, May 1, I arose early. The morning was wet and stormy, and I was detained by the muleteer demanding extra pay for the horses, which, however, the Bey decided against him. We accordingly started about seven, and soon after began to ascend the Pass through the mountains, having on the left the miserable village Lerna, seated on the swampy shore of the gulf, anciently called Hydra; and hence was derived the Lernean hydra or monster, celebrated by Virgil. This was no doubt nothing more than the malaria fever personified, which prevails severely in this district. In fact, the whole of this country had a wretched, forlorn, deserted appearance. The population was scanty, oppressed by poverty, misery, and tyranny, their countenances for the most part pale and haggard, expressing despair and dejection, to extricate themselves from which appeared hopeless; whilst the surrounding country, naturally fertile, aided by the influence of a genial climate, only required the industry of man to make it yield the richest reward. Nothing was done; all lay dead and desolate. Nature seemed to have done everything for man, while man had in everything neglected nature; a not uncommon occurrence, and one which, as far as my experience goes, has produced more misery than almost anything else.

We began to ascend the Pass over a rude causeway, about eight or ten feet wide, in many places cut up into deep gullies by the torrents; the ascent steep, rugged, and difficult, fit only for mules and pack-horses. The evening was fine, however, the air fresh and invigorating, and the scenery beautiful, which raised my spirits after the melancholy caused by the depressing influence of the sight of so much desolation. Having got to the summit of the Pass, we had a magnificent view on both sides. Looking back, we had the Gulf of Nauplia, with the picturesque fortress at the entrance, the plain of Argos and its ruins, also those of Mycene and Tiryns, backed by a rugged outline of hills; on the west the fertile plain, with its capital Tripolitza, and the ruins of Mantinea, backed by a bold line of hills, with the magnificent Taygetus rising 7000 feet, flanked on both sides by the blue waters of the Mediterranean. We reached Tripolitza just after dark, and, as usual, put up at the best Greek house where we could find admittance; and as I had a good sumpter mule, Demetrius soon prepared a capital supper, which I enjoyed much after a hard-working day of nineteen hours, during which I had been fourteen hours in the saddle. I slept soundly, notwithstanding the attacks of numerous fleas, to which by this time I had become tolerably well accustomed.

Tripolitza, although the capital of the Morea, was but a poor place, with about 4000 inhabitants, and the Turkish garrison of about 300 men. The Governor's palace was merely an assemblage of some low, insignificant buildings enclosed within a wall. The rest of the town consisted of irregular, crooked, narrow, dirty streets, some of them hardly wide enough for a carriage, for which, indeed, there was no need, and there were none; with a place or square near the centre, surrounded by cafés, where the Turks and better class of natives, such as they were, sat idly smoking their long pipes and nargillas with that stupid indifference and repose which characterized them, perfectly regardless of any but themselves, and quite unconscious and indifferent to the misery around them. There were one or two bazaars, where all the business of the place was transacted; in these cloths, furs, shawls, leather articles, such as saddles, bridles, &c., silks, arms, pipes, tobacco, and provisions of all kinds, were sold, but scarcely any articles of European manufacture. The town was surrounded by a crenellated wall, unfit for artillery, and incapable of resistance to European troops; and there were two or three mosques, with their minarets rising above the mass of insignificant buildings, and a few Greek churches, the interiors of which were decorated with rude and almost grotesque representations of saints. These edifices, towering above the rest, had a picturesque and imposing effect when viewed from a distance, but this vanished the moment you drew near and entered, when little else but dirt and misery met the eye.

The following day after my arrival I walked about the town, amusing myself with the little which was to be seen; and the next day I rode, accompanied by my servant Demetrius, to the ruins of the ancient Mantinea, formerly a city of considerable importance, and well fortified according to the style of that period. It was surrounded by a strong lofty stone wall, about 25 feet high, flanked at intervals by circular turrets, and in front of each entrance there was a kind of lunette or advanced work to protect it, so that an enemy, in endeavouring to enter, was exposed to flank attacks from the garrison. These walls were surrounded by a wide ditch filled with water, and the total surface enclosed by them is considerable. The ruins within consist of the remains of a hippodrome, theatre, and other private and public buildings, concerning which, from their scattered and dilapidated condition, it was difficult to define anything accurately. In the neighbourhood of Tripolitza there are numerous kalavatha, or subterranean conduits for carrying off the water for drainage, irrigation, and supplying the inhabitants. The river flowing through the plain, if such it can be called, is an insignificant stream; it is connected with the subterranean conduits described above. The surrounding country was tolerably cultivated, and produced corn, pulse, oil, and wine, both red and white. This wine was made in a very rough way, and, in order to preserve it, was mixed with a certain quantity of turpentine, which gave it a strong bitter flavour, by no means agreeable at first; and until you become accustomed to it, you cannot drink it with any degree of relish, but by degrees you like it, and it is very wholesome.

Whilst at Tripolitza I became acquainted with the late Mr. Hodgets Foley, afterwards M.P. for Droitwich, a very amiable, good-natured person. As an instance of Turkish civility, Mr. Foley was taking his usual ride in the evening, accompanied by his servant, and on his way met with a party of the delhi, or cavalry, playing at football, a game in which they much delighted. As they passed he civilly pulled up to get out of their way, when two or three of them at once, without the smallest provocation, attacked him, laid hold of his horse, and otherwise maltreated him, and he with some difficulty extricated himself from their hands, and rode back to Tripolitza as quickly as possible, to lay his complaint before our Vice-Consul, who was a Greek. He made his complaint to the Governor, who promised redress, but (Turkish fashion) never gave it. Foley's mistake was this, that instead of getting out of the way, he should have ridden right into the midst of them, laying about him on both sides with his whip as hard as possible; then they would have feared and respected him, and have never offered him any violence.

After having seen all that was worthy of attention in the town and neighbourhood, I started for Misitra, the site of the ancient Sparta, some miles from Tripolitza, the route to which lay through the plain. We reached Misitra, a miserable village, towards evening, and obtained tolerable quarters in a respectable Greek house, where we were received very kindly, and most hospitably entertained, which of course I paid for by a handsome present. I sat down to supper with the family, consisting of the husband, his handsome wife, daughter, and son, before a low table, placed upon the floor; the dishes consisted of soup, a species of stewed mutton, vegetables, and cakes and fruits. After dinner the servant came round with a basin and ewer, and poured water over our hands, and then presented a napkin to wipe them: we next adjourned to another room, where coffee and pipes were served, and the daughter sung very prettily some Greek airs, accompanied by the mandoline; and thus I passed a very agreeable evening.

The following morning I rose early and proceeded to the ruins of ancient Sparta, some miles distant. Here nothing is to be seen of this ancient celebrated city, but some mounds, fragments of walls and buildings, the extent and designation of which it is extremely difficult to make out. It was in vain, therefore, to attempt more, and so I next determined to prosecute my journey into Maina, said to be inhabited by the descendants of the Spartans. This was considered at that time to be a journey of considerable difficulty and danger, and, indeed, impracticable, on account of the lawlessness of the natives, who were said to be nothing more than a set of pirates and robbers, at war with the whole world and each other. Lord Byron wished to go there, but was dissuaded from it, although no person ever accused his lordship of want of courage. My janissary advised me not to make the journey, and said that he would not accompany me, as we should all be murdered; and added that the Turks, who had frequently endeavoured to conquer the country both by land and by sea, had always failed. The more they tried to dissuade me from going, the more I was determined to do it, being stimulated also by doing that which few persons would attempt; therefore I discharged my janissaries and extra horses and servants, and sent them to meet me at Calamata, the other side of the peninsula, and decided to undertake the journey with Demetrius alone, who was well acquainted with one of the captains or chiefs of the district of Marathonesi, and who was anxious to go himself, and said that the difficulties had been greatly exaggerated, and that we should get through very well. I resolved, therefore, to start on the following morning.

Before proceeding farther I will endeavour to describe the country which we were about to visit. The district of Maina is situated at the southern extremity of the Morea; it is bounded on the north by Mount Taygetus, and its subsidiary range of mountains, and by Cape Matapan on the south, so that it forms a triangle, jutting out into the Mediterranean; and on the east and west sides it is bordered by that sea, so that on all sides, except on the north, it is surrounded by water. As to the aborigines, it is difficult to find any well-authenticated account; it is supposed that they at one time formed part of the Spartan republic. After the conquest of the Greeks by the Romans, and subsequently by the Turks, they took refuge here, and being of a warlike and restless temperament, disdained all subjection to any one chief, but divided themselves into separate communities, and chose their chief for the time, who was the most able and successful warrior amongst them. These communities or tribes were wholly independent, and were frequently at war with each other, but always united to defend themselves against the common enemy, and they drove out the Turks, who frequently, but in vain, attempted to bring them under subjection, and latterly had quite given it up. Such was the state of the country when I visited it. The whole population was considered to be about 20,000, divided into capitanates or beylichs, with a supreme chief, to whom, however, the allegiance was merely nominal, and only accorded when the whole country was invaded by a common enemy. On other occasions, each tribe or community made war upon the others upon the least provocation, or upon the surrounding countries and seas as it suited them; in fact, they were a horde of pirates, at war with all mankind, who treated them accordingly. It was no uncommon event at that day for a British ship of war to bombard its seaports and to land a strong body of men to chastise the natives whenever they could lay hold of them; but as the coast is intersected with numerous creeks, backed by innumerable inaccessible mountain fortresses, to which they escaped when overpowered, it was almost impossible to root them out. Every petty chief had a stronghold or fortress, where he lived with his most intimate followers. In some cases this fortress was surrounded by a strong loop-holed wall, to which no access was given except by a drawbridge or ladder, which could be drawn up when attacked; and the access to the castle within was the same. In others it was a simple square or round tower, with the entrance about 10 feet above the ground, approachable only by a ladder, which could be removed at pleasure, and was always drawn up at night. The country for the most part was uncultivated, and although extremely fertile, capable of producing corn, wine, oil, fruit, and silk in abundance, served little more than to supply the deficiency when booty was scarce. Such was the general state of Maina, which I determined to visit. We accordingly left our heavy baggage behind, and I started early in the morning, accompanied by Demetrius and one baggage horse. Shortly after leaving Misitra we crossed the frontier, came to a wild, desolate country with nothing but brushwood, occasionally mixed with lofty valonia trees, the bark and fruit of which form a valuable article of export.

At midday we halted as usual to give our horses, as well as ourselves, rest and refreshment; and the ever-active Demetrius, in a comparatively short time, provided us with a tolerable dinner, and after a couple of hours' rest started again on our journey, and passing through the same kind of wild country we reached the sea-shore and the ruins of the ancient Marathonesi, the greatest portion of which had been submerged by an earthquake, and we could distinctly discern the buildings through the clear water. A few miles farther we reached the neat little seaport town, the modern Marathonesi, the capital of the tribe of which Demetrius's friend was the chief. In the course of our day's journey we were not a little inspired and gratified that we had not been attacked or even molested, notwithstanding the dismal accounts and forebodings which had been made to us. It is true we met a few rough, straggling, armed fellows on the way, and they looked at us with a fierce, independent scowl, but seeing us well armed and prepared to meet them, passed on without further remark.

We arrived, soon after sunset, at the house of the chief, and were most kindly received by him, and he showed us into some clean comfortable bedrooms, furnished, to my surprise, in the French fashion. The fact is, the chief was a great merchant and trader with the Ionian Islands, Italy, and France, and possessed considerable wealth, and was far more civilized than the rest of his countrymen. He was a fine handsome courteous young man, about twenty-eight, of the middle size, and extremely well-proportioned, sinewy, active, and vigorous. His countenance was oval, of the true Greek form, with dark piercing eyes, black eyebrows and lashes, well-defined nose, small mouth, with compressed lips, and beautiful white teeth; with long moustaches, curling several inches below his chin, and his head covered with a profusion of black hair hanging down in natural ringlets; he wore no beard, his chin being closely shaven; in fact, he was a very dashing personage. He gave us an excellent supper, and entertained us with numerous anecdotes and accounts of the country and its inhabitants; he asked me how I could think of venturing among such a lawless people, whom no stranger ever visited, and strongly recommended me to go back, as after having learned that I proposed to cross the country and return by the opposite coast, he said the journey would be attended with considerable danger, and he feared that he could not protect me. Both Demetrius, who sat at table with us, and myself told him that it would be disgraceful to go back, and if he would furnish us with an escort of some of his followers they should be properly paid, and we had no doubt but that we should get through our journey without difficulty. Finding that we were determined to proceed, he opposed our intentions no more, but said he would do the best he could for us. Being rather tired with our day's ride, we went to bed and slept soundly until soon after sunrise the next morning, when we were up and stirring. Our friend, the chief, told me that he would provide me with an escort of twenty-five well-armed men, whom I might safely rely upon. We accordingly breakfasted, and went to look over the town, which was the neatest and cleanest I had seen in Greece. There were several vessels taking in and delivering their cargoes; in fact, the little town was bustling with activity and industry. All my escort, consisting of five-and-twenty stout, active, wiry fellows, were mustered outside of the town waiting for me, and a fierce, rough, savage-looking set they were, armed to the teeth with guns, pistols, and yatagans, clad a good deal like Albanians, with kilts, jackets, hose, and red skull-caps, with long moustaches, close-shaven chins, and long hair falling down their shoulders. Most of them had several scars from wounds received in their various fights. Before starting I examined their arms to see if they were properly loaded and flinted, as it frequently happened that they were in bad order and would not go off when they pulled the trigger, and often burst, so that they did more harm to themselves than to their enemies. I found the arms, however, better than I expected, and giving some of them fresh flints and good powder and ball, they were properly loaded. I then said a few words of encouragement, which their chief, who was with me, interpreted, and told them to behave like men, and take care of me, and they would be properly paid. The men answered one and all, that there was no fear but that they would take good care of the English Milord, as they called me, and off we started in high spirits for the opposite side of the peninsula.

The view from Marathonesi was extremely beautiful. On the north the noble Mount Taygetus, with its numerous peaks, deeply-indented gorges, ravines, and its sides covered densely with woods, rose towering above the clouds. At its base there was a torrent running east and west through a valley bordered by precipitous cliffs towards the sea. On the south side of the torrent was a narrow road, and the perpendicular cliff on the south side of the road was full of caverns admirably adapted for the concealment of brigands, pirates, and bad characters of all kinds, of which so many abounded in the country. We had to pass through this gorge, and at the time there was a great number of these vagabonds concealed in the caves, waiting to intercept the merchants going backwards and forwards to and from Marathonesi. I was warned of this, and determined to drive these rascals, if there were any, out of the caverns before we passed. I accordingly told about eight or ten of the most active of my escort to cross the river and climb the base of the hill, and to fire into the caves, whilst I and the main body proceeded along the road; and as we searched each cave as we advanced, if any robbers were concealed in them, they could not escape; and if they attempted to do so, they were exposed to two fires. We were thus proceeding cautiously through the Pass, and had scarcely well entered when my men on the opposite side of the valley commenced firing into the caverns, and at first they were sharply replied to by the concealed bandits. This did not last long, for the robbers, who were not equal to us in numbers, found that they were out-manœuvred, and if they remained in the caves must all be taken, and they expected little mercy from their captors. They therefore left the caves, and fled as fast as possible, upon seeing which my people on both sides of the torrent followed them with equal speed, firing at random whilst they went forward. This rude and desultory mode of fighting was fortunately attended with no slaughter, although several flesh wounds occurred. My party being the stronger, we overtook the brigands, and made three or four prisoners. At this my people were delighted, and at the same time exasperated against them, and wished to shoot them. I at once interposed, but found that my men were so excited with the fight and subsequent pursuit, that it was extremely difficult to control them. However, after a little while I pacified them by saying that it was unmanly and unworthy to kill an enemy after he had surrendered. This did not seem to have much impression on them, as they considered the conquered their lawful prize, and that they might do what they liked with them. I ultimately quieted them by saying that I would pay a good sum per head for each of the four taken, which they willingly accepted, fearing that if they refused I should report them to their chief. I therefore released the captives, who upon their knees thanked me for their lives. Having reached the end of the Pass, and the road being tolerably clear to the end of the day's journey, I dismissed my escort with about ten shillings each, besides a capital dinner, and a good sum for the captives, with which they were pleased, and left me with many thanks and cheers, expressed very cordially in their peculiar and rough manner.

Having parted with my escort, I rode leisurely forward with Demetrius and the baggage horse, until we reached the castle of a capitano, to whom I had a letter of introduction from my excellent friend the Captain of Marathonesi. On our arrival, which was just as the sun was setting, we found the chief surrounded by numerous followers seated on stone benches fixed to the gate of the outer wall by which the castle was surrounded. The old chief and his attendants were quietly enjoying their pipes and coffee, whilst the cowherds were driving in the cattle for the night, just as described in Homer. The chief and his suite arose to meet me as I advanced and presented my letter of introduction, which he received with much dignity, and welcomed me most kindly to his castle, and certainly a wild, rough place it was. Within the outer wall was placed the keep or castle, consisting of a great square tower massively built, containing a few rough apartments with little or no furniture except a few mattresses and cushions, which were removed during the day. The space between the main tower or residence was filled with numerous outhouses and sheds for the cattle and servants and guards of the chief. After a short conversation we adjourned to the inner castle, and were shown into the principal room, about 30 feet square, with divers rows of low cushions placed all round against the walls. Directly after our arrival, a low table was brought in, and the chief, his attendants, myself, and servant Demetrius, to the number of about a dozen, sat down cross-legged to supper, which was immediately served. This consisted of stewed mutton, with vegetables, roast game, sweets, and fruits, which for the most part we ate with our fingers. Before each guest was placed a bottle full of wine, but there were no glasses on the table, so that one had no alternative but to drink out of the bottle, which was refilled several times with good red wine. After dinner we took to our pipes and coffee, during which we had a good deal of agreeable conversation, which was chiefly devoted to inquiries about England, in which they appeared to take the greatest interest. After having well supped, and having smoked and conversed as long as we felt it agreeable, one by one we dropped off to sleep upon our mattresses, armed, and without undressing, ready for a surprise at any moment. I soon fell asleep, and awoke at daybreak thoroughly refreshed. We all got up, shook ourselves, washed, and took coffee and a pipe, and then strolled about till breakfast. This castle is situated on a fine bay on the sea-shore, near the base of Taygetus. Outside the walls of the castle there is a straggling village. Besides fishing-boats, there were several others of a more suspicious character, being sharply built, and having the appearance of pirates. We got a capital breakfast with the old chief; afterwards rowed in the bay; and then taking a hearty farewell of my worthy host, started for Kitriai, the castle of the Bey, who was the nominal chief of the whole of Maina. We reached Kitriai a little before sunset, and were most kindly received by the Bey, a fine old warrior, about 5 feet 10 inches high, well built, square, and muscular, with an open, well-developed Greek countenance, thick moustache, and flowing white beard. There was something mild and gentlemanlike in his manner, very different to any other we had met. We were shown to our apartments, which were very comfortable, overlooking the sea and surrounding coasts. I spent the next day very agreeably in looking about the adjacent country, which was very pretty, but not so wild and grand as that nearer to Taygetus. Besides the Bey, there were several of his counsellors, his secretary, and one or two Greek priests, all accomplished persons, who spoke the Greek in its purity, much to my gratification and instruction. The town of Kitriai is about the size of Marathonesi, but not nearly so neat and clean, neither had it so much trade.

On the morning after, we started at daybreak, and proceeded along the coast to Calamata. This town ranked about the third in the Morea, and was then doing a good trade; but, like all the towns in the Morea, was miserable and dirty, with narrow streets, low buildings, from one to two stories high, two or three insignificant mosques and Greek churches, and a square near the centre, with several cafés, filled, as usual, with idle Turks and Albanians, smoking their pipes and sipping coffee and sherbet, whilst numerous Greek merchants and Jews were transacting their business. The bazaars were close by, and crowded with purchasers. The harbour was full of vessels and sailors of various kinds—Hydriote Greeks (amongst the best sailors on the Mediterranean), Austrian, French, and a few English vessels, besides a number of boats called caiques, and fishing vessels, altogether a busy scene. I took up my quarters at a Greek house, as rough and dirty as usual, and found my heavy baggage, servants, and janissary whom I had sent from Misitra.

Next day I started at daybreak through a wild, undulating, partially-cultivated country to visit the two Doric temples of Messenia, where Mr. Cockerell discovered the celebrated Phigalian marbles. It is said that these marbles were sent to Malta for sale, of which due notice was given by advertisement, and a certain day fixed for the auction, to which commissioners were sent by the British and other Governments. The sale, however, by some mistake, took place at Athens. The British agent, not being aware of this, remained at Malta. There was therefore no person present to bid on the part of the British Government, and they were knocked down to the Bavarian agent. When this was announced, the British Government refused to deliver up the marbles, upon the plea that the sale ought to have taken place at Malta. The matter remained in abeyance until the Peace of 1814, when the Bavarian sovereign was in England. Certain negotiations then took place, and the British Government ultimately agreed to give the marbles up to Bavaria, upon condition that a complete set of casts was sent to the British Museum. By this unfortunate affair England lost a valuable prize; and this loss was the more annoying, because, having already obtained the Elgin marbles, the Phigalian marbles—which represented an earlier stage of Greek art—would have rendered the English collection the most complete in the world. I spent the greater part of the day at Phigalia in measuring and sketching the temples, which are situated upon a hill commanding a fine view of the picturesque surrounding country.

I took up my quarters in a Greek house, got supper as soon as possible, and went to bed, with the customary annoyance of vermin. The country around was naturally rich and fertile, but left, as usual, almost in a state of nature. Undulating, with clumps of wood scattered about, it resembled a good deal the park-like scenery of England. There were numerous flocks of sheep and goats and a few herds of cattle feeding upon excellent herbage, guarded here and there by wild, shepherds, armed with guns and yatagans. The peasantry appeared very poor and ragged, but, notwithstanding, many of the women had skull-caps made of gold coins strung together, called mahmoudies, about the size of a sovereign. This composed their whole wealth; and it is a curious fact that these head-dresses were respected, and never stolen. The men also, and in some cases the women, had the girdles round their waists fastened together by large discs of silver about the size of a small plate. Thus there was a strange contrast of wealth and poverty exhibited in the same person.

We started early next day, and at noon on the second day reached the Alpheus, the most considerable river in the Morea, and which flows through the famous plain of Olympia. The river Alpheus here was 150 to 200 yards broad, but fortunately the waters were low, although we had some difficulty in crossing without swimming our horses. In the Olympian plain there were scarcely any remains worth noticing, though numerous fragments of marble, brick, and pottery are strewed about. We got some very rough quarters for the night in a village near Olympia, and next day, passing over a similar kind of country, reached Patras before sunset, at which we greatly rejoiced, as we were tolerably tired with our rough journey in the Morea. I succeeded in getting, as I thought, clean, comfortable quarters in a respectable Greek house. I was shown upstairs into a large well-furnished room, with apparently a delicious-looking bed covered with a snow-white counterpane. They gave me an excellent dinner and some good wine, after which, being very tired, I threw myself on the outside of the bed for a nap. I slept most soundly for about five or six hours, and when I awoke it was quite dark; I felt all the torments of the damned, for I was almost devoured by immense bugs. I jumped out of the bed in an instant, rubbing my face, which was covered with blood and bugs, and struck a light, when I found that the bed too was almost covered with them. I very soon caught some dozens, I may almost say hundreds; I undressed, and cleared my clothes of them, then dressed myself, and waited until daylight, being resolved not to trust to a Greek bed again, but to my own mattress.

I spent the next day at Patras quietly, and then started off on my return to Athens, taking the circuitous route by Delphos, Livadia, Eubœa, Negropont, and Marathon. We coasted along the south shore of the Gulf of Lepanto, passed our old friend the ruined convent near Vostizza, and dined at the house of a Greek merchant near, who kindly invited and treated me most hospitably. After dinner, just before I departed, our worthy host said he would show me a curiosity, and without more ado he brought out a large tub full of black snakes each about four feet long; at the word of command all these reptiles leaped out of the tub and began dancing about the table, at which I got considerably alarmed, when my host burst out laughing, and said they would do me no harm unless I irritated them, which of course I had no idea of doing. I therefore looked quietly on, and in a short time he called to them and they quickly leaped back into the tub. He said these snakes abounded in the cornfields which had just been reaped, but they were timorous, and never did any harm unless provoked; their bite was not dangerous, although for a time it occasioned a good deal of inflammation. I parted from my worthy host with many thanks, and then hired a large boat to take myself, servants, horses, and baggage across the gulf to Scalo, on the opposite side. We reached the landing-place about sunset, after a four hours' voyage, and proceeded to a caravanserai close by for the night. This was a most wretched place, in the middle of a marsh filled with large frogs and mosquitoes innumerable; and what with the loud croaking of the frogs, and the bites of the mosquitoes and bugs, it was impossible to sleep with any degree of comfort. There was a party of Turks and Albanians at the caravanserai who suffered as much as we did; being unable to sleep, they revenged themselves on the kanghè or master of the caravanserai, and belaboured him with their sticks most unmercifully, and left him without paying for anything. The poor fellow complained to me bitterly; but unfortunately I could give him no redress. As soon as it was daylight we started also; I paid him handsomely, although we had scarcely anything from him, but lived on our own provisions. He was very grateful, and hoped to see me again, but I told him that was very improbable.

As we cleared the marsh and got amongst the splendid mountain scenery, in the midst of which the classic Mount Parnassus reared his majestic head, towering above the whole, the fine fresh air completely revived us. We passed Salano, surrounded by olives, and breakfasted under the trees, much to our comfort. A little before noon on the 26th of May we reached the village of Kastri, the site of the celebrated Delphos, situated at the base of Mount Parnassus. Enhanced as it is by classical recollections, the magnificence of the surrounding scenery it is impossible to describe; let it suffice to say I shall never forget it. Of the temple and spring scarcely anything remain, and a good deal must be left to the imagination. I descended into a small plastered well, said to be the ancient Castalian spring, which is situated in a cavern at the base of the mountain; here the oracle was said to have been placed, and although I was not inspired, nevertheless I could not divest myself of the sanctity of the place and of the wonderful influence which the oracle once held over the ancient world. The surrounding scenery, with Parnassus towering above the clouds, added much to my enthusiasm, which was increased by the sight of seven or eight fine eagles soaring over our heads and screaming wildly. I was delighted with the scene, and only regretted that I had not time to ascend to the summit of Olympus.

After a halt of three or four hours, during which I dined, I started in high spirits for Livadia, where we halted for the night. This was a rather large town, with extensive cotton manufactories, seated on the side of a lofty hill overlooking the Lake Copais and the fertile plain of Thebes. We got tolerable quarters here, not forgetting the ever present bugs and mosquitoes. Next day I visited the ancient Orchomenus, the remains of which are very inconsiderable, consisting of the treasury, similar to that of Mycene, but elliptical instead of circular; also of the citadel. I also visited the ancient lion of Chæronæa, which is tolerably perfect, cut out of the solid rock. Returned to Livadia to sleep.

On the 28th May I examined the cave of Trophonius. The weather was very hot, as it was the latter end of the month. I then travelled on until, passing through the miserable modern town of Thiva, for of the ancient celebrated Thebes there are no remains worthy of notice, we took a route to the north-east, and entered a defile, in the centre of which there is the beautiful tranquil Lake of Copais about four miles long, upon which some wild swans and ducks were enjoying themselves.

The next day we were off at daybreak, and on our way examined the kolavothora or outlet at the lower end of Lake Copais, which was a rough, unfinished tunnel cut through the sandstone rock; by means of this tunnel the surplus waters of the lake are discharged into the sea, otherwise the lake would overflow and totally submerge the surrounding plain. This is a considerable work, but is not carried far enough; if it were the whole lake would be laid dry, a large quantity of valuable and fertile soil would be gained, and the surrounding district rendered much more healthy. From there we proceeded through an arid, stony, undulating country to Negropont. Upon reaching the shores of the arm of the sea which divides the island of Negropont from the main, and which is scarcely 200 yards wide at the narrowest point, although very hot, I stopped to make a sketch of the beautiful scene; having done this we proceeded across the bridge which connects the island with the mainland, to the fortified town of Negropont. The town is surrounded by a strong wall with circular towers at the angles; these towers mount some heavy guns, particularly towards the sea front, where there is a casemated battery, incapable, however, of resisting the broadside of a heavy frigate. This town at the time of my visit was garrisoned by fierce, bigoted Turks, who had the utmost contempt for Christians, nevertheless with a certain respect for Europeans, particularly the English. The chief inhabitants were Greeks, whom the Turks kept in the most perfect subjection. The interior of the town consisted of narrow streets with low, mean houses, and a few mosques, Greek churches, bazaars, cafés, &c. There was very little trade, although the place is well adapted for it, as the island is rich and fertile. We got into a Greek house and made ourselves as comfortable as we could, and one of my Greek servants for greater protection requested one of my old hats, which I gave him, and of which he was very proud, although it contrasted strangely with his picturesque Greek costume.

There was nothing to be seen in the town, and as we did not appear to be welcome guests, the sooner we left the better, so the next day I started for Marathon, having sent the heavy baggage down to Athens, which I expected to reach the same evening. This renowned battle-field interested me exceedingly; it is about two miles wide, and is bounded by the sea on the east, and the ranges of Mounts Hymettus and Pentelicus on the west. I fancied in my enthusiasm that I could discover the positions of the different commanders, their manœuvres, and in fact the whole battle from beginning to end. I remained there several hours, until near sunset, and then left with Demetrius at a hand-gallop, expecting to reach Athens in the course of a couple of hours or so. We had no guide, and thought that our most direct course lay through the wood on our right, where there was a convent, which would serve as a halting-place if necessary. Off we started, and were soon in the wood, which became thicker and thicker as we advanced; the road, which at the first was only a horse-track, now became fainter and fainter; and at last we lost it altogether. The sun had now set, and the short twilight had disappeared. After floundering about the wood for some time, our horses came to a dead stop, and neither whip nor spur would make them budge an inch farther, but they backed, snorting violently, and rearing occasionally so as almost to unseat us. We could see nothing, it being quite dark, and could not imagine why our horses had come to such a dead stand; we therefore threw the reins on their necks and allowed them to have their own way; and most fortunately we did so, for they had more sense than their masters. When left to themselves they immediately turned round and went back for some distance, then turned to the left and commenced a very rapid descent, occasionally sliding for a considerable distance; sometimes losing their footing, then slipping and sliding again, so that we could hardly keep our seats. At last we came to the bottom, cleared the wood, and reached a grass plain. It was now about midnight; we did not know where we were or where the convent was; in fact, we were lost, so that there was no alternative but to picket our horses and lie down on the grass till morning. Although very tired and hungry we had nothing to eat; we accordingly wrapped ourselves in our cloaks and lay down and slept until sunrise. When we woke we found ourselves at the bottom of a perpendicular cliff above 80 feet high; we now saw the cause of our horses' halt the night before; we were doing all we could to force them over this precipice, but by a merciful Providence their instinct saved us. We were sincerely grateful for our escape, and shuddered to think how narrowly we had missed a terrible catastrophe. We then looked for our horses, but they were nowhere to be seen; we at once concluded that they must have been stolen by robbers, who were said to be numerous in the neighbourhood. After searching everywhere for about a mile round, we at last observed them quietly feeding under a hill, and had very little difficulty in catching and mounting them. We rode to the convent which was close by, and which for the darkness we could not find the previous night. Here we got a capital breakfast, and rested, and the good monks having heard our story, congratulated us on our escape, not only from the precipice but from the robbers. We dined here also, and rode to Athens in the cool of the evening, where I took up my old quarters. Before leaving the convent I went out and took a sketch of Athens, which is about four miles distant, and the surrounding country. The view is extremely beautiful; in the front you have Athens, with its Acropolis, rising out of the plain, backed by the island of Ægina in the midst of the Ægean Sea, and beyond that by the mountains of the Morea; whilst as a foreground there is Mount Hymettus, with the convent rising out of the woods, and Pentelicus on the right. This I think is the best view of Athens.

Being again settled in Athens, I set about finishing my studies of the antiquities, and taking sketches of the neighbourhood; as the weather was very hot, being now the middle of June, I started off one morning before sunrise with Demetrius to take a sketch of the Isthmus of Corinth and the surrounding scenery, and found an excellent position which commanded the whole, near a marsh with a small lake in front. I was so absorbed in my sketch that I did not think of the malaria, which was very fatal in that place. The sketch occupied about three hours, and although I had taken some breakfast in the open air, and Demetrius the same, we both felt very ill and returned to Athens, and sent directly for the doctor, when I found that I had the malaria fever in the form of an ague. Demetrius was much worse, and towards night he became delirious, and remained so for some time. I rallied in three or four days, but remained very weak. Demetrius was confined to the house for above a fortnight. Although I could go about, still the fever clung to me more or less, and I could do very little. At this time, a singular person, a Mr. Scott, a friend of Lord Byron's, called upon me and offered every attention; he was a shy, kind, and well-informed man, living quite shut up by himself, seldom stirring out till dusk, when he was to be seen galloping round the walls of Athens quite alone. There were also at Athens, Mr. Hodgetts Foley, Mr. Beaumont, Mr. Waddington (the late) Dean of Durham, and the Rev. Mr. Hanbury. The latter was a tall, fine-looking man, of fair complexion, with long light brown hair hanging over his shoulders, and a long beard, altogether a very striking figure; he was perpetually thinking of firearms; so that it might well be said, that instead of being a man of peace, he was a man of war.

I was now introduced to the Maid of Athens, so celebrated by Lord Byron; she certainly was a handsome and elegant young woman, about twenty, with a very pleasing manner and lively and intelligent in conversation. She had a younger sister, a very agreeable person also, but not so handsome. They lived with their parents, who made their house a very pleasant resort for strangers. By this time, the beginning of July, I was a great deal better, and was recommended to take a sea voyage for change of air. The above-named gentlemen had hired an Hydriote brig to take a month's cruise amongst the islands, and I agreed to join them. We accordingly started about the end of the first week in July, the weather still very hot. We left the Piræus in the morning, and as there was very little wind we were becalmed off Cape Colonna, which forms the eastern promontory of the Gulf of Athens. Here we cast anchor, lowered a boat, and went on shore to visit the celebrated temple.

As we landed we were much struck by the appearance of the beautiful ruin, perched upon the summit of the promontory in solitary grandeur, and overlooking the surrounding coasts and islands, altogether forming a very imposing and charming scene.

Whilst in the midst of silent admiration at this beautiful temple, all of a sudden Hanbury cried out, "There goes an old hawk," and fired at it immediately. This dispelled the charm, and we all burst out laughing. Having finished our exploration, we returned on board our vessel to dinner. A light breeze soon afterwards sprang up; we lay-to for a short time off the island of Syra, and the following morning got under way for Paros, a fine land-locked port, which we reached early in the day, and landed after breakfast. We then proceeded to explore the quarries from which the celebrated marble was extracted.

These quarries are situated near the top of a hill, about a mile distant from the port, with a steep broken road for an approach. The ancients, instead of opening out a good face, so that the quarries might be worked to any extent with advantage, excavated caverns in the side of the hill, and having arrived at the good sound rock, cut it out by wedges and picks in such masses as they required, so that the inside of the quarry presented a long gallery from whence the stone had been taken, worked in a very regular manner in steps one above the other, without any appearance of waste; in fact, the rock is so solid that when worked by pick and wedge no blasting is necessary; indeed, if blasting were adopted, it would entail considerable waste. At the same time, it appeared to me that if the solid rock was bared from the surface and a good length of face opened out, the pick-and-wedge system might be worked to almost any extent, with much greater advantage; and if inclined planes with railways were made to the port, and a proper embarking jetty, with cranes, carried out into deep water, so as to enable large vessels to come alongside at all times to receive their cargoes, this fine marble might be quarried and exported at a very moderate cost, infinitely below the price now charged for Carrara marble. As the Parian marble is of a beautiful white cream-colour, almost free from veins, and of an even, close texture, the sale would be immense and yield an excellent profit; and considering the enormous price of fine marble adapted for sculpture, amounting to one and two guineas and upwards per foot cube, according to the size of the block and fineness of the quality, it is astonishing that the value of these quarries has not been recognized before, and that capital has not been forthcoming to work them. The harbour is spacious, with ample depth, and well protected against all winds.

The following day we devoted to the examination of the celebrated stalactite caverns of Antiparos, a small island on the north-west of Paros, and only separated from it by a narrow strait. We accordingly went there after breakfast, and being accompanied by a sufficient number of guides with torches and wax lights, commenced the exploration. We entered by a lofty arch, and after proceeding some distance came to a magnificent chamber, from the roof of which depended the most magnificent stalactites, many of them eight and ten feet long, reflecting the light of our torches like so many diamonds. From this hall issued several galleries, some of which had not then been explored; we pursued our way through all which had been investigated, and returned, after a fatiguing walk, to the surface. This island, like that of Paros, is composed of fine, close marble, which might be developed in the same manner with considerable profit. We next set sail for Naxos; and here the whole party started in a native open boat to visit some ruins, which turned out to be not in the least worth seeing. We were first becalmed for many hours, and on our return were overtaken by a violent storm, and only escaped, as the entire coast was one wall of almost perpendicular rock, by one of the sailors accidentally remembering the vicinity of a small creek, into which the captain, a brave and skilful man, managed to steer us through the breakers. We ultimately returned to Naxos, after an absence of thirty-nine hours, by far the greater part of which we passed without food, as the ruins being only eight miles from Naxos, it was considered a mere morning's sail. The day after our return we invited the English Consul, a Greek, and several of the principal inhabitants to dinner, and had a jovial party, for the Greeks like good cheer, and are certainly not water drinkers. The wine of Naxos, of which a considerable quantity is made, both for home and foreign consumption, is excellent; it is chiefly white wine, resembling a good deal brown sherry, and if well made is equally good. Besides wine they export Velança bark, wool, figs, currants, and other articles.

Having nothing further to see, we set sail from Naxos, with a lively and lasting recollection of our visit to the *celebrated* antiquities, and then steered for Scio, the ancient Chios, passing amongst numerous islands, many of them very beautiful; some were covered with woods and verdure to the water's edge, others consisted of bold barren rocks rising perpendicularly from the water to a great height, and terminating in lofty peaks. The scene was ever changing, and we enjoyed views of wood, water, plain, and mountain, combined in the most charming manner, and of endless variety.

It was amusing also to see how well the Greek sailors managed our ship. At this time of the year strong northerly winds prevail during the day, against which we had to heat. In tacking about and manœuvring the vessel they showed considerable skill and activity.

Mitylene, where we next arrived, is a poor little place, surrounded by steep, lofty hills covered with wood almost to their summits. There is some trade of wine and oil, and wood and bark. I was much amused by a fight between my man Demetrius and a Turk who insulted him; fortunately they had neither firearms nor knives, but they used their hands in the most clumsy manner possible, and after a little time I parted them before they had done any mischief.

We spent about three days in the island, which is well worthy of a visit on account of its rich and beautiful scenery, and then returned to Scio, where our party separated.

I took lodgings, and rested a few days, for I was still very weak, and liable to attacks occasionally from the malaria fever. I had several enjoyable rides about this island. A ridge of lofty hills, extending from north to south, added greatly to the picturesqueness of the scenery. The soil is extremely rich and fertile, and a large quantity of excellent wine, besides oil, silk, fruits, grain, wood, &c., is produced. A great trade is carried on with different parts of the Mediterranean, particularly Constantinople and Smyrna. Perhaps, with the exception of Sicily, it is the richest island in the Mediterranean, and has a great number of wealthy merchants and proprietors. The port, which is formed by a small creek, defended by two stone piers, contained numerous vessels, some of considerable size, and bearing the flags of most European nations. The town was better and more substantially built than any I had hitherto seen, and there were some good bazaars, rich with the wares of the East and with European articles. The square in the centre was surrounded by handsome cafés and houses, and also one or two large mosques and Greek churches. The island was governed by a Turkish officer of rank, with a small body of janissaries and cavalry. The following year, 1821, this island suffered terribly during the Greek revolution, when the inhabitants attempted to overcome the Turks, who were too strong for them, and showed them no mercy; many were massacred, and the island was devastated. Having taken a sketch of the port, and seen everything of interest, I hired a Greek open boat, and embarked with Demetrius and all our baggage for Smyrna, where I got tolerably comfortable lodgings in a Greek house.

Smyrna is situated at the south-eastern extremity of the gulf of that name, and extends northward along the coast for about a mile and a half. The shore of the bay in front of the city is lined with a quay-wall, formed partly of wood and partly of stone, with small wooden jetties projecting from it at irregular intervals. Between this quay-wall and the houses is a road, which runs nearly along the whole sea-front. From this road innumerable narrow, crooked streets lead to the different parts of the town; many of these are so narrow that two horses can scarcely go abreast. The buildings for the most part consist of two and three stories. The bazaars were numerous, extensive, and rich, filled with all the commodities of the East, such as rick silks, plain and brocaded, shawls, jewellery, arms, leather articles, and pipes of all kinds, rough iron and pottery ware, besides tobacco, fruit, and provisions. All these bazaars were laid out in different sections, with a particular trade assigned to each, and were kept in tolerable order; the Jews and Armenians were the bankers, and had their counters or stalls, with offices behind, where they transacted business to a large amount. All the Consuls' houses were in front of the quay, and were substantial stone buildings of considerable extent, and, with the flags of the respective nations which they represented, had a very imposing appearance. In front of the quay were moored at times several hundreds of vessels of all nations, and the immense number of boats and barges passing continually between them and the shore, formed a most busy and lively scene. There were several fine and extensive mosques, which no Christian was ever permitted to enter; also many Greek churches, with their gorgeously-bedaubed paintings representing innumerable saints. I attended the Greek service at different times, but generally came away anything but satisfied; their strong nasal tone in saying prayers, and their indifferent music and singing, greatly disappointed me. Towards the land the city was surrounded by a high wall, and at each gate was a strong guard. At the south end there was a kind of citadel on a hill, in which the Governor resided, and about half a mile from the city were the cemeteries, deeply embosomed in cypress trees. Many of the tombs were very pretty, of white, well-sculptured marble, with inscriptions from the Koran engraved upon them, and they were tastefully ornamented with garlands of flowers.

The exports from Smyrna are various, and its trade with Europe is very extensive. The articles of export are figs, raisins, silk, oil, bark, grain, &c.; figs and raisins, however, are the chief, and these are exported to a vast amount, and certainly they are the finest in the world. There are two crops, those which are ripe about the end of July, which are mostly consumed in the country and neighbourhood; they are most delicious, and extremely wholesome. The second crop is ripe about the end of September. These are gathered, dipped in water, then carefully packed in boxes and exported, chiefly to England, France, and the north of Europe, and by the time of their arrival they are covered with a rich coating of crystallized sugar, and are fit for use. At the season of shipping the figs, which I witnessed, the greatest activity is visible everywhere—the producers selling, the merchants buying, the packing-case makers splitting the wood and making up the boxes—the packers carefully stowing the figs—the men loading them into lighters, by which they are transported to the fast-sailing vessels waiting for them, which are generally clipper schooners of about 120 to 170 tons—the Consuls' offices besieged with numerous applicants for their clearance papers—the whole combined to form a most active and industrious scene, not omitting the numberless dinners and social parties at the different hotels and coffee-houses, which are thronged with natives of all nations. I walked about and witnessed this busy scene with much delight and satisfaction. I must not omit the melons, both sweet and water, and the grapes, which were the most delicious I ever tasted. The finest raisins are those called sultanas grown at Scala Nova, a small port to the south-west of Smyrna, which I visited. The grapes from which they are made are especially delicious. The town is a poor place as regards the buildings; but, notwithstanding, there is a considerable amount of wealth. When I was there the town had recently been visited by a severe attack of plague, and many of the houses were shut up, all the inmates having died, and, having no heirs, the property belonged to the Government.

Whilst I was at Smyrna the British Consul was Mr. Werry, a fine old gentleman of about sixty-five, with considerable vigour and talent, just the man for the place. All British subjects were under his protection, amongst whom were a considerable number from the Ionian Islands, who were the most troublesome and daring vagabonds, committing all Sorts of crimes, and mixed up with every row in the place; these rows daily occurred, and bloodshed and murder not unfrequently was the consequence. Old Werry was the arbiter, and delivered his judgments with great impartiality and justice. The Turks also were very troublesome, and frequently attempted acts of great barbarity. Werry, however, was equal to the occasion, and at such times went direct to the Governor and claimed an audience; he was at once admitted and seated at the head of the divan; then he claimed protection for his British subjects; and whenever the Governor demurred, Werry drew his sword, threw his hat down on the floor, and threatened him with a British frigate to blow up the town before his eyes. The Governor would try to pacify him by offering pipes and coffee, and soothing words, which Werry treated with contempt, and nothing would do but releasing the prisoners, which was soon effected. Then Werry strode off in triumph, scarcely condescending to return the Governor's salute. The Turks looked upon him as a madman; but they have always entertained the greatest respect for him, and allowed him to have his way, invariably calling him the *mad Consul*. Nevertheless, he was a most kind and worthy man, greatly respected and liked by his countrymen, the other consuls, and by the whole population. He received me with the greatest kindness and hospitality. I was a frequent visitor at his house, both in town and at his country seat, Bridjar, about nine miles distant to the south-east.

Whilst at Smyrna I made acquaintance with my countryman, Mr. (afterwards Professor) Donaldson, a M. Parke, and another French architect, who had just returned to Smyrna after a very successful exploration of some of the most remarkable ruins of the ancient cities of Asia Minor, and their portfolios were filled with drawings of these most interesting relics of antiquity. I passed several happy days in their company, and Donaldson most kindly made out for me a map by which I might be enabled to visit these remarkable ruins; I therefore determined to follow it out as nearly as possible without delay, and took leave of them with many thanks.

On the next day, having obtained the usual firman, I started, in company with my faithful servant Demetrius, to explore some of the ruins of Ionia, with a Turkish surgee, or guide, a janissary, and a couple of baggage horses, leaving my heavy baggage at Smyrna. After passing through a wild, rich, and almost uncultivated country, we reached Aiasolok, on the Meander, near the ruins of Ephesus, in the evening, and took up our quarters in a Turkish caravanserai. Demetrius soon prepared a good supper, laid our mattresses, and we both slept soundly until the morning. Every inmate was up transacting his business soon after daybreak, and I at once sallied forth to examine the ruins of Ephesus. This celebrated city was situated on the left bank of the Meander, at the base of an elevated ridge of hills. The only remains which we could distinctly make out was the amphitheatre, of the rest little could be discovered; but numerous fragments of columns, cornices, blocks of masonry, pottery, brickwork, &c., lay scattered about. I made a sketch of the whole from the best place I could find. The ruins of Ephesus formed the foreground, with the Meander and its numerous windings and the marshy valley flanked by the mountains on each side, whilst in the background was the miserable town of Aiasolok, with its fortress and mosque on a conical hill above it, backed by a wild range of mountains of considerable elevation, which completed the view. The scene was most picturesque, replete with solitary grandeur and desolation. The stillness was something remarkable; nothing was seen moving, except at rare intervals a long line of camels laden with merchandise, led by their guide, smoking his pipe, and mounted on a donkey; or here and there a solitary heron fishing amongst the reeds of the Meander, and flocks of wild ducks, which took to flight at the slightest disturbance. It was singular and melancholy to remark the ruins of the three most celebrated of the Churches of the world before me, namely, those of the ancient Greeks, Romans, and Christians, dominated over by the Mahometan mosque. I remained alone lost in contemplating this solemn and desolate scene for two or three hours, reflecting upon the vicissitudes and instability of human grandeur, and walked slowly back to the caravanserai at Aiasolok, where I took my breakfast, and then started on my journey. We slowly wound our way through the Pass between the mountains at the back of Aiasolok, and descended into the plain of the ancient Magnesia, the site of one of the seven Churches, through which two streams flow sluggishly to the sea, and reached the ruins about a couple of hours before sunset. I picketed my horse here whilst I examined the ruins, and sent Demetrius forward with the other horses and baggage to a village, where I proposed to pass the night, ordering Demetrius to have supper ready by my arrival.

The principal building is the hippodrome, which was tolerably perfect, and I took some pains to measure it accurately, which occupied me about two hours. There was nothing else worthy of remark, although there were numerous fragments of buildings scattered around. I met a Turkish Aga, handsomely accoutred and well mounted, attended by a servant, who watched my movements with considerable attention, and politely accosted me. Having finished my work, I mounted and galloped off to the place of my rendezvous as quickly as possible. Passing through the adjacent forest I observed something like hammocks slung to the trees about 10 or 12 feet from the ground; these, I learned, were the sleeping places of the shepherds who attended their flocks in the neighbourhood, and as there were no habitations near, they slept in these hammocks to keep themselves out of the influence of the malaria, which is very fatal in these parts.

About sunset I reached my destination, which was situated at the head of a valley surrounded by lofty mountains. There was a cluster of miserable mud huts, but no place fit to sleep in, so Demetrius was obliged to take possession of an open shed, with a bare mud hut adjoining, in which he placed the baggage, spread my mattress in the shed, and laid out the table attached to my canteen in the open air in front, upon which he had got ready a tolerably good supper. This canteen, it should be mentioned, was a very smart affair; all the utensils were plated on the outside and gilt on the inside, and, being quite new, had a very stylish and attractive appearance: it was much finer than I wanted, and had been sent out from England by my brother. No sooner had I sat down to supper, than I was surrounded by a considerable number of Kurds, wild-looking fellows, armed to the teeth, who had come from their own country with their flocks, for the rich pasture which this country afforded. Their encampment was close by. They appeared very friendly, and admired my turn-out with great satisfaction, thinking, no doubt, that it was real gold and silver. I treated them very courteously, gave them wine, coffee, and tobacco, for which they appeared very thankful, and we parted, as I thought, excellent friends. It was a fine summer evening, and soon after became dark. The Kurds retired with many thanks, and being tired I lay down for the night on my mattress without taking off my clothes, whilst Demetrius and the other servants were in front, and the horses picketed before them, having only their girths slackened, so that we were ready for a start at any alarm. I examined my pistols, my men did the same, and we all retired; but I was very restless and feverish, and could not sleep, although excessively weary. I was continually roused by the barking of the great mastiffs which were kept by the shepherds to drive off the wolves and panthers in the vicinity. At last, towards midnight, I fell fast asleep. I awoke all of a sudden to find two fierce-looking Kurds by my side, one with a pistol close to my head, the other flaring a lighted torch in my face, with one hand thrust under my pillow, to steal my purse and valuables. I sat up directly and secured my pistols; seeing this, the fellow with the pistol pulled the trigger, but fortunately it only flashed in the pan. I at once fired. I think I hit him somewhere, for he uttered a loud cry, and instantly took to flight with his companion. My men awoke at the noise, jumped up directly, and fired also. By this time we were surrounded by the Kurds, and had a hand-to-hand fight for a few minutes, when I called to my men to mount and gallop off as quickly as possible. I did the same, and off we went at full speed, leaving the baggage behind. We could do no other, for by this time the whole tribe was on the alert; they came running to the help of their companions, and commenced firing at us. Fortunately it was hardly daylight, and they could not see us clearly. In order to avoid the shots that whistled past us, we crouched down upon our saddles, and soon got out of their reach. They did not follow us, either being deterred by the stout resistance we had made, or being satisfied with the booty which they had got. To have remained longer would have been madness, for we should soon have been overpowered, and probably all murdered, as there were at least thirty or forty of them, and we were only five. We never pulled bridle until we had made about six or eight miles; it was then broad daylight, and the sun was just rising above the horizon; we continued, however, at a smart pace, passing through a rich and partially cultivated valley, until, about seven o'clock in the morning, we reached a small town, where the Aga, or Turkish Governor of the district resided. Here we stopped at the caravanserai, tolerably tired with our morning's ride of about seventeen miles, and heartily thankful for our providential escape. After a good wash, and breakfast, I sent my janissary, a gallant fellow, to demand an audience of the Aga, which he granted at midday. I went accordingly, showed my firman, and then related my story by means of an interpreter. The Aga was a fine-looking man, about forty, surrounded by his divan and a number of well-armed Tartars and guards. He listened to my tale with the utmost gravity and attention, and then burst out into a violent rage at the indignity which I had suffered, and said that such an insult to an English gentleman had never occurred before in his district, and he would lose no time in bringing the offenders to justice. He was most particular in inquiring as to the place where the offence had occurred, and the number and description of my assailants, and then ordered a strong body of well-armed and well-mounted Tartars to the place, with strict orders to bring them to him immediately; in the meantime he ordered coffee and pipes to be served, and offered me any money I required. He also asked me where I was going. I told him that I was going southwards for two or three days. He then said that upon my return he should be happy to see me again, when he would have all the culprits in custody for me, and if I could identify them he would have their heads taken off directly. I took my leave, with many thanks for his courtesy; to which he replied, with the greatest civility, that he was too happy to be of any service to an Englishman. I left soon afterwards, and crossed another mountain ridge, passing through a wild, desolate country, and descended into a fertile valley, through which a small river wound its tortuous course. On the way I examined the ruins of the ancient Priene, seated on a hill on the right side of the valley. Here I saw considerable remains of some temples of the Ionic order, besides the relics of numerous other buildings, which it was very difficult to define. This city was surrounded by a strong wall of massive masonry, considerable remains of which were visible. We crossed the river by a ford, and took up our quarters for the night amongst the ruins of Miletus. There were only a few scattered wretched hovels in the vicinity, but none of them fit for our accommodation; I therefore preferred the open air, and determined to bivouac amongst the ruins; as I had already provided myself and my people with mattresses and coverlets, and a good supply of provisions, I felt myself comfortable and independent. I was roused about daybreak by something tugging hard at my pillow; upon jumping up I descried a large wolf close by, with several more near him. I immediately laid hold of a stout stick, and dealt him a heavy blow on the head. This aroused Demetrius and the servants, who fired several shots at the other wolves, and they all scampered off as quickly as possible, so that we had no further molestation.

Early in the morning I set about examining the ruins, of which there were but few; the principal was the theatre, but even of this, except some massive walls, there was little remaining. I made a sketch, with the Lake Bofi and the mountains behind it for a background, the whole forming a very beautiful picture. We then went on to Yirondi, which we reached about four o'clock in the afternoon, and got comfortable quarters in the caravanserai. The next morning I proceeded to examine the ruins of the celebrated Temple of Apollo Didymæus. This is one of the largest and most magnificent specimens of the Ionic order, and well worthy of the attention of every traveller who takes an interest in architecture. I was much pleased with and instructed by it. Of the front columns several were still standing, and numerous fragmentary blocks of pillars, entablature, pediment, and substructure lay scattered all around, the whole being of the most elaborate workmanship. When entire it must have been a very magnificent work of art, as the remains evidently show.

After finishing my examination, I determined to return to Smyrna as quickly as possible, having much to do before my tour was completed. I accordingly started soon after midday, although the weather was very hot, and the next day I sent my janissary to request an audience of the Aga, which he granted me at midday. He then said that he had captured all the Kurds, and ordered them to be brought from the prison. Accordingly the whole of them, amounting to about thirty in number, were produced, and marched before me slowly, when the Aga asked me if I could recognize any of them, in which case he would take care that justice should be administered. I was pretty well aware what this justice would be, namely, that their heads would be taken off if I recognized them. I certainly remembered several of their faces, but having been robbed only of a few articles, worth about eighty pounds, I did not consider that the punishment of death could be put in comparison with my loss, although it is true they would have murdered me if they could. I therefore resolved to save them, as they had already suffered sufficiently by being confined in a miserable dungeon with scarcely any food for four days. They were accordingly discharged, having been, as I understood, pretty well punished, in addition to their imprisonment, by heavy fines, and perhaps sundry stripes into the bargain, so that it was not likely that they would attack an Englishman again. The Tartars had recovered several of my lost things; for this I made them a handsome acknowledgment, with which they were much pleased, and gave me many thanks. I also rendered my best thanks to the Aga, and we parted upon the most amicable terms.

Having returned to Smyrna, I there met my friend Scott, whom I had previously known at Athens, and we resolved to proceed at once to Constantinople by land. The first night we reached the modern Magnesia, a large town situated on a plain, surrounded by rude crenellated walls, inhabited chiefly by a Turkish population, and containing nothing worthy of remark. The following morning we proceeded through an open undulating country, very fertile, but, as usual, only partially cultivated. Here and there were extensive plains, some only tenanted by shepherds tending their large flocks of goats, cattle, and sheep, others growing cotton, maize, corn, beans, and tobacco, occasionally interspersed with vineyards; the inhabitants looked poor and miserable, and the villages and towns wore the same appearance.

After a long, hard day's ride of eighteen hours, we came in sight of Broussa, the ancient capital of the Turks. The surrounding country was rich and beautiful, and covered with luxurious gardens, intermixed with comfortable villas and houses; the city, with its numerous mosques and towering minarets, lay in the foreground; whilst the magnificent Mount Olympus, above 7000 feet high, rearing his head above the clouds, formed a noble background. When within a mile of the city I felt very tired and thirsty. Just then a countryman approached, and offered us some bunches of magnificent grapes, which I accepted with much gratification, and gave him a handsome present. I devoured them voraciously. We reached the caravanserai at sunset, and as there were a great number of travellers, we got but poor accommodation. I awoke about three in the morning with a violent diarrhœa, which continually increased, until I became so exhausted that I could scarcely move. It turned out to be dysentery, and my friends considered that it must end fatally. Before leaving England I had consulted my friend, the late distinguished Dr. James Johnson, as to what was to be done under similar circumstances, and he recommended large doses of calomel as the only and best means of cure. I had provided myself with this, and took it accordingly, with the assistance of which I rallied considerably, and we then determined to proceed at once to Mondania, on the shores of the Sea of Marmora, where we should find much better accommodation. Off, therefore, we went, without being able to see anything further of the beauties of Broussa. I was much fatigued with the journey, and immediately after our arrival took to my bed. Here the dysentery came on more violently than ever. I again had recourse to a large dose of calomel, and ordered Demetrius to get some chicken broth as soon as possible. This the good fellow got ready without delay. I swallowed as much as I could, and then fell back on my mattress perfectly exhausted. I was delirious for four or five hours; but at the end of that time I broke out into a violent perspiration; the calomel had done its work effectually, and I became conscious and tranquil, although very weak. The disease, however, had been arrested, and I recovered rapidly.

I reached Constantinople at last by sea, and after a day or two of repose I regained my strength and sallied forth with a janissary to examine the Moslem capital. I first paid a visit to our ambassador, Mr. Frere. He received me most kindly, and asked me to dinner the next day. The entertainment was sumptuous, and I spent a delightful evening. I was afterwards frequently invited to this most hospitable house, and always received the same kind attention, ever returning from it both amused and instructed. Mr. Frere was a very accomplished, unassuming gentleman, ever ready to protect his countrymen; he discharged his onerous duties in such a manner as to gain the respect of the Turks (which was very difficult in those days), as well as that of the *corps diplomatique* and foreigners of every nation, and the respect and affection of his own countrymen. I also frequently visited the hospitable house of our worthy and most excellent Consul, General Cartwright. He was a man above the common order, and no one was better calculated to deal with the Turks; frank, open, courageous, honest, and decided, understanding thoroughly the people he had to deal with, and never flinching from his duties, firmly upholding his countrymen when right, but no less inflexible in refusing his support when wrong. He was hospitable and sociable to a degree, yet withal never neglected the least of his duties, and was universally respected and beloved.

Constantinople has been often described, and is now so well known that I need not weary my reader with any detailed description. But as I was there at a time when the reforms of the Sultan Mahmoud had not long commenced; when the janissaries were still in existence; when the old Turkish bigotry, insolence, and fanaticism was at its height; when the fierce vigour which had formerly carried out its conquests was nearly extinct, and European ideas and civilization had not yet dawned; when, in short, the Government was most corrupt, and the whole Turkish population sunk in indolence and sensuality, a few rough notes of what I then saw may perhaps not be unacceptable.

Amongst the sights during my short stay was the marriage of one of the Sultan's daughters, which was made the occasion of great fêtes, and amongst others a grand tournament or display of djerid. This consisted of a number of horsemen extremely well mounted, each armed with a short blunted dart, which they hurled at one another, and those who received the greatest number of hits were declared to be losers The game was very exciting, and the display of horsemanship and the activity and skill of the riders were worthy of the highest admiration.

There was also another sight equally new and beautiful, the sultan proceeding in state to the mosque of St. Sophia during the Bairam. It was a very imposing spectacle, all the actors in it being clothed in the ancient picturesque costume—long flowing robes of endless variety of brilliant colours, furs, and turbans of every shape, those worn by the Sultan and his great officers mounted in gold and silver and studded with precious stones. The Sultan's body guard consisted of about one hundred fine-looking men wearing dresses of the most brilliant colours, richly ornamented with gold embroidery, and having on their heads helmets of finely polished brass, surmounted by a crescent of the same metal, being three to four feet long, on the top of which were fixed plumes of the finest white ostrich feathers, flowing on all sides. The Sultan, clothed in a splendid dark sable pelisse, with a green turban ornamented with the plumes of a bird of paradise, set in a most costly diamond aigrette, rode in the midst of them, mounted on a magnificent pure white Arabian charger, covered with housings richly embroidered with gold. He was a very handsome, stern, dignified-looking man, about fifty, with sharp, piercing black eyes, moustache, and beard; his nose short and well developed, and a medium-sized, well-defined mouth; his whole appearance was very stately, grave, and solemn, expressing majesty, firmness, and courage. As the procession moved slowly, silently, and majestically along, through a dense mass of spectators, everyone was impressed with awe and admiration, and certainly it was one of the most unique and finest sights of the kind I had ever witnessed. I afterwards saw a review of several thousand of the Ottoman troops, consisting of infantry, cavalry, and artillery. These, though inferior in discipline and mechanical contrivances to European armies, nevertheless expressed a degree of fierce and enthusiastic courage, mixed with a thorough contempt of Christians, which inspired them with the confidence that they could overwhelm all their enemies, and plant the crescent over the cross with triumphant success.

Constantinople proper was inhabited chiefly by Turks imbued with the most bigoted hatred and contempt for Christians, and it was attended with not a little danger to go amongst them. As I walked boldly along I was frequently saluted by fierce scowls and curses loud and deep, and sometimes with a small shower of stones, which but for the presence of the janissary, whom they feared, might have been attended with serious consequences. These janissaries, a certain number of whom were attached to every embassy for their protection, were called pig-keepers by their comrades, who considered them as an inferior class. The streets for the most part were narrow, crooked, and dirty to a degree; here and there was a fine stone mansion, inhabited by some Turkish grandee, but this presented only a dead stone wall to the street. The population of the city generally looked poor, miserable, and oppressed. I tried to visit the mosques, but the fanaticism of the Ottomans was so great at that time that no Christian was permitted to enter, and I was more than once, when trying to penetrate the outer court of one of them, driven away by a shower of stones. On the arrival of a fresh ambassador from any Christian court a special firman, allowing him to visit the mosques, was issued, when the different strangers in Constantinople at the time were allowed to accompany him. The ambassador and suite on these occasions were always accompanied by a strong guard of janissaries, for protection. It happened that a little time previous to my visit the new Russian ambassador, Count Stroganoff, arrived, and obtained a special firman to visit the mosques, when he was accompanied by two or three hundred of his own countrymen and other strangers, attended by a strong guard of janissaries. Whilst in one of these edifices, a Frenchman, it is said, spat upon the pavement; immediately a cry of horror was uttered by the priests and assistants, the mosque was defiled; the alarm was raised, which spread like wildfire, and the place was quickly surrounded by a vast multitude of angry Turks, many of whom rushed into the mosque, shouting for vengeance on the infidels who had desecrated their temple, and attacking them with the most savage ferocity. The ambassador and his followers were obliged to defend themselves and fight for their lives. The doors were shut, and there they were obliged to remain until the Government sent another strong body of janissaries to rescue them from their perilous situation, which they did with the greatest difficulty; and whilst they were escorted back through the streets, they were assailed with yells, curses, and missiles of every description; many were seriously hurt, and they had some difficulty in escaping with their lives.

At Bouyukderé I spent several days at a comfortable Greek hotel, and was much amused at seeing two or three grave Turks enter the hotel one evening and engage a private room. For a time everything went on quietly. Then came a most tremendous noise, and shouts of all kinds; in fact, they were getting very drunk. Then all was silent, and I imagined they were senseless; but shortly after, to my great surprise, they sallied forth and entered their boat apparently quite sober. I afterwards asked the Greek waiter whether the water they had drunk had rendered them so noisy. He replied, laughing, "No; they had two bottles of rum apiece." He said, farther, that it was their constant practice, especially during the Ramadan, to go prowling about in search of some quiet place where they might indulge. Knowing this, some of the proprietors of Greek inns, after closing the front door, left a private one open, and allowed none but Turks to enter, not even appearing themselves when they did. The Turks, seeing the coast clear, would enter, and presently find a room in which was a table covered with good things, both to eat and drink, whereupon they helped themselves liberally, and after leaving a handsome donation on the table, walked out as quietly as they came in, satisfied with having cheated their religion and satisfied their appetites without anyone being the wiser.

Constantinople is supplied with water from five reservoirs, situated on the western base of the Little Balkan Mountains, and communicating with the city by stone conduits. They were originally built by the Byzantine emperors, and their preservation was enforced by repeated imperial edicts, some of which are still preserved in the archives of Constantinople; a heavy fine was imposed, amounting to a pound of gold for every ounce of water taken from these reservoirs by any individual without express permission. The water for Pera is supplied from the southern reservoirs, near Bagtche, by means of a conduit, upon which are placed at certain intervals hollow stone columns, called *sous terrasi*, which rise to the level of the source from which the water is taken; the water in the conduit rises up these pillars nearly to the same level, and thus acts as a safety valve, overcoming the friction of the water in the conduit, forcing it forward to the level required at Pera, and so relieving the pressure upon the upper surface of the conduit, which otherwise would be blown up through such a long line, had it not a vent to relieve the pressure. This ingenious idea was the invention of the Turks, and was adopted for the purpose of saving the expense of the lofty stone aqueducts used by the Romans and their successors. Some expedient of this kind became inevitable, as they had no iron pipes large or strong enough. In modern times a similar device, called the stand-pipe, is occasionally employed to relieve the pumping engines; but such is not necessary along the line of the conduit, because the material of the conduit is made of cast iron of such dimensions that it is strong enough to more than amply resist the pressure of the highest column of water which it may be necessary to employ.

The whole of the sides of the mountains where the reservoirs are placed used formerly to be covered with dense woods, and when I visited the reservoirs in October, 1820, nothing could be more beautiful; but after the massacre of the janissaries, about a year and a half subsequent to my visit, a remnant took possession of the forest, and committed intolerable acts of brigandage, until the Ottoman Government, thoroughly roused, surrounded the entire forest with troops, and destroyed it and its inmates together. There is also a fine aqueduct, but there can be very little doubt that proper means are not taken during the rainy season to preserve and store the water nature then places at their disposal. If this were done, the serious evil of drought would be avoided; and means are now being taken to ensure a more constant and better supply. This scarcity of water, for the same reason, has been felt in London and other great cities. We shall continue to suffer from these droughts until men come to understand that only a certain and known quantity of water falls upon the earth, and that at certain periods; and that it is necessary to store the surplus waters to supply the deficiency of the dry season. This great truth is fortunately now beginning to be fully appreciated in the civilized world, and it is to be hoped that we shall no longer hear the cry of scarcity of water.

On leaving Constantinople I took passage in a native vessel for Alexandria. Passing down the Hellespont, at its narrowest point I remarked the positions selected by Xerxes for erecting his famous double bridge, or rather two bridges, one taking a north-west and the other a north-easterly direction. An ancient author, Polyænus, says "that they connected together a vast number of ships of different kinds—some long vessels of fifty oars, others three-banked galleys—to the number of three hundred and sixty, on the side towards the Euxine Sea, and three hundred and thirteen on that of the Hellespont. The former of these were placed transversely; but the latter, to diminish the strain upon the cables, in the direction of the current. When these vessels were firmly fastened to each other, they were secured on each side by anchors of great strength—on the upper side, towards the winds which set in from the Euxine; on the lower side, towards the Ægean Sea, on account of the south and south-east winds. They left, however, openings in three places sufficient to afford a passage for the light vessels which might have occasion to sail into the Euxine or from it. Having done this, they extended cables from the shore, stretching them upon large capstans of wood. For this purpose they did not employ a number of separate cables, but united two of white flax with two of byblos. These were alike in thickness, and apparently so in goodness; but those of flax were in proportion much the more solid, weighing not less than a talent to every cubit. When the passage was thus secured, they sawed out rafters of wood, making their length equal to the space required for the bridge. These they laid in order across upon the extended cables, and then bound them fast together. They next brought unwrought wood, which they placed very regularly upon the rafters. Over all they threw earth, which they raised to a proper height, and finished the whole by a fence on each side, that the horses and other beasts of burthen might not be terrified by looking down upon the sea."

This bridge of boats, for such it may be properly called, resembled materially those constructed by the moderns. The addition of the suspension cables, which connected both ends of the bridge with the shore, must have contributed greatly to its strength. The floating bridges used in modern warfare, however, have this advantage over those constructed by the ancients (of which that devised by Xerxes was a very favourable specimen); the boats or pontoons upon which such bridges are now erected are specially constructed for the purpose, and an army on the march can carry these pontoons, like other baggage, and when necessary a bridge can be built in a very short time. When the army has passed the river the bridge can readily be taken to pieces and the materials transported elsewhere, to serve the same purpose again. An enemy pursuing, unless provided with similar appliances for constructing a bridge, would find his farther progress barred; but in the floating bridges used by the ancients it was possible for the pursued to erect a bridge that would be used also by their pursuers. The floating bridge by which Darius crossed the Bosphorus is said to have been similar to that contrived by Xerxes, although no precise record exists to enable us to ascertain exactly how it was fashioned. Alexander adopted a different course; taking advantage of such vessels as he could obtain on the spot, by rowing or sailing he crossed the particular river or strait which opposed his march. We have a remarkable example of a floating bridge on the river Douro opposite to Oporto, but the bridge erected by Xerxes seems rather to have resembled that thrown over the Adour.

The only incident that occurred on our voyage to Alexandria was furnished by a Turkish Aga, one of the passengers, who came on board in considerable state. His attendants spread his carpet on deck, that he might attend to his devotions. However, the sea was so rough that directly he appeared on deck he lost his balance and rolled over and over, which greatly hurt his dignity. He was very much enraged, and, to revenge himself, immediately set to work belabouring the unfortunate captain with his cane, saying that it was entirely his fault for giving them a foul wind, though all the time it was blowing most favourably, though rather too fresh for the Aga's comfort. I burst out laughing, but soon checking myself, interceded for the poor captain and got him off, while the Aga retired to his cabin and delivered himself over to seasickness.

The situation chosen for the city of Alexandria, which Alexander the Great founded as the emporium of the Mediterranean after the destruction of Tyre, was peculiarly favourable, for the seven mouths of the Nile at that time existing were ill adapted for the safe passage of large vessels; all of them were more or less obstructed by bars, upon which during the stormy seasons there was a heavy surf, so that it was extremely dangerous to attempt their navigation. Alexander therefore resolved to adopt a situation totally independent of the Nile, and accordingly selected Alexandria, as being the farthest point from the mouths of that river, and the least affected by the alluvium brought down by it, and that which is carried along the shore by the western littoral current.

At the time the present situation was selected there was a small island not far from the shore. This island he connected with the mainland by means of an embankment faced with masonry, thus forming a double harbour, namely, one on the eastern side, which was protected by the promontory bounding the bay; and one on the western side, protected by a reef of rocks running in a westerly direction from the original island, and the projecting point of the bay to the west, so that the western harbour was tolerably well shielded on all sides, with the exception of the entrances between the rocks on the north; but through these only a comparatively small amount of swell can penetrate. It contains space for all the vessels that are likely to frequent this port, and the accommodation may be further increased, to almost any extent required, by making docks inland. It is true that in order to connect this port with the Nile above the bars and the interior country, a canal or other means of communication was requisite. A canal at that time was the only effectual method known, and this was accordingly adopted, and was made to the Canopic branch of the Nile, skirting along the shores of the Lake Mareotis for a considerable distance; but it was frequently filled up by the sand blown in from the desert and the adjacent shores. This sand was as constantly removed, and the communication between Alexandria and the Nile more or less imperfectly preserved. In 1851 the Pacha completed a line of railway between Alexandria and Cairo, and thus got rid of the uncertainty and expense of maintaining the canal and the river navigation. This railway was made under the direction of the late eminent engineer, Mr. Robert Stephenson, and now a perfect and economical communication is kept up by this means between Alexandria, Cairo, and the interior.

The eastern harbour of Alexandria, the water being very shallow, is now scarcely used except for the small coasting vessels; but it does not appear that in either the eastern or western harbours there is any material tendency to shoal.

When I was there, Alexandria, although possessing considerable trade, was but a poor place compared to the present city, and all the business of shipping, transhipping, and storing of goods was carried on in a very rude and costly manner, which is now materially changed for the better.

Besides Pompey's Pillar, the other important relic is Cleopatra's Needle, then as now lying prostrate. This needle was given to the British army, and a large subscription, amounting it is said to about 20,000*l.*, was raised to transport it to England, and there to erect it in some conspicuous place in the metropolis, as a trophy of the signal success of the British army. For some reason or other, never properly explained, this has never been done, although seventy years have elapsed since the money was subscribed. When I returned from Upper Egypt, in the month of March, 1821, I found that the English Government, after repeated applications, had sent the 'Spry,' sloop of war, commanded by Captain Boswell, with my old friend and schoolfellow, Captain Wright, of the Royal Engineers, to report upon the cost, and the best plan for bringing Cleopatra's Needle to England; but this all ended in nothing being done, although Captain Wright's report was very favourable, both as to the cost and feasibility.

On reaching Cairo I had an audience with the celebrated or notorious pasha, Mehemet Ali. He received me with great dignity and civility, and said that the English were his best friends, upon which I could not help saying to myself, Thank God we are not his subjects! He was no doubt a very remarkable man—cool, determined, able, and courageous. He reduced the turbulent rulers and Bedouin tribes of Egypt to subjection, and procured a degree of tranquillity and peace for the oppressed inhabitants which they had not enjoyed for many generations; and if he plundered them himself, he would allow no one else to do so. Nevertheless, one cannot but be horrified at the atrocities with which all this was accompanied and accomplished. While here I saw a few Mamelukes who had escaped the massacre of 1809, and who were permitted to reside here. I was much struck by their martial air, their richly-embroidered costumes, and superbly mounted pistols and scimitars. As a body they were nearly exterminated, the survivors being allowed to subsist on the little property saved from the wreck. Egypt, like Constantinople, has since been so well described, and is now so well known (though at that time my journey to the second cataract was deemed extraordinary), that it is unnecessary for me to go over the same ground.

I hired a khangé, or small boat, manned by a reis, or captain, and eight rowers, with a tolerably comfortable raised cabin behind, divided into two parts, in neither of which, however, could I stand upright. But by this time I had become pretty well accustomed to the usages of the country, and my limbs being tolerably supple, I did not much mind it. I engaged the khangé at so much per month, I finding provisions for myself and servants, the reis agreeing to do the same for himself and his crew. Having stocked my boat with everything we were likely to require for four months, I embarked at Bouloe, accompanied by my servant Demetrius and a janissary. We came to anchor every night near the most convenient village, and started next morning soon after daybreak. At midday I halted for an hour, and sometimes more, in order to allow the captain and his crew to take their dinner comfortably, whilst I and my servants took ours. When the wind was unfavourable the crew were obliged to row or track the boat against the current; at such times I used to land and walk along the bank, gun in hand, exploring the adjacent antiquities or the surrounding country; and I found that I could easily keep pace with the boat. Where there was any object particularly worthy of remark, I had the boat moored as near to the bank as I could, whilst I went ashore, examined, sketched, and measured the objects in question at my leisure, and in this way I took measurements of all the edifices of any note. When there were none, I used to indulge in a sporting excursion, and found ample amusement. I not only procured many interesting objects of natural history, but shot numerous wild fowl, that were a very welcome addition to my table. Thus I passed a very agreeable time. Never idle for a moment—always employed, either for instruction or amusement; and my whole time was my own. I carried my house with me; and when there was nothing interesting on the route, and I had filled my book with sketches, if my journal was in arrear, I used to remain in the boat posting it up, whilst we were sailing, tracking, or rowing.

On my way up a sad accident deprived me of the services of my faithful attendant—I may say friend—Demetrius. We met a boat, having on board the O'Conor Don, Captain Groding, and another, coming down. They hailed us, and as they were short of provisions, I gave them a liberal supply, and invited them to dinner, the two crews regaling themselves and firing *feux de joie* at the same time. Suddenly a tremendous explosion, followed by a loud cry, was heard, and hastening on deck I found poor Demetrius covered with blood from two deep wounds in the throat and breast, his pistol having burst in his hand. At first I thought it would have been fatal, the effusion of blood was so great. I succeeded, however, in tying up the wounds and stopping the hemorrhage. It was then a question what should be done with him, as I was not surgeon enough to cure him, and no advice was to be had. In this emergency my new friends were so kind as to offer to take him to Cairo, where he could be properly attended to. I gladly accepted the offer, and supplied him with a sufficient sum of money, and requested them to place him in charge of the Consul, with directions to procure the best medical advice, for which

I was of course responsible. Independently of the accident, I was much grieved to part with him, and was at first greatly at a loss for his services; but my janissary, an Italian turned Mussulman, and who spoke Arabic very well, and was not a bad cook, cheerfully consented to do what he could; and the reis also, a very civil fellow, as well as his crew, did all in their power to compensate me for the loss of Demetrius, so that after a short time we managed between us to get along pretty well.

I will only add one more incident, because it seems to show that the serpents used by serpent-charmers are not always deprived of their fangs, as is usually supposed, but that there is some real secret which renders them harmless or powerless. I was measuring the Temple of Edfou, when I saw a peculiarly venomous serpent come out of its hole, whilst an Arab boy who stood by fixed his eye steadily upon it the moment he saw it, the reptile fixing his eyes on the boy. The lad began waving his hands gently up and down, humming a peculiar tune in a low, monotonous tone. The serpent seemed to be charmed, and lay perfectly still, listening to and keeping its eyes attentively on the boy, who, finding that he had charmed it, was about to secure it; but at this I was so horrified, that I took up a large stone and killed the reptile. The boy was very angry, and assailed me with the most vehement gestures and imprecations, at which I laughed heartily. I afterwards learned that he was the son of a serpent-charmer, and was collecting these reptiles for his father.

In proceeding along the banks of the Nile, I observed that the land inclined from the margin of the river to the base of the adjacent hills. This is nothing more than might have been expected, as it is usual under all similar circumstances, and is caused by the periodical inundation. Thus, whenever the river rises above the margin, the current naturally diminishes in velocity as it encroaches on the banks, and to a certain extent becomes stagnant, and then deposits the alluvial matter with which the waters are charged; and as the water spreads farther on both sides from the main body of the river, it becomes clearer, and contains less alluvial matter. This is a wise provision of nature, for it enables the waters to extend a long way, and thus to irrigate a great extent of land. In order to ensure this irrigation more effectually, it is only necessary to keep open sufficient channels, which may be done with facility. But suppose this was not the case; suppose the land farthest off silted up first, then it would be necessary, for the purpose of irrigation, to raise the water by artificial means at considerable extra cost, to irrigate those lands farthest from the river. However, in process of time, as the land rises both at the sides of the river and the parts more remote from it where they have attained the utmost level of the floods, recourse must be had to art to irrigate the lands, otherwise their fertility and cultivation must cease, as the quantity of rainfall in the lower valley of the Nile, as it passes through Egypt, is comparatively trifling; in fact, the fertility of the country depends almost entirely upon the floods.

It is very probable that these waters might be utilized to a much greater extent by establishing large reservoirs in the adjacent valleys, which would be filled during the rising of the floods; and when these latter have subsided, the stored-up water could be discharged during the dry seasons for irrigation, navigation, and numerous other purposes. According to the present system, a vast quantity of water is allowed to waste, and the means of cultivating a large additional tract of country, now a desert, is lost. This object might be carried still farther by improving the channel of the Nile up to the great lakes of Albert and Victoria Nyanza, for the most part now a marshy, pestiferous district; this will very probably be done, as the subject becomes better understood; indeed, it is surprising how little the advantages which nature offers us in this respect are turned to account.

The delta of the Nile, like other rivers of the same class and magnitude under similar circumstances, advances outwards into the Mediterranean, and in proportion as it moves forward, the depth of water increases, and the width of the delta becomes greater, so that it requires a larger quantity of alluvium to maintain its progress, which becomes necessarily slower—that is, as far as concerns the alluvium brought down by the Nile. But then it must be observed, that as the delta proceeds outward, the stagnation produced by the protrusion of its apex into the Mediterranean causes a greater accumulation of alluvial matter to be deposited on both sides of the apex, and consequently two great bays are formed, one on either side, although the shores of these bays necessarily do not advance so rapidly as the centre portion. In proportion as the several branches of the Nile advance seaward, so their courses become lengthened, and consequently the total fall or inclination of the current becomes diminished, so that it cannot keep them all open; and hence, out of the eight branches or mouths of the Nile which existed in ancient times, only two now remain—namely, the Damietta and Rosetta mouths, and these are slowly deteriorating.

Whilst upon this part of the subject, it may be advisable to make a few remarks about the Suez Canal. This great work consists of an open cutting or trench from the Bay of Pelusium (Port Said) to the Red Sea at Suez, a total length of 99 miles, 196 feet wide at top, 72 feet wide at bottom, and 26 feet deep, with side slopes of 2 to 1. At the Mediterranean end there is a rise of tide or variation in the surface of the sea of from 1 foot to 2 feet, and at the Suez end from 2 feet at neaps to 6 feet at springs.

On the Mediterranean the entrance to the canal is protected by an artificial harbour composed of two piers carried from the shore. The western pier is carried out 2400 yards in a straight line, pointing towards the north, it then inclines slightly to the east for 330 yards, so that the total length of the west breakwater or pier is 2730 yards, or 8190 feet. The eastern breakwater or pier is carried out from the shore at a distance of 1530 yards from the commencement of the western pier, and is extended in a northerly direction 2070 yards, where it terminates at 760 yards from extremity of the west pier, which constitutes the entrance. Thus the two piers enclose a space of 500 acres, with a depth within of 26 feet. This harbour is said to be well protected against the prevailing or north-west winds.

This outer harbour, called Port Said, is connected with extensive quays and basins within, from whence the canal proceeds across the isthmus. At 52 miles from Port Said there is Lake Timsah; also Lake Ismaila and the Bitter Lakes, at 57 miles from Port Said. These Bitter Lakes cover a surface of about 100,000 acres, and will always ensure a considerable draught or current from the Red Sea, to compensate for the large amount of evaporation which is constantly going on, particularly during the summer season, and is said to amount to about 250,000,000 cubic feet daily. In order to supply fresh water to Suez, Ismaila, and Port Said, a considerable channel has been made from the Nile, at Cairo, to Suez and Ismaila, and a double line of cast-iron pipes between Ismaila and Port Said, with pumping engines of the requisite power at the former place.

The entrance to the Suez end of the canal is formed by an extensive double embankment through the shoal water, increasing gradually from a width of 72 feet at bottom, to 980 feet, where there is an open tidal dock, with 26 feet depth at low water.

This is no doubt a very extraordinary performance, rendered remarkable for the vast amount of capital which has been raised by a single individual—not an engineer—and the wonderful energy and perseverance with which he has accomplished it, opposed by innumerable obstacles, political and financial, which would have daunted and overwhelmed any person of ordinary physical powers. Although as an engineering work it is encountered by no unusual difficulties in the execution—being simply a matter of digging and dredging upon a vast scale— yet it certainly entitles M. Lesseps and his officers to *the greatest credit*.

Having now generally described the canal, let us consider how far natural obstacles exist which should cause any doubts as to its being possible to maintain the canal at such an expense as will enable it to produce something like a reasonable profit upon the capital expended in making it.

These obstacles may be enumerated as follows:

Firstly. The alluvial matter brought down by the Nile, and that from the prevailing littoral westerly current in the Mediterranean.

Secondly. The sands driven by the north-westerly winds into Port Said.

Thirdly. The sands driven into Suez by the southerly winds.

Fourthly. The sands driven into the canal from the surrounding deserts by the kamsin, or south-east winds.

Fifthly. From the great evaporation which will take place, and the consequent requirement of a corresponding supply of water both from the Mediterranean and the Red Sea.

Sixthly. Whether the expenses which must necessarily be incurred in overcoming these obstacles will amount to such a sum as will render the canal practically useless, that is to say, that it will not be worth the while of the Company to maintain it.

Before considering these important questions, it will be right to investigate the natural causes which have formed the Isthmus of Suez. I think we may conclude that Africa was originally an island, and that by degrees the waters of the Red Sea, driven in by the southerly winds, and those of the Mediterranean, driven by the northerly winds, brought with them a great quantity of alluvial matter; at the junction of these waters the currents would be destroyed, and the alluvium with which the waters were charged would be deposited and form a bank or bar, which by degrees rose above the ordinary level of the sea. This bank, once formed, would continually increase, not only from the alluvium brought in by the seas, but also from the sands blown in by the northerly and southerly winds from the surrounding deserts; and thus, in the process of time, the present isthmus would be formed. I think that the practical evidence of this is undoubted. The same operations are still in existence, and it is simply a question of time as to the increase.

Having discussed the cause of the formation of the isthmus, I will now proceed to consider the objections or obstacles above mentioned. With regard to the first, the waters of the Nile are constantly bringing down alluvial matter, but whether the quantity brought down now is the same as formerly, or greater or less, is a question which nothing but experience can decide. But as far as experiments have already been made, it appears that the accumulation which has already taken place is considerable, and if it proceeds in the same ratio as hitherto, it must shortly become a very serious question whether it should be removed or not, and whether it might not be remunerative to do so.

Secondly. With regard to the sands driven in by the northerly winds; these must be very considerable, and not being able to escape, they must accumulate and tend to fill up the harbour; this will necessitate constant dredging to keep it open.

Thirdly. The sands driven by the tide and the southerly winds into the Suez end of the canal. These also must be very considerable, as is already evinced by the great extent of shallow water at the northern extremity of the Red Sea. This also must be reduced by dredging.

Fourthly. With regard to the quantity of sand which may be expected to be driven into the other parts of the canal from the surrounding deserts, during the winter and spring prevailing kamsin, or southeasterly gales. This quantity has been proved by one year's experience to be not less than 310,000 cubic yards, and at times it may possibly be much more. It is proposed to check this by planting the sides of the canal with trees. Still a great deal of dredging must be constantly required.

Fifthly. The evaporation from the Bitter Lakes, and parts of the canal adjacent, is said to be 250,000,000 cubic feet of water, which is equivalent to about three-quarters of an inch daily. This water will have to be supplied chiefly from the Red Sea; and as it will have to pass through such a narrow channel, the velocity of the current will probably amount to two or three miles per hour, and if the banks of the canal are not well secured by paving, or similar works, they will be liable to be seriously affected. The constant indraught of the current will impede vessels coming from the Mediterranean, whilst it will facilitate the passage of vessels coming from the Red Sea; and in the same manner there will be a constant current from the Mediterranean, but by no means to so great an extent. These currents will also very probably bring in a considerable quantity of alluvium. They will not, however, very materially interfere with the passage of steam-vessels, although, if the latter are permitted to go at full speed, the waves produced by them will scour away the sides considerably, unless well protected by stone paving, fagoting, or similar works.

Upon the whole, viewing the difficulties above mentioned, the question naturally arises, whether they are of such a character as to be insurmountable; and to this I think we may safely say that they are not. What has been done once, as has been proved by the completion of the canal, can be done again, and will be maintained with much less difficulty. So far, therefore, viewed simply as a work of engineering, it resolves itself into a question of cost, or in other words, will the work pay as a commercial speculation, seeing that it has already cost 20,000,000*l.*, and a great deal more is still required before it can be said to be quite complete, besides a very large sum for annual maintenance, and what this last item will be it is very difficult to decide, and nothing but experience can prove; still the more the canal becomes known, the more in all probability will it be used, and therefore the more money will be available for keeping it in repair. In addition to the dredging, it is very probable that the piers, both at Port Said and Suez, will have to be extended considerably.

The whole of the valley of the Nile, from the head of the delta below Cairo, is bordered by ridges of sandstone hills a few hundred feet high, with generally a plain monotonous tableland above, intersected by numerous ravines; no granite appears until we reach Assouan. In the vicinity of this place there are numerous quarries of fine granite, chiefly red; the masses are so compact that blocks of almost any size may be obtained from near the surface. From this district all the granite for the obelisks, statues, and columns of the various buildings, temples, and pyramids on both sides of the valley have been taken. Transported in flat-bottomed boats and rafts to the places where they were intended to be used, they were landed by means of inclined planes of wood, with rollers, and wedges, assisted by numerous rough capstans or windlasses, worked by countless gangs of men. This work appears to have been done with considerable skill, and the necessary combined operations were carried on simultaneously, by means of well-concerted signals. These operations are very clearly explained by the sculptured figures, and have been published by Sir Gardner Wilkinson and others.

Whilst examining the granite quarries near Assouan, I observed several imperfect blocks, which the Egyptians had commenced quarrying, but found them defective, and ceased working them. The mode of quarrying seems to have been nearly the same as we employ at the present time, namely, by wedges, levers, and pickaxes. Even if gunpowder had been known it would have been of very little use, for it would have in most cases destroyed the blocks, and the waste would have been enormous.

It does not quite clearly appear of what metal the tools were composed with which they worked the granite, whether of iron or bronze; if of the former, it must have been case-hardened. The polishing would be done by attrition with emery or sharp silicious sand. As for the stone, being sandstone, it was easily worked, and softer tools only were required.

As soon as I got back to Alexandria my first inquiries were for poor Demetrius. To my great delight I found him quite recovered, and overjoyed at seeing me; but he said he had been very ill for six weeks, and had several times considered himself dying. I immediately set to work making arrangements for my journey overland to Palestine, when I was suddenly interrupted by letters from my father, saying that he was very unwell, and wishing me to come back as soon as possible. I instantly dismissed Demetrius with a very handsome gratuity, discontinued my preparations, and, through Mr. Lee, our Consul at Alexandria, hired a passage to Malta in a merchant vessel. I was, however, laid up for several days at Cairo with fever, which was not pleasant, as the plague was then raging. As soon as I was sufficiently recovered I proceeded to Alexandria, where I found that the merchant-man had already sailed. However, as I have said before, Captain Boswell, with H.M. sloop, the 'Spry,' was there, having Captain Wright on board, and the latter having finished his report on Cleopatra's Needle, the sloop was about to return to Malta, and they very kindly offered me a passage. This I gladly accepted, and on arriving at Malta found that the plague had broken out on board the merchantman in which I had engaged a passage, and that several of the crew had died. She was then in quarantine, and before long everyone on board died of the plague, and I saw the ship burnt. I humbly thanked an all-merciful Providence for its inscruta-ble dispositions.

At Malta, though better, I still retained the fever, and consulted Dr. Groves, the head of the naval medical department. He looked very grave and said little, but recommended me to go to England as soon as possible, while he told a friend of mine that he did not think I should recover, the fever having taken such hold of my constitution. However, by the aid of bark and port wine I became tempo-rarily myself, and as soon as I was able I proceeded to pay my visits. I called on the Commander-in-Chief of the Mediterranean squadron, Sir Graham Moore, who gave me every assistance in his power, and introduced me to Admiral Woolley, chief of the dockyard, to whom and his amiable wife I owe my gratitude for their kind hospitality. They introduced me to Sir Manly Power, the commander of the military forces, and from him and the various messes of the regiments composing the garrison I met with the most friendly reception. I here made the acquaintance of Mr. Strutt, one of the sons of Mr. Strutt, of Belper, and with him I explored both Malta and Gozo. He shortly after left for Greece and the East, where he died of malaria, which I much regretted, as he was a remarkably fine, intelligent, and spirited young man.

While I was here His Majesty George IV.'s birthday was celebrated with the greatest *éclat*. At twelve o'clock the entire garrison was reviewed by Sir Manly Power, with the usual salutes from the forts and ships of war. At sunset all the guns on the works were fired. This was a magnificent sight, and, that I might see it properly, I was invited to dine at the artillery mess at St. Elmo, from whence the first gun was fired; this was taken up by fort after fort until the entire fortress seemed in a blaze, and the whole was terminated by a magnificent bouquet of rockets.

I embarked for Naples on board a cutter commanded by Mr. Thurtle, one of the oldest midshipmen in the navy. He was a very peculiar and rather eccentric character, rough and sarcastic in manner, an excellent sailor, and a kind-hearted man. His long service and great experience rendered him well known throughout the navy, and his wit and good-humour made him a favourite wherever he went; in fact, he was a privileged person throughout the fleet, and enabled to assume a character which few officers of far higher rank thought of attempting. Unfortunately he was the brother of that daring and dissipated criminal who shortly afterwards murdered his companion, Welsh, at Elstree, for which he was hanged at the Old Bailey. This was a sad blow to the poor midshipman, and one from which I afterwards heard that he never recovered. My voyage with him was most agreeable. He was very kind and attentive, full of fun and humour, yet never for a moment neglecting his duties, and he kept his crew and ship in the most perfect order.

On landing at Naples I got apartments in the Gran Bretagna (then, and I believe now, the principal hotel), with more ease than usual, in consequence of the rapid advance of an Austrian army, thirty thousand strong, commanded by General Baron Pirmont, at the request of the King of Naples. I saw them enter, and very fine troops they were, especially the Tyrolese yagers. Later on, when I wished to make an excursion to Pæstum, I obtained, through the kindness of the colonel of the regiment, two Tyrolese corporals as escort. They sat, rifle in hand, on the box of my carriage, and mounted guard while I sketched. We saw several ill-looking fellows, armed, prowling about, but none dared venture within reach of the rifles. Letters from my father reached me here, saying that he was a good deal better, and that there was no necessity for my hurrying; I therefore, while still hastening home, thought I might allow myself a few days on the road to visit those objects of interest which I had previously omitted.

While at Naples I also examined an open pier, of which there are numerous specimens about the bay. This pier is constructed in a peculiar manner of pozzolana mixed up with irregular-shaped pieces of brick, marble, and tufa, or volcanic stone. The piers were formed by enclosing the space in a wooden dam or box, then filling it with the materials above mentioned, which after a short time set under water, and became a solid mass; the cofferdam was then removed and the pier left standing; then another was constructed in the same manner, until the required number were completed; upon these piers arches were built, and upon the top of them a roadway and parapet were constructed. The piers were about the same thickness, or rather greater, than the span of the arches. The object of the openings and piers was to provide a barrier which should be just sufficient to break the swell, but not large enough to obstruct the current, as the latter prevented any sediment from accumulating on either side of the pier. This is a very ingenious and novel mode of constructing piers, and it is strange that a like method has not been adopted in England, where the vast quantity of alluvial matter carried by the currents along our coasts accumulates round solid piers, and frequently fills up the harbours, rendering them almost useless, unless the accumulated matter be removed by dredging at continual expense.

Whilst breakfasting at the hotel at Pozzoli I felt an attack of my old enemy, the malaria fever, coming on; I therefore got back to Naples as soon as possible, and was obliged to go to bed, and there I remained for several days perfectly helpless.

Thinking it would leave me, I did not send for medical advice. However, finding myself gradually getting worse, I was recommended to send for Dr. Roskilly, formerly an army surgeon, who had now established himself at Naples as a physician, and from his well-known skill and courteous manners had obtained considerable practice, particularly amongst the English. He came, and I found him such as described; he examined me carefully, and, after a considerable time of grave consideration, he said that I was in a very bad way, that the fever had got such a strong hold of me that it was continually undermining my constitution, and that, if not speedily arrested, he could not answer for the consequences; indeed, I might not live to see England. The worthy doctor, with whom I was much pleased, then left, and promised to return the next day at the same time, which he did. He then said that he had given my case his utmost consideration, and he was more than ever convinced that some decisive measure was necessary. My constitution was still good, and strong enough for the measure he would propose; and if I would submit to him, he felt confident that he could carry me through. He explained his remedy, which was nothing more or less than that I should take blue pill, as much as I was able; this, he said, would thoroughly eradicate the fever. I accordingly consented, as my case was desperate. It succeeded completely after about a fortnight's trial. I felt very weak, but I was entirely a different man. I soon began to recover strength and appetite, and in about another week I was enabled to go out, and became myself again. No doubt the remedy was severe, but I must in justice to the able doctor say that he perfectly cured me; for although in after life I was frequently exposed to malaria influence in the fens of Lincolnshire, Walcheren, the worst parts of Holland, Tunis, and elsewhere, I never had an attack of the malaria fever.

On my way northwards I stopped two days at Mola di Gaeta, where the Apennines approach the sea, and where are numerous traces of the villas of the old Roman aristocracy; it was well I did so, for the brigands had mustered in great strength in the adjoining mountains, and committed many atrocities; amongst other things, they had just carried off a number of pupils from a neighbouring school. The Austrians, who were there in great force, at the request of the local authorities came to the rescue. A very strong patrol of cavalry kept the main roads, while a still more numerous body of Tyrolese swept the valleys and penetrated the recesses of the mountains. The brigands at first defied their pursuers, but finding themselves hard pressed, they killed their unfortunate prisoners. On hearing this the Austrians attacked at once, and having slain a great number, made the rest prisoners. I saw about thirty chained together, who, I believe, were all shot the next day. The road being now clear, I resumed my journey.

At Florence I carefully examined and measured the celebrated bridge of the Most Holy Trinity, by Ammanati, across the Arno, near the Gran Bretagna. This beautiful structure, a masterpiece in the art of bridge-building, consists of three arches; the arches may be termed semi-elliptic, slightly pointed at the crown; perhaps they might be more appropriately termed Gothi-elliptic. The slight pointing at the crown may have been done to allow for sinking, which did not take place to the extent calculated upon. Each of the arches is surrounded with a moulded archivolt of equal thickness throughout, with an ornamented scroll or shield in the centre of the spandril walls supporting the roadway, which is bounded by a solid panelled bridge. The piers are terminated by acute angular buttresses carved up to the top, and panelled also. The whole is built of marble, and is certainly one of the lightest, most elegant, and most scientific structures of the kind in existence. Some doubts as to its stability have existed at various times; nevertheless, it still exists without a flaw, and as a model to engineers and architects.

I visited the magnificent picture gallery of the Palazzo Pitti, replete with choice specimens of the great Italian masters. The palace itself is a fine example of the simple, massive, rusticated style, devoid of ornament. The effect is imposing, and shows what may be done by bold, well-defined masses, without resorting to that exuberant over-ornamented style, intermixed with all sorts of coloured marble, so much the fashion of the present day, particularly in England, where the climate is peculiarly unsuited for it; in any case it can only be termed a vitiated taste. If the building is well designed and properly grouped in effective masses, so as to give well-defined portions of light and shade, there can be no need of extraneous ornaments, as they only detract from the general effect; and where a building is not well designed no ornament can render it effective.

The port of Leghorn consisted of several solid stone piers, which did not appear to be laid out in the most scientific manner, and considerable improvements were in contemplation when I saw it. The great difficulty it has to contend with is the quantity of alluvial matter brought down by the Arno, which spreads along the coast for some distance both to the north and south. This might be obviated by judiciously adopting the principle of open piers, as invented by the Romans, so as to allow the littoral current full scope; this would sweep away the alluvium as fast as it was deposited, and the requisite protection from westerly winds might be easily obtained if the works were constructed on a proper system.

Passing by Carrara, I was much struck by the waste consequent on the clumsy method of working the quarries, and of transporting and loading the blocks. It occurred to me at the time how easy it would be to construct an iron tramway from the quarries to the shore, with a suitable pier furnished with powerful derricks at its termination, and this, with proper machinery for quarrying, all of which would have cost but little, would materially reduce the price of the marble, and consequently increase the demand.

As I passed the magnificent and well-protected Gulf of Spezzia, I thought, "What a splendid site for a naval arsenal!" This idea has now been at least partially carried into effect, and if the works are properly designed and executed, it ought to be one of the finest in Europe.

The morning after my arrival at Genoa I arose early, quite refreshed, and went first to the harbour, which I had always heard was one of the finest works of the kind in the world, and as such my excellent father expected that I should give him a complete account of it. I therefore examined it very minutely, and took great pains in tracing, sketching, and measuring it, when I could do so without being observed, for there was a good deal of jealousy about it. At first sight I was very much struck by the extent and magnificence of the bay, with the fine old town rising like a vast amphitheatre of palaces round it, surrounded by a huge circle of forts which crowned the summits of the hills; and by the great extent and massiveness of the two outer moles, the depth of water enclosed within them being capable of receiving at all times the largest vessels of war. But when I began to examine more narrowly I found that there, was a great error in the design of the main or outer moles. The southern, which was the longest, consisted of two arms, the outer one inclining inward at a considerable angle, whilst the northern mole consisted of a single straight line projecting from the shore in a south-south-west direction, so that the entrance pointed south-south-west, and was consequently exposed to the full effect of the severe gales which blow up the Gulf of Lyons during the winter and autumn months; the consequence is that during severe gales from this quarter a heavy swell sets into the entrance and produces such a strong current throughout the interior of the harbour that vessels in front of the town and under the southern mole can scarcely ride with safety at their moorings. This is a serious defect, and it might be easily remedied, although up to the present I have not heard that anything has been done. In the construction of the works there was nothing particular to find fault with, but there was a serious error committed in the disposal of the sewage; the whole of this was discharged into the harbour, so that in hot weather a most disagreeable effluvia arose in front of the lower part of the town; moreover, it caused the accumulation of a considerable quantity of alluvial matter. This had to be removed by constant dredging, in which convict labour was employed. The alluvium was emptied into lighters and sent out to sea to be discharged. The old arsenal was situated at the south-east corner of the inner harbour; it was quite unequal to the requirements of modern times.

While I was at Genoa an accident occurred by which my old acquaintance the sloop 'Spry' was nearly lost. Captain Boswell, having served his time, had been succeeded by another captain, who, contrary to strict rule, had his wife on board. The evening on which the captain had given orders to prepare for sailing there happened to be a grand ball in the town, and the lady persuaded her husband to take her, and defer sailing to the morrow. By this time the 'Spry' had left the harbour, and lay at single anchor outside the north mole. During the night a strong gale sprung up from the south. As the ship could not regain her old position the officer in command let go two more anchors, but two cables having parted she dragged her third anchor, and was driven close to the rocks. No assistance could be given from the shore, and, as her position seemed hopeless, the officers and crew took to their boats, and managed to get off in safety. To the astonishment of everybody, however, the last cable held, and the recoil of the waves, so close was she to the rocks, actually prevented her from striking, and so she remained until the gale abated, and the ship was saved. The odd part of it was, that when the crew were about leaving the ship my old friend Dr. Biggar was asleep in the cabin, and when aroused flew into a violent rage, and threw a bootjack at the head of the midshipman who woke him. After several ineffectual attempts, the midshipman was obliged to leave him, and he turned round and slept soundly till the morning. Judge of his surprise when he awoke and found what had happened during the night!

After passing rapidly through France, I embarked from Boulogne, September 23, 1821, in a small packet, during a very strong gale, with about fifty other passengers.

The following day we started with the morning's tide with about seventy passengers, of whom Colonel Hylton Jolliffe, a friend of my father's, was one. We left Boulogne about eleven, and reached Dover at 2 P.M.; I directly started by coach for London, and reached my father's, 27, Stamford Street, the same night. I found him in very bad health, lying upon the sofa in the principal front bedroom; he was glad to see me.

He continued in the same weak state, although in perfect possession of all his great faculties, dictated to me several letters on business, and talked of sundry new works that he was about to undertake, particularly the new London Bridge, and the removal of the old one, which had been for some time under discussion in Parliament; a Bill for this purpose had actually been introduced during the past session, and my father had been requested to prepare a design for it, which he did, and it was very similar to that since executed by myself. My father's bodily health appeared to decline gradually; he was confined almost to the sofa, and could do little more than walk across the room; in this manner he continued until the 3rd of October. He went to bed as usual, perfectly sensible and composed, and hoped that he would soon be better, as he was most anxious to return to business and make up for lost time. I went to his room on the morning of the 4th of October, and found that a considerable change for the worse had taken place; he seemed to be in much pain, and was walking about the room, evidently scarcely knowing what he was about. I got him into bed, and immediately sent for his physician, Dr. Ainslie, who had known him all his life, but he was unfortunately out of town. The apothecary, Mr. Welbank, came, and we consulted together as to whom we should send for, and ultimately summoned Dr. Roberts, who, although in good practice, had never seen my father, and consequently knew nothing of his constitution or complaints. He did the best he could, but evidently thought the case was very serious. My father lay in bed all day, almost unconscious, although I thought he knew me. I remained with him nearly the whole day, and about five o'clock in the afternoon he appeared to be sinking fast, and breathed very heavily, which alarmed us all excessively. In a short time this ceased, his features began gradually to relax, and he breathed his last at half-past five on the afternoon of the 4th of October, 1821, in the sixty-first year of his age.

The disease which killed my poor father was that of the kidneys and liver, as far as we could ascertain. All my brothers and sisters were assembled round his deathbed. It was a sad, sad sight, and afflicted us most severely; we had, however, the melancholy satisfaction of having done all in our power, though of no avail, to arrest the fatal event. He was universally known and respected; the news of his death spread immediately throughout the town, the public papers were filled with leading articles giving accounts of his public and private life, and everybody deplored his loss. One of the most powerful and touching articles was written by his talented friend Perry, the proprietor of the 'Morning Chronicle,' who was then at Brighton in bad health, and died there three months afterwards.

It seemed to be the universal wish that he who had rendered so many services to his country and was so generally beloved, should be buried in St. Paul's, and arrangements were made accordingly. The funeral took place a few days afterwards, at eleven o'clock, and he was attended to his last resting-place by a vast concourse of literary, scientific, and private friends. The late talented John Wilson Croker, Esq., then First Secretary of the Admiralty, wrote the epitaph, which was composed in the most feeling and scholar-like manner for which that able gentleman was so particularly well qualified.

CHAPTER III.

It was some time before I could recover from the shock. I had been absent abroad nearly two years and four months, and had passed through so many different scenes, that when I returned to England everything seemed perfectly new to me; being deprived of my father so unexpectedly threw me almost into a state of despair, so that I scarcely felt myself equal to undertake the responsibility of following his noble career, which I could never expect to equal. After giving way to my melancholy reflections for about a month, I determined to rouse myself to the utmost and to do my best, and with his brilliant example before me, and cheered on by his numerous attached friends, I felt that if I had no chance of attaining the same degree of celebrity as my dear father, I might still do something, and although *lungo intervállo*, I might still keep up the name. I determined therefore to set to work in right earnest and endeavour to obtain some of my father's numerous appointments. My first ambition was to succeed him in his numerous great works then being carried on by the Admiralty, such as the Plymouth Breakwater, and the new Chatham and Woolwich dockyards.

That most excellent and able man, the late Lord Melville, was at the head of the Admiralty; the distinguished and gallant Sir George Cockburn, one of Nelson's officers at the Nile and elsewhere, was the First Naval Lord; John Wilson Croker and Sir John Barrow were the Secretaries; and there never has been such a galaxy of talent at the Admiralty since. All these great and good men have since passed away from us, not without, however, leaving behind them indisputable monuments of their skill and the great benefits they conferred upon their country. As for myself, I owe them my deepest gratitude, and shall never forget their kindness. I was appointed by the Admiralty to succeed my father as their engineer. This high honour at my early age (for I was only seven-and-twenty) filled me with the greatest thankfulness, although I felt it was due to no merit of my own, but rather to the regard and respect which they entertained for my father; I therefore resolved to do everything in my power to render myself worthy of it, and set to work with right good will.

The next appointment I obtained was as drainage engineer to the Eau Brink Commissioners. This was at that time the greatest work of the kind, at the head of which were the late General Lord William Bentinck, afterwards Governor-General of India; the late Sir Andrew Hammond, Bart.; the late Sir Charles Browne, physician to the King of Prussia; and the late Thomas Hoseason, Esq., of Banklands, in the district of Marshland, near Lynn. These able and distinguished men formed the Committee for carrying into effect the Eau Brink Cut, for the improvement of the drainage of the great level of the fens, called the Bedford Level, amounting to about 300,000 acres of valuable land. This work consisted of a cut for altering the channel of the Ouse, by means of which nearly two miles of the navigation of that river would be saved, and an additional fall for the drainage of five feet perpendicular would be gained. This great work had been planned nearly a century before, but had always been opposed by the inhabitants of the fens, as being in their opinion inadequate to effect the desired object. At length, after great opposition on the part of the townspeople, who alleged that it would ruin their harbour and trade, the plan was decided on, and an Act of Parliament was obtained, in the year 1781, to carry it into effect, and to lay a tax of 4*d.* per acre per annum upon all fen lands which it was supposed would derive benefit from it; certain guarantees were given to Lynn Harbour and the interior navigation interests, as well as to the owners of the banks of the Ouse, that they should be indemnified for any damages they might sustain in consequence of the Eau Brink Cut being executed. Under this Act it was decreed that there should be two engineers, one appointed by the drainage interests, namely, the late Robert Mylne (the architect of Blackfriars Bridge), and Sir Thomas Hyde Page, R.E., as the engineer for navigation. These two gentlemen were to decide the direction and dimensions of the proposed Eau Brink Cut, which was to commence below German's Bridge and to terminate a short distance above the boat wharf at Lynn. They, however, differed so materially that it was necessary to call in an umpire to decide between them, and the late scientific Captain Joseph Huddart, of the Trinity Board, and the inventor of the celebrated patent cable machinery, was appointed arbitrator. Captain Huddart made his award; but when it was determined to carry on the works, it was found that the whole of the funds appropriated for that purpose, which amounted to about 80,000*l.*, had been expended in litigation and the cost of obtaining the Act of Parliament, so that the whole matter fell to the ground.

Meanwhile the defective state of drainage of the great level of the fens still continued, and everybody was convinced that the only remedy was to carry into effect the Eau Brink Cut as awarded by Captain Huddart. It was therefore resolved that a new Act of Parliament should be obtained for this purpose, increasing the tax upon the lands proposed to be benefited sufficiently to cover the costs according to the estimate of my father, who was appointed engineer-in-chief of the drainage interests; and the above-mentioned gentlemen, Lord W. Bentinck, Sir A. Hammond, Sir Charles Browne, and Thomas Hoseason, Esq., were appointed as the executive committee.

The Act having passed, Messrs. Jolliffe and Banks tendered for, and received and executed the contract.

The effect of this work greatly exceeded the most sanguine expectations of its supporters. Immediately after it was opened the low-water mark at the upper end of the cut fell five feet, and the drainage waters were carried off with a degree of rapidity which astonished the whole country. The autumn and winter of 1821-2 was characterized by an unusual quantity of rainfall, and if it had not been for the opening of the Eau Brink Cut the whole, or the greater part, of the level of the fens would have been under water, and therefore the fenmen were very well pleased with the result. At this time I was appointed to succeed my father as chief drainage engineer, and the late Mr. Telford had been previously appointed the chief engineer for navigation. Immediately after my appointment, which was in the month of December, 1821, I went to Lynn to examine the works, and was much astonished to find the great effects which had been produced by the Eau Brink Cut. Instead of the circuitous old shallow course, full of shoals and obstructions of every kind, there was a fine straight, deep channel, two miles shorter than the old one, of the proper width, bordered by strong banks of the full height; the floods passed off without difficulty, and the navigation was so much improved that the lighters and barges going up the river from Lynn saved several tides. It is true that upon examining the country between Denver Sluice and Cambridge, there was a great deal of water out in several places, but this was attributed to the interior drains and rivers not being properly defended and embanked, so that they could carry off the water to the main outfall below. I also examined the new steam pumping apparatus, which had lately been erected for draining Soham and other fens. This, although proposed by my father in 1786, was the first of the kind that had been erected. It consisted of a scoop wheel, with a perpendicular lift, worked by a condensing engine. It answered its object completely, and has since been imitated by numerous others with equal success in different parts of the fen and lowland districts. Yet in many places it has been found very difficult to induce the fen proprietors to combine together in order to effect a natural drainage, which would be better and less expensive; they prefer to act independently of each other, and adopt the steam wheel. Still, even with this, the main outfalls must be improved to their fullest extent, otherwise the water cannot run off; and when the floods in the adjacent rivers rise so high that the banks are endangered, the pumping must be discontinued, otherwise the banks will break, and then a greater injury will ensue. Nevertheless, the steam pumping apparatus is an immense improvement on the old windmill, which could only work when there was wind.

My next appointment was to succeed my father as engineer-in-chief to Ramsgate Harbour.

This harbour was established by special Act of Parliament for the purpose of affording shelter to vessels of 300 tons lying in the Downs during south-west gales. When these winds prevail that anchorage is crowded with all classes of vessels, and the smaller ones, not being so well found with ground tackle as the larger, are liable to be driven from their moorings and fall foul of the larger vessels, causing them to go adrift also, and thus creating considerable confusion and damage; but by having Ramsgate under their lee, the small vessels can always get under way, run for it at the commencement of the flood, and reach it in safety long before high water, at which time the Goodwin Sands are covered, and a heavy sea rolls into the Downs. In fact, Ramsgate Harbour was made for clearing the Downs of small vessels, so that the large ones may ride in safety, and so far has effectually answered its object.

There were a certain number of trustees appointed under the Act, who were selected from the principal merchants and shipowners of London, and the Deputy Master and three or four Elder Brethren of the Trinity were members of the Board of Trustees *ex officio*. At the time I was appointed engineer, the celebrated Sir William Curtis, Bart., member for the City of London, was chairman, and the worthy Deputy Master of the Trinity House, Sir John Woolmore, represented that Corporation with three others of the Elder Brethren. As this was one of the oldest and most important harbour trusts in the kingdom, I felt great honour in being appointed their chief engineer, the more so as the celebrated Smeaton and my father had previously occupied that position. It was here that Smeaton followed out the idea that had been originally proposed at Dundee, of establishing an inner basin with sluices for the purpose of scouring away the mud which continually accumulated in the outer harbour, owing to the great quantity of alluvial matter brought in from the adjacent coast and waters, which otherwise in a very short time would have filled it up and rendered it useless. Ramsgate Harbour was also celebrated for being the first place of the kind where the diving bell was introduced by Smeaton for the purpose of laying down moorings and removing obstacles under water. The diving bell was afterwards, in 1813, much improved, and rendered, for the first time, applicable to building masonry under water with as much security and accuracy as building upon dry land. In such an exposed situation it was more economical and expeditious than the old cofferdams; it was in this manner Mr. Rennie rebuilt the east pier-head in 17 feet at low-water spring tides, which was originally made by Smeaton, and which failed. I also succeeded my father at Sunderland, Donaghadee, Port Patrick, and Kingstown harbours, the West India Docks, besides other places soon afterwards, so that I had a large business, and was daily getting more.

The most difficult and anxious work, however, at that time was the new dockyard at Sheerness, designed and partly carried into effect by my father. He originally, in the year 1807, recommended that the old dockyard, which was composed only of some old wooden slips imbedded in the mud, a few storehouses, a wretched basin, lined with wooden walls, and some timber jetties, should be abolished. He said that it was on the lee or wrong side of the harbour, that the foundation for new works was of mud and quicksand, that the space, on account of the buildings in the old town, was very confined, and, therefore, that to make a good dockyard there would be very expensive, and he thought it would be far better to make a new complete establishment at Northfleet, just above Gravesend, and to get rid of Woolwich, Deptford, and Sheerness altogether. Mr. Pitt, then Prime Minister, decided that it should be done; but when he died the matter fell to the ground, as I have before said, although the land was bought for the purpose.

However, after the great war had terminated, in 1815, the nation was naturally anxious to reduce the expenditure as much as possible, so that the House of Commons would not listen to the idea of expending any large sums upon great new works either for the navy or the army; and it was only after considerable difficulty that the House of Commons would grant money for the repairs of Sheerness Dockyard, and, like most extensive repairs, it was found, when too late, that it would have been far better to have abandoned Sheerness altogether.

Nevertheless it was absolutely necessary to do something to the Thames and Medway dockyards, to keep in repair the large fleet of vessels which was there laid up in ordinary. The total number of pennants flying at the close of the war was about 1000, and the last vote for seamen in 1815 included 127,000 sailors and marines. As there was greater depth of water at Sheerness than at any of the dockyards, and as the harbour immediately contiguous was capable of accommodating with ease any number of large vessels, Sheerness was decided upon as the place where the greatest repairs and improvements should be made, and it ended in an entirely new dockyard being built. The works accordingly commenced in 1815, and the late Lord Melville laid the first stone at the north end that year. It ended in expending nearly 3,000,000*l.*, 1,700,000*l.* of which went to the engineering department, and the remainder in the purchase of ground, buildings in the town of Sheerness, in storehouses, mast and timber ponds, smithery, admiral's and officers' houses. The engineering works were of the most difficult kind; the foundations were composed of nothing but soft mud and loose quicksands to an almost interminable depth, so that my father was obliged to invent an entirely new system of hollow walls faced with granite in front and brick behind. This system of walls, which was entirely new, by giving a greater superficial area of bearing surface with the same weight of materials, rendered them thoroughly secure. He had adopted this kind of construction with perfect success at the docks at Great Grimsby, in Lincolnshire, and they succeeded equally well at Sheerness, although exposed to much greater difficulties. The dockyard as completed consists of one basin of nearly 3 acres, at the east end of which are three docks for first-rate vessels of war, with a depth of 9 feet at low water of spring tides, the basin being of the same depth, so that with a rise of tide of 18 feet at spring tides the largest vessels can always be docked at those times. There is also another tidal basin of about the same size, together with large storehouses, smithery, mast ponds, boathouses, admiral's and officers' houses, chapel, &c., and ample space for timber and other materials. This dockyard, therefore, as was the intention, is well adapted for keeping in repair the numerous vessels lying up in ordinary in the adjacent harbour, or for executing any repairs which vessels on the northern stations may require; it never was intended to be a building yard, and it answers its original purpose well. It should be mentioned, that here Mr. Rennie first introduced cast-iron gates for the dry docks; these fitted to the granite quoins so nicely by polishing the two surfaces, iron and stone, with emery, that they worked together perfectly, and were completely water-tight; and although they have now been in use forty-five years, they are as good as on the day when they were made.

At the time that these works were proceeding, it was decided to make a new large granite dry dock at Chatham, similar to those at Sheerness; and some improvements in the line of river wall, which partly interfered with the free circulation of the tide there, were being constructed when I was appointed, and I finished them.

The breakwater in Plymouth Sound, which was designed by my father and commenced in 1815, had made considerable progress; in fact, the great mass of rubble stone had been deposited throughout its entire length, so that vessels of war as well as merchantmen could safely lie under its protection during the heaviest gales from the westward. When I was appointed engineer to the Admiralty, the late excellent Mr. Joseph Whidbey, who was distinguished for his scientific acquirements, and who had sailed round the world with Vancouver, was the superintendent of the breakwater, so that I had but little to do with it until the latter end of 1824, when, on the 22nd and 23rd of November of that year, a violent storm occurred from the south-south-west, the most dangerous wind, and its effect upon the breakwater above low water of spring tides was to disarrange nearly the whole of the superstructure, and to transfer a very large portion of the stone from the south to the north slope. The effects of this severe storm were considered to be so serious as to create considerable doubts as to the security of the breakwater, and even Mr. Whidbey was alarmed. I was accordingly sent by the Admiralty to make a detailed report as to what had occurred. I carefully examined the whole work, and had numerous transverse and longitudinal sections taken to show its exact state, and reported that the main body of the work remained as substantial as ever, but that the rubble from above low water to the top, on the south or sea slope, had only been laid at an angle of 3 to 1, and the waves during the storm above mentioned had in a great measure disturbed it, and had transferred a very large portion from the south to the north slope, increasing the south slope to 5 to 1. Now it is singular that my father, when it was decided to raise the breakwater from the level of half tide (which was the original intention) to above high water of spring tides, always said that the outer slope should be laid at 5 to 1. After his death, however, Mr. Whidbey, with an idea of economy, reduced the south slope to 3 to 1, so that the effects of the storm had been to confirm my father's views; and not only had no real damage been done, but it had consolidated and strengthened the breakwater materially, and had given a practical example as to the best mode of completing it, and I recommended that the outer slope should be finished at 5 to 1, and the inner at 2 to 1. The Admiralty, however, feeling their responsibility, thought it advisable, in addition to myself, to consult three other experienced engineers, and the late Mr. Telford, Mr. Josias Jessop, and my eldest brother, George, were appointed for that purpose.

We proceeded to Plymouth in the month of March, 1825, and spent several days in examining the breakwater. We finally recommended that the outer slope should be finished at 5 to 1 and the inner at 2 to 1 (as I had previously suggested to the Admiralty), that the outer slope and top should be paved with rough square blocks set closely together, and that the inside slope should be paved with rubble. Mr. Whidbey was, unfortunately, of a different opinion, and recommended a nearly vertical wall of solid masonry on the top, of which the Admiralty did not approve, and adopted our plan, which was ordered to be carried into effect. Mr. Whidbey was so much annoyed that he resigned his situation and retired into private life near Taunton. This was much regretted, as he was a most able and honest public servant, and beloved and respected by everybody who knew him. I was then appointed chief engineer, and upon considering the subject again, I thought that, as the base or toe of the outer slope was the most exposed part of the work, it would be better to strengthen it by benching, which would effectually break the force of the waves before they could reach the main body of the work. This was approved by the Admiralty, and has since been carried into effect with complete success.

I may here relate a curious anecdote in connection with the death of Mr. Perceval. Messrs. Fox, Williams, and Co., the great mining contractors in Cornwall, took the first contract for blasting the rock and depositing the stones on the breakwater. In 1815 Mr. John Fox, a Quaker, having come to town on business, breakfasted with my father and several others, including myself. The conversation happened to turn on the death of Mr. Perceval. Mr. Fox said in a simple, unaffected manner, "I remember it very well; it is a curious story, and now I will tell it you. I was then visiting my friend Williams at Redruth. I went to bed as usual, and awoke in a most restless state, having had an extraordinary dream. I dreamed that I went to the House of Commons, where I had never been before, and having no admission into the interior of the House, I sat down quietly on one of the benches in the lobby, expecting a Cornish member who had promised when I came to London to give me a ticket of admission to hear the debates. Beside me on the bench sat a tall, muscular man (describing Bellingham most exactly), who appeared to be very restless, and continually asking whether Mr. Perceval had come to the House, and every now and then putting his right hand into his left breast pocket. At length, after waiting some minutes, there was a bustle, and several persons near me said that Mr. Perceval was coming; and shortly after Mr. Perceval made his appearance (Mr. Fox describing the exact dress he wore, namely, a blue coat with gilt metal buttons, white cravat and waistcoat, with nankeen shorts, white stockings, and shoes, according to his usual attire in the summer). Immediately after Mr. Perceval made his appearance, the man who sat next to me got up, and, advancing close to Mr. Perceval, drew a pistol from his left breast pocket, fired, and Mr. Perceval fell at his feet. This occasioned great commotion. The man who fired the pistol was at once seized, and I rushed out and asked what had happened, and the bystanders told me that Mr. Perceval had been shot by a man named Bellingham, who was the identical individual who had been a few minutes before sitting by my side. When my dream had come to this point I awoke in the greatest agitation. I could not account for it. I had never seen Mr. Perceval, nor his murderer, Bellingham; I had never been in the lobby of the House, and I had been in no way connected with Mr. Perceval, either by correspondence or otherwise, still I was so much affected by the dream that I felt convinced that Mr. Perceval had been murdered. I passed the remainder of the night in great restlessness. I could not sleep, but was always thinking of the dream, being thoroughly convinced that it was true. I came down to breakfast at the usual hour, in the most anxious and nervous state, which I in vain endeavoured to conceal as much as possible; but my friend and partner Williams and his whole family observed it, and said that I looked very ill, and kindly asked me to explain the cause. After much pressing, I told my story. Friend Williams and the whole of his amiable family said that it was nonsense; that I had been unwell, and still was so, and said that they would send for their family doctor. I said no; I felt perfectly convinced that my dream would, unfortunately, prove but too true, and that the mail, which would arrive in the evening, would bring a confirmation of it. They tried to laugh me out of it, but nothing would do; I therefore went about with my friend Williams, transacting our mining business, being convinced that the arrival of the mail in the evening would confirm the truth of my dream in all particulars. We returned to dinner at five o'clock; at nine the mail arrived, and confirmed every particular of my dream. I was afterwards taken to the House of Commons, where I had never been before, and I correctly pointed out the whole particulars of the melancholy transaction exactly as they occurred, to the astonishment of my friends and the bystanders. The whole story seems so strange that I cannot account for it. I relate it to you just as it occurred to me."[2]

This is certainly one of those marvellous instances of foresight which baffles all comprehension. John Fox was generally considered by his numerous friends and acquaintance to be a most honest, plain, straightforward, business man, and incapable of stating anything but what he believed to be true. I heard him relate the dream, and my father and all present believed it. The curious part of the story is how he should have dreamed such a thing, being in no way connected with it.

About this time (1825) the several victualling departments of the navy at Plymouth were very inefficient, and divided into three or four establishments—one at Southdown, opposite to Devonport, another at Cremill Point, near Stonehouse, a third at Plymouth, and the fourth in Plymouth town—being several miles from each other, so that the extra expense and delay in provisioning vessels of war was considerable. Upon this being represented to the Admiralty by the Victualling Board, of which the late General Stapleton was the chairman, they determined to make an entirely new victualling establishment, concentrating the whole of the several departments into one, upon a well-organized plan, so that every operation should be carried out with the greatest dispatch and economy.

After much discussion Cremill Point, being nearest to the dockyard at Devonport, and being in other respects, as to depth of water, &c., possessed of peculiar advantages, was finally selected as the best place for the new establishment, and I was ordered by the Victualling Board to prepare the necessary plans, specifications, and estimates, and to see them carried into effect.

Whilst these works were proceeding, a proposal was made to the Admiralty to apply the Cornish system of engines to the Admiralty steam-vessels. Up to this time steam, according to the principle of Watt, not exceeding 5 lb. pressure upon the square inch, was only used, whereas in the Cornish condensing engines steam of the pressure of 37 lb. per square inch was then usual, with much greater effect and economy. I was accordingly sent by the Admiralty to Cornwall to investigate their scheme. I took with me my principal assistant, the late Mr. William Lewin, a very zealous, able, practical engineer. We examined the whole system very carefully at different mines, measured the actual work done and the quantity of fuel consumed, and came to the conclusion that the Cornish system of high-pressure

---

[2] I have always heard this story in Cornwall; and a pamphlet on the subject, now very scarce, was published at the time. Mr. Fox, I have been told, was taken before a magistrate, and made an affidavit concerning it the morning after the occurrence, and before the mail came down.—C. G. C. R.

condensing might be applied to the navy with the greatest advantage, and reported the same to the Admiralty. It was not adopted at the time, although high-pressure condensing engines have since been introduced into the navy; at the present time the pressure has advanced to 27 and 28 lb. per square inch, and in the non-condensing engines to 100 and 120 lb. Watt, although he had tried steam at every temperature, yet generally preferred low pressure; his reason for this must have been the imperfect means then at command for controlling it, and perhaps he was right at the time, for the manufacture of wrought iron had not then made sufficient progress to render it capable of resisting with safety the great power of high-pressure steam. He pointed out the way, and others, through the improvements in the manufacture of iron, have profited by his discoveries. I must not omit, in justice to the very talented Mr. Perkins, his views as to the value of high-pressure steam, upon which he made numerous experiments; amongst other things, he proposed to use it as a destructive power in war, by means of his celebrated steam gun, which created much sensation at the time; and it is rather singular that this has not been taken up in modern times, when every invention which can add to the methods of destruction is most greedily adopted. There can be no doubt that the "*steam gun*" may be used with the most terrible effect in fixed batteries either for musketry or artillery, and probably the day will come when we shall see batteries worked by steam, with a great diminution of manual agency; and the men employed to work them will be so protected as to render the destruction of human life comparatively trifling.

Amongst the other legacies which my dear father had left to me, was that of building London Bridge. He had shown, to the satisfaction of the Committee of the House of Commons, the impracticability of keeping up the old bridge by any amount of repair or alteration; that it could not be rendered fit for the improvement of the river Thames, nor for the continually increasing traffic which must pass over it, at any reasonable expense, and that therefore an entirely new bridge would be by far the most satisfactory. The House of Commons was so fully convinced by his arguments that it unanimously condemned the old bridge, and refused to comply with the recommendation of the City of London that it should be altered, but resolved that a new bridge should be built according to the design made by my father. An Act of Parliament was accordingly passed to this effect, the late talented and energetic Mr. Holme Sumner being the leading member of the Committee. The Corporation demurred to this, saying that there were other engineers equal to Mr. Rennie, and demanding a public competition. The advertisement for designs was accordingly issued, and a great number were sent in and referred to the Committee of the House of Commons and the Government, according to the previous Act. After considering the various rival plans, it was finally decided that that of my father was the only one which complied with all the required conditions, and it was therefore adopted. At this resolution great discontent was manifested by the different competitors. The Corporation of London also objected to this decision, as they considered that they ought to have been the sole judges. But finding that the Government adhered to its decision, they submitted, at the same time urging that they, who were to pay the expense of the new bridge, ought to select the engineer that was to execute the work. The Government agreed to this, subject to their approval; and finally the Committee appointed by the Corporation to carry into effect my father's design, fixed upon me as the engineer-in-chief; Mr. Richard Lambert Jones was appointed the chairman of the executive committee of the Corporation. A very able and efficient chairman he was, and conducted the whole to a conclusion, to the great satisfaction of the Corporation and of the Government.

The design, as I have already observed, was made by my father, but no detailed working drawings, specification, or estimate, had been prepared by him; it fell to my lot, therefore, to do this. As I had acted under my father during the construction of the Waterloo and Southwark bridges, I had become so thoroughly acquainted with his system, that I had no very great difficulty about it.

The design consisted of five semi-elliptical arches, the centre being 150 feet span, the two next arches 140 feet span each, the two side or land arches 130 feet span each; the two centre piers were 24 feet thick at the springing, and the two side piers 22 feet each; the whole was to be built of the best grey granite. The width of the roadway was originally designed to be 48 feet, but was afterwards increased to 54 feet wide, at an extra cost of 46,000*l.*

It was intended by my father that the new bridge should be built on the site of the old one, which was to be pulled down in the first instance, and a temporary wooden one was to be built above it to accommodate the traffic whilst the new bridge was building. It was considered that as soon as the fall of 5 feet occasioned by the old bridge should be removed, the river would be restored to its natural state, and there would then be less difficulty and expense in making the cofferdams and founding the piers and abutments; the old approaches to the bridge would then be preserved, and thus a less quantity of valuable property would be required. The wish, however, of the Corporation to preserve the old bridge during the construction of the new one was so strong, that there was no possibility of resisting it. I therefore yielded to their desire, and agreed to build the new bridge immediately above the old one, and as near as practicable to it; notwithstanding, I felt at the time that there would be considerable difficulty, risk, and extra expense in so doing, on account of the great depth of water in which the piers must be founded, namely, 28 feet at low water of spring tides, and the strong current and fall through the old bridge both during the flood and ebb, particularly during the latter. It should be observed that the old bridge stood as it were upon a hill, the foundations of the piers being from 28 to 30 feet above the bottom of the river immediately above and below it, occasioned by the great fall and scour produced by the contracted waterway; thus it was necessary to secure the piers by large projecting starlings, and to throw considerable quantities of stone continually round them, in order to prevent the old bridge from being carried away. However, there was no alternative but to build the new bridge above the old one, and I accordingly set about taking every possible precaution in order to prevent accident.

As the loose stone thrown round the piers of the old bridge was continually washed into the holes immediately above and below, it was in vain to attempt driving the piles for the cofferdams of the piers until this stone was removed, which was done by dredging. The cofferdams for the piers were elliptical in form, this shape being the best adapted for resisting the strong current in which they were placed; they consisted of two main rows of piles each 14 inches square, each pile being properly hooped and shod with wrought iron, and driven 25 feet into the bed of the river. These piles were connected together in the horizontal direction by three rows of braces 15 inches square, namely, one at the level of the lowest tides, another at the level of half tide, and the third at the level of high water. At every 10 feet the two rows of piles were fastened together with wrought-iron bolts 2½ inches diameter, which passed through the horizontal braces or walings, as they are technically termed, and were secured outside and inside with additional wooden cleats 15 inches square and 8 feet long, so as to cover the joints where the main horizontal braces met; outside of these were large iron plates, and as the bolts were screwed at each end, they could be tightened up to the full bearing without crushing the timber. On the outside of these two main rows of piles was a third row of the same dimensions, and driven the same depth into the bed of the river at a distance of 6 feet in the clear from the two main rows, and connected together with a tier of horizontal braces, and to the two main rows of piles with bolts, cleats, and plates of the same dimensions as those already described. When the piles had been driven to their proper depth, and had been properly secured to each other as above described, the joints between every pile, which had been previously fitted to each other, were well caulked with oakum, and the outside joints were covered with melted pitch, so as to render them water-tight; the spaces between the three rows of piles were then filled with strong well-puddled clay.

In addition to the above three rows of piles, there was a fourth row on the inside, driven down in the form of a parallelogram, corresponding with the exact size of the foundation of the piers, and to the same depth as the outer piles. Every tenth pile, and those at the corners or angles, extended up to the level of low water. Upon the heads of these piles longitudinal and transverse braces were fixed across the inside of the dam, at the level of low water, half tide, and high water; so that the dam was braced internally and externally in every direction to resist the pressure of the water, like a well-made cask. There was a powerful steam engine, with the requisite pumps, attached to each dam, to remove any water which might either rise from the foundation or from the outside. Each dam was provided with a trunk secured by a valve 3 feet diameter, laid at the level of low water, so that in the event of any unusual pressure of water coming against the dam, these valves were opened, and the dam was then filled with water, and all mischief was prevented. The first pier cofferdam on the Southwark side was completed, the water pumped out, and the earth was excavated to the depth of 30 feet, going below low-water mark of spring tides; the bottom consisted of the stiff London clay. Piles, consisting of Baltic fir, elm, and beech, 22 feet long, and 12 inches diameter in the middle, properly hooped and shod with wrought iron, were then driven 20 feet into the solid ground, or until, with a weight of 12 cwt. falling 18 to 20 feet, they would not move above an inch at a blow. These piles were driven 3 feet 6 inches from centre to centre, both in the longitudinal and transverse direction. After having driven them, their heads were cut off and accurately levelled. The loose earth between their heads was then removed, to the depth of 12 inches, and the spaces filled in with stone bedded in concrete; all the rows of piles were then connected together in the transverse direction by Baltic fir sills or beams 14 inches square, well fitted to each pile head by jagged wrought-iron spikes 20 inches long and three quarters of an inch square, driven through the sills into the pile heads below; the spaces between the sills were well filled in with stone and brickwork; another row of sills was then laid in the transverse direction above the pile heads and spiked down to the lower sills in the same manner; the spaces between the upper sills were then filled in with stone and brickwork. The whole surface of the foundation was covered with elm plank 6 inches thick, closely jointed together and bedded in mortar, and well spiked down to the sills below with jagged spikes 10 inches long and half an inch square. Upon this platform the masonry was built, each course diminishing in length and width by a series of offsets 12 inches wide, until they reached the shaft of the pier, when it was carried up solid to the springing of the arches. The whole of the exterior masonry was of the best whitish-grey granite, and the interior stone was of the best hard Yorkshire grit stone from the quarries of Bramley Fall. The abutments were constructed upon piles and masonry of the same character as the piers.

The first stone was laid with considerable ceremony on the first pier coffer-dam from the Surrey shore by His Royal Highness the late Duke of York. The dam was fitted up with great taste like an amphitheatre, with seats all round, the whole being covered at top with a handsome coloured canvas awning adorned with numerous flags of all nations. The Lord Mayor, assisted by the Aldermen, Common Councilmen, and Committee, with Mr. Jones, the Chairman, attended in great state, and everything went off well. After this pier had advanced nearly to the level of high water, one day whilst examining it, standing upon one of the cross beams, my foot slipped, and I fell headlong into the dam upon the top of the masonry; fortunately, my left foot caught in a nail in the beam, and I hung by it for a few seconds. This somewhat broke and changed the direction of my fall, and I pitched upon an inclined plank, upon which I slid until I struck my head against a stone; my hat deadened the blow; as it was, however, I was cut about the forehead and half stunned. The after effects of this fall were very serious; my whole system got such a severe shaking, that I did not recover thoroughly until nearly ten years afterwards, and I carried on my large professional business with the greatest difficulty. The works made satisfactory progress, and the centres for the first and second arches from the Surrey shore were soon fixed.

Each centre was composed of eight ribs, framed upon the truss principle, resting upon a continued series of wedges in one piece, laid horizontally and resting upon tressels or legs formed by the piles of the cofferdams, which had been cut off and levelled for that purpose. The mode of setting and fixing the ribs was the same as that adopted at the Waterloo and Southwark bridges. A large lighter was constructed especially for this purpose. In the centre was placed a strong framing, which rested upon eight screws, four in each row, working in a strong cast-iron box, to which levers were attached, by means of which the screws and framing above them could be gradually raised and lowered at pleasure; at one end of the framing there was an upright scaffold. The centres, I have already said, consisted of eight main ribs framed together separately. As there was no room to frame these centres near the bridge, a special workshop and wharf were provided by the contractors at Millwall, in the Isle of Dogs; when ready they were launched in one piece, from a properly-prepared platform, into the river, and towed to the Southwark end of the bridge, where the lighter, with its apparatus of powerful sheers, crabs, and tackle, was in readiness to receive them; by these means they were gradually hauled up an inclined plane, and then raised upright upon the platform, supported by the frame and screws beneath, and firmly braced to the scaffolding in the lighter; two centre frames were thus placed upon it at one time, and adjusted by the screws to an extra height of 2 feet, so as to allow for any deficiency in the rise of the tide. Two ribs having been placed upon the framing, the lighter was hauled off from the shore and placed in front of the opening in which the centre was to be, the lighter being moored 100 yards from it, about half an hour before high water; upon the turning of the tide it was gradually allowed to float down with the ebb current to its place. By the time that the lighter with the centre ribs arrived in its exact position there was always 2 feet to spare, in order to allow for any deficiency of the tide; as the tide fell the two ribs were securely deposited in their places upon the framing and wedges below them. It should be observed, that upon the wedges there was an additional framing so as to reduce the weight of the main ribs of the centring. When the main ribs had been deposited upon the framing wedges, they were securely braced together until the whole number of ribs required for each centre was fixed, when they were all braced firmly together longitudinally, transversely, and diagonally. This mode of fixing centres for arches of any span was most successful and economical, and I believe that my father was the inventor of it, if it may be termed an invention. My excellent and talented friend, the late Robert Stephenson, adopted the same method for fixing the tubes of the Conway and Menai Straits bridges. He told me that he was not aware that my father had proposed it before him; but in the 'History of the Britannia and Conway Bridges,' edited, I believe, by the present Mr. Edwin Clarke, who was employed under Mr. Stephenson at the Conway Bridge, it is expressly mentioned that my father had previously employed the same plan.[3]

But to return to London Bridge. The works proceeded successfully; the fifth or last arch on the City side was completed in 1829. The centres of the first, second, and centre arches having been removed, it was found that they had subsided only 3 inches each, which was the exact distance that had been allowed for, with an extra half inch for the centre arch. Upon examining the arches and piers after it was supposed that they had subsided, it was found that there had been an unequal sinking, that the two centre piers had subsided on the east side slightly more than on the west side. I was much puzzled at this, and could not for some time account for the difference. Three or four of the quoin arch-stones of the second arch from the City shore had been fractured for about 8 or 9 inches; this, however, was of no

---

[3] I may mention here that before Telford built the suspension bridge near Bangor, Mr. Rennie proposed a cast-iron bridge of several arches, on the site of the present Britannia Bridge, which would ultimately have carried the railway, but the scheme was considered too bold at the time. The possibility of the construction of the present magnificent bridge, which marks an era in engineering, is of course due to the great improvements in the manufacture of iron since Mr. Rennie's time. Thus every age has its specialty, and what cannot be done at one time becomes practicable at another.

consequence, for it is always difficult in such large arches to get the workmen to bed the quoins accurately. The same thing had occurred in the Waterloo and other bridges; I was therefore led to investigate the subject more thoroughly during the construction of the new bridge. It was found that the cofferdams for the piers in several instances were made in front of the openings or arches of the old bridge, which could not be avoided. These dams necessarily still further obstructed the waterway through the old bridge; I therefore felt that it was absolutely essential to find relief for the ebbing and flowing tides, detained both by the old bridge and the cofferdams of the new bridge, and accordingly recommended that, on the south side of the main arch of the old bridge, which was only 80 feet wide, two arches should be thrown into one, and that the intermediate pier should be removed, by which means a single wide opening would be made facing the space between the cofferdams of the arch and the centre on the Surrey side of the new bridge, but the difficulty of doing this arose from the fact that the traffic over the old bridge could not be interrupted for a moment. I soon, however, found an expedient; I ordered that the requisite number of logs of the best Baltic fir timber 15 inches square should be prepared sufficiently long to extend over the two arches and piers of the old bridge which I proposed to remove. Having got these ready, I stopped up one half of the roadway, leaving the other half open for the traffic, and working night and day, I laid these whole timbers spanning the two arches to the adjoining piers close together, bolted them to each other, and secured them to a longitudinal bearer of the same dimensions imbedded in the masonry of the adjoining piers, so that the timbers which were to form the increased opening rested firmly upon them. Having done this, I removed the masonry of the intermediate arch by degrees from beneath the timber girders, placing a strong diagonal strut or support under each girder as I proceeded, at the same time connecting together all these diagonal struts. Having completed one half of the temporary arch or opening, the traffic was diverted over it, and the other half was completed in the same manner; the whole operation was accomplished within ten days, and the traffic was not stopped for one hour. The intermediate pier of the old arch was then removed entirely. As the work advanced to the fourth pier on the City side another similar opening was made. These alterations relieved the river materially, and enabled the works to be carried on much more securely, and greatly diminished the fall through the old bridge at low water. Nevertheless, there was increased scour against the dams where the openings were made, which occasioned the slight unequal subsidence before mentioned. The last or fifth arch was completed on the City side, January, 1829.

It has been mentioned that Lord Liverpool's Government had always taken the greatest interest in the construction of new London Bridge, and gave the Corporation of London every support in their power, not only for the accommodation of the great traffic across it, but for the improvement of the Thames, which the removal of the old bridge would effect; and amongst the men most zealous in Lord Liverpool's Government were the late Marquis of Salisbury, then Lord Cranbourne, and the late Earl of Lonsdale, then Lord Lowther; both these noblemen had considerable talent, and, fully alive to the advancing ideas of the day, were mainly instrumental in forwarding this great work. That amiable, able, and conscientious nobleman, the Earl of Liverpool, had succumbed to the effects of a paralytic stroke, and the Duke of Wellington was now Premier; he took the greatest interest in the promotion of London Bridge and everything connected with it; so that the Corporation of London, who had hitherto been radically inclined, or had rather been opposed to the Tory Government, turned rather Conservative than otherwise, and the Duke became most popular with them; he invariably, whenever he could, accepted their invitations to Guildhall and the Mansion House, and was always received by them with the greatest respect and attention. Richard Lambert Jones, the Chairman of the London Bridge Committee, was his particular favourite, and he always shook Jones by the hand, a favour which he did not accord to everyone.

At this time the bridge had made considerable progress towards completion, and the important question arose, what was the best plan for the approaches? It was originally intended, in order to save expense, that the old line of Fish Street Hill, on the City side, should be adopted, pulling down such of the houses on this line as might be necessary to make the incline not steeper than 1 in 30; but inasmuch as the great traffic of Upper and Lower Thames Street interposed materially with the main body of the traffic coming north and south, it became most important to consider how this might be avoided, and the old idea of making an arch over Thames Street was revived, and was favourably received by the London Bridge Committee; the question was accordingly referred by them to me. I had always felt that the old approach by Fish Street Hill was a very great difficulty; but, restricted as I was to the old line of approach, I felt that I could not get out of it without some pressing necessity; I was therefore glad to have the opportunity of reconsidering the subject, particularly when proposed by the Committee. It was quite clear, that in any case an arch over Thames Street to separate the great cart traffic of that quarter from the main coach and passenger traffic coming from the City, Southwark, and the northern and southern parts of the town, was absolutely necessary, and I should have proposed it myself in the first instance, if there had at that time been any chance of its adoption, and I am quite sure that my father would have done the same. But if it had been proposed to divert the traffic from the old line of Fish Street Hill at the first, it is most probable that it would never have been carried; I was obliged, therefore, to confine myself to the old approaches, leaving the future to develop itself. The Committee, however, having taken up the idea of an arch over Thames Street, I was only too happy to fall in with it; but as nothing could be done without the sanction of the Government, it was determined by the Committee to bring the subject before them at once; the more so, as it would require a much larger sum than originally calculated to make the approaches, for which the Committee had no funds, and a new Act of Parliament would be required. Plans and estimates were accordingly prepared for the new approaches, and submitted to His Majesty's Government. The Duke of Wellington took the greatest interest in the subject, and investigated it to the fullest extent; he visited the place himself, he interrogated the Chairman of the Committee and myself most closely, and at length, being fully convinced that it was necessary, gave the consent of the Government, provided that the funds could be found. In order to meet this difficulty, the Committee proposed to increase the coal tax, which, with the necessary sinking fund, would pay off the whole sum necessary to make the new approaches, which were estimated at 1,400,000*l.*, in a given number of years. The Government consented to this, and the requisite notices, plans, &c., were given and deposited in the month of November, 1828.

Early in 1829 the Bill was brought into Parliament, and was most strongly opposed by the great northern coal-owners, Lord Durham, Lord Londonderry, Lord Lauderdale, and others, as well as by a considerable body of Londoners, and after a hard fight the Bill passed the House of Commons and got into the Lords; but here the opposition was more violent and powerful than ever. The Duke of Wellington, however, having been thoroughly convinced of the necessity and justice of the measure, determined that it should be carried if possible, whilst the Opposition were equally determined to throw it out. The Committee accordingly met in the Painted Chamber of the House of Lords, and the extraordinary number of forty peers, including seven cabinet ministers, assembled, the Duke of Wellington being in the chair. The Opposition comprised, amongst others, Earl Grey, his son-in-law the Earl of Durham, the Marquis of Londonderry, the Earl of Lauderdale, &c. Such a committee upon a private Bill has never since been seen in the Lords, and perhaps never will be again. The brunt of the battle fell upon me; I was the leading witness, and had to establish the whole case. I never felt more nervous in my life; I was to be prepared upon all points to defend everybody else's errors as well as my own. I knew there were several weak points, and though I had an excellent case upon the whole, I still felt the greatest difficulty about it; I knew also perfectly well, that if I broke down, my career as an engineer was ruined for ever, for the Government had pinned their faith upon me; I therefore had made myself thoroughly master of the subject, and determined to sink or swim with it. Mr. W. Montague, then surveyor to the City, was a very sensible, practical man, and of great experience in the valuation of property, and possessed considerable influence with the Corporation; but whether it was jealousy at my being so much younger than himself, or whether he thought the post of honour should have been given to him, I cannot say, but he did not act cordially with me. That very remarkable man, Richard Lambert Jones, the Chairman of the Bridge, with his usual tact and sagacity, at once saw this, and thought that if he was examined there might be some discrepancy which our enemies would take advantage of; he therefore, with the concurrence of his Committee, determined that I alone should fight the battle of the estimates, upon which the whole fate of the Bill depended. It was well, both for Montague and myself, that this course was taken; for Montague, when the first Bill of 1821 was carried, had made the surveys of the property to be taken for the approaches on both sides of the bridge, which was confined to 180 feet above the old bridge. This limit ought to have been taken in a direct line, at right angles to the old bridge; but unfortunately it was taken according to the line of the shore, which near the old bridge receded considerably, whilst the old Fishmongers' Hall projected beyond it, making the direct line, if taken, as it ought to have been, at right angles to the old bridge, 20 feet longer than if measured according to the line of the shore. This difference of 20 feet rendered it necessary that old Fishmongers' Hall should be removed, and negotiations immediately commenced between the London Bridge Committee and the Fishmongers' Company with respect to the purchase of this piece of land. The Fishmongers' Company behaved very fairly;

they said that they did not wish to build a new hall, as the existing structure answered their purpose very well; but if they were compelled to part with this 20 feet of ground, they must build a new hall, which they did not want to do. However, as they had no wish to impede the construction of the bridge and approaches, they were willing to sell the strip of land on which part of their hall stood, namely, 20 feet in width, for 20,000*l*. 1000*l*. per foot at first sight appeared a very large sum, although at present it would be considered remarkably cheap. The Committee of the bridge, looking at the matter fairly, resolved to pay the Fishmongers' Company the required sum in full compensation for everything, and the Fishmongers' Company might, if they thought fit, build a new hall at their own expense. This agreement was then settled, provided that the Bill for making the new approaches by an arch across Thames Street should pass the legislature. Knowing this fact, I was very anxious that this error of Mr. Montague's should not come out before the Committee of the House of Lords, as it was no doubt a great mistake, and if it had transpired, it might materially have injured our case. I determined, therefore, during my examination, to keep it out of view as far as possible; at the same time, if it was fairly put to me, of course I felt myself bound to give every honest explanation. Perhaps Mr. Montague was not anxious to be examined; anyhow, he was not, and the whole burthen fell upon me.

The opposition in the Lords' Committee was headed by the Earl of Durham, a very able and intelligent man; he would have made an excellent lawyer if Providence had so designed it, and in this instance he conducted his case admirably. I got through the examination in chief very well, and the opposing counsel commenced his cross-examination, and made nothing of it. Then Lord Durham got up and for three days I underwent as severe a cross-examination as I ever experienced, either before or since. He seemed to be aware of all the facts, and omitted nothing to render his case triumphant. I always feared that he was coming to the mistake about Fishmongers' Hall, but he never did, and I had to lead him away from it as far as possible; at last he got to the frontages in the different streets of the respective parishes which were proposed to be taken, and the new frontages of the new streets which were to be erected. Here I showed very clearly that the lineal frontage according to the proposed new streets would be greater than the frontages taken away. This, however, did not satisfy his Lordship, for he contended that some of the parishes would lose a great deal more frontage than they would obtain. I had some idea that this would be the case, and therefore did not think it necessary to take the individual frontage gained and lost by each parish. I thought it was quite sufficient to know that upon the whole a greater line of frontage would be given by the new approaches, than taken away from the old. The parishes which had petitioned against the Bill on this account, argued that in some of them the rates would be greatly diminished, and that in others they would be greatly increased, which would cause an unequal and unjust distribution. I still kept to my point, and said upon the whole, without going into detail, the parishes would be the gainers, and it was for them to adjust the rates amongst themselves. Lord Durham was very indignant at my obstinacy in maintaining this point, and tried in every way to make me confess that I had made a mistake; I nevertheless stuck to it, and said that if I had tried to equalize all the frontages, my survey might have extended to the Tower, and there would have been no end of the expense. By this time he was losing his temper, and said that if I was not very careful I might go to the Tower still. At this the Committee smiled, and his Lordship, being fairly baffled, sat down, and I, having been told that they had no further questions, left the witness-box with the greatest alacrity.

It was considered that I had made out the case for the Bill so completely, both as regards the estimates and the absolute necessity, in a public point of view, of carrying the new approaches into effect, that no other professional witness was put into the box, except the present able chamberlain, Mr. Scott, then chamberlain's chief assistant, who gave such clear and straightforward evidence with regard to the funds which were disposable by the Corporation, and the way in which they were administered, that the Committee of the Lords passed the Bill for the London Bridge Approaches with but few dissentient voices. The Duke of Wellington, and five or six Cabinet Ministers with him, attended every day, and in fact kept Parliament sitting to pass the Bill. It was curious to observe that he never for a moment interrupted the opponents of the measure; he gave them full scope, and never said a word until they had had their say, then he put the question, and carried it without difficulty.

After the fifth arch, the first on the Surrey side, was keyed, the sinking was observed to be 4 inches, or about 1 inch more than the others. I could not very well account for this. I also observed that two of the quoins on the south-east end of the fourth arch from the Surrey shore had splintered off at the soffit, but no crack could be observed in the spandril walls; but upon levelling the piers, it was found that the east end had subsided from 10 to 14 inches more than the upper or west end. This I could account for in no other way but that there had been a greater scour here than at the upper end, and that the piles had to some extent been laid bare. I levelled the arches and piers constantly after this for several months from a fixed standard gauge, but could find no alteration; I therefore felt satisfied that the whole of the pier abutments had come to their final bearing, and the works were continued as fast as possible towards completion.

In 1830 the Duke's Government retired, and he himself became as unpopular as he before had been popular; yet he never deserted London Bridge, and was more frequent in his visits than ever. I often used to attend him at five and six in the summer mornings; he generally came on horseback, and remained from half an hour to an hour, and sometimes more if necessary. At length the whole bridge and the approaches were completed, and His Majesty, King William, at the special request of the Corporation of London, condescended to fix a day on which he would open the bridge in person. Earl Grey, who had strongly opposed the Bill for the improved approaches two years before, now, as Premier, accompanied His Majesty to inaugurate the opening of the same. Perhaps, as a spectacle of the kind, it was the most brilliant of any that had taken place for fully a century; and the whole Corporation, including the Bridge Committee, did everything in their power, for the honour of the City of London, to render the pageant as splendid as possible. The whole of the space at the north or City end of the bridge was covered with a magnificent tent, several hundred feet long, decorated in the greatest taste with the flags of all nations, and with ancient and modern arms grouped round the standards forming the supports of the tent, under which were arranged tables for 1400 guests, for whom a splendid collation was provided. His Majesty, King William, came in the royal barge in state, accompanied by all his ministers, and upon his arrival was greeted with a salute of twenty-one guns from the Tower. All the piers and arches were decorated with lofty standards displaying the national emblems; the whole of these, as well as the great tent and decorations, were under the direction of Mr. Stacey, of the Ordnance department of the Tower, and the greatest credit is due to him for the admirable taste which he displayed. The ceremony consisted in King William walking over the bridge, accompanied by his ministers, the Lord Mayor and Corporation, and the Bridge Committee. When His Majesty arrived at the Southwark end a balloon ascended, carrying Mr. Richard Crawshay; the Tower guns then sent forth another salute, and King William and his *cortège* returned to the tent at the City end of the bridge, where they partook of lunch with the usual ceremonies, and returned by water as they came, with another royal salute from the Tower. The day was remarkably fine, the river was covered with boats filled with gaily-dressed people; the wharves, warehouses, and bridges were thronged with spectators; in fact, it was a great metropolitan holiday; everything went off well, and all appeared to be satisfied; I was particularly so. I had been very hard worked, I may say almost night and day, for some time past, to get things ready, and was of course rather tried; nevertheless, the success which attended the whole rendered me completely unmindful of myself, and I forgot all my fatigues, for I was amply rewarded for all my troubles and anxieties.

A few days after the opening of London Bridge, Jeffreys, the cheesemonger of Ludgate Hill, presented a petition to Parliament through the well-known Henry Hunt, stating that the new bridge was coming down. Jeffreys was very much annoyed because he had received no recompense for his repeated proposals with regard to the new approaches, though he was never regularly employed; but he was one of those active, intelligent persons, who are always interfering in matters which do not concern them. If he had devoted himself to his own business he might have done well; but, unfortunately, he neglected this, and fancied himself a great engineer, a post for which he was absurdly incompetent. He mistook his vocation, and in attempting to do that for which he was wholly unfit, he neglected the business of a tradesman, for which he was thoroughly suited. The petition ended in its being referred to a Commission, consisting of J. Walker, Telford, and Tierney Clark, who examined the bridge carefully and made their report, which was colourless and came to nothing, as the subsidence spoken of had taken place two years and a half before, and had not increased, nor has it done so up to the present day.

In all works of great magnitude, and particularly in such a difficult situation as that in which London Bridge is built, it is impossible to be certain of attaining absolute perfection, but the Committee, being perfectly satisfied of the stability of the new bridge, determined to remove the old one forthwith, and I received orders accordingly. The removal was contracted for by Messrs. Jolliffe and Banks, for the sum of 10,000*l.*, they having the benefit of the old materials, except in so far that they were to fill up the holes in the river below both bridges, to the extent of 14 feet below low water of spring tides, which was rather more than the average depth of the river in the vicinity; they were also bound to remove the whole of the foundations of the piers, starlings, &c., of the old bridge, to the same depth. The whole of these operations were completed in the year 1834, when the river, after a lapse of 658 years, was restored to its natural state.

The history of old London Bridge is replete with interest, and forms a very curious epoch in the annals of bridge building before the embankment of the river Thames by the Romans. The Southwark side, which is in many places considerably below the level of high water of spring tides, was frequently flooded, and numerous creeks were formed in it, so that the river must have been very unequal in its depth, and filled with numerous shoals, and fordable at low water in several places near London; and there was evidently an appearance of a ford at the site of old London Bridge, as in many cases the piers were founded on the original ground, which must have been dry, or nearly so, at low water; these piers were in many instances wider than the adjoining arches, so that they offered considerable obstruction to the free flow of the tidal and fresh waters through the bridge. These obstructions necessarily increased the velocity and scour of the current, and threatened to carry away the old bridge. Great starlings, or timber casings of piles, were erected round the bridge piers, and the spaces between them were filled in with chalk. These starlings still further narrowed the openings of the arches, so that at low water some of them were little more than 8 or 10 feet wide, and the obstructions became so great, that the fall at low water increased to 5 feet perpendicular. Five openings on the south end and one in the north end were occupied by water-wheels for pumping water for the City. The obstruction caused by these works was so great that the celebrated Smeaton was employed by the Corporation of London to take down the two arches near the centre, and replace them by a single one of 80 feet span.

The original bridge is said to have been built in the year 1176; but between that time and the period of its removal in 1834, it underwent so many alterations and changes that it may almost be said to have been rebuilt several times. It was originally covered with houses, as everyone knows, leaving a narrow passage between for the traffic. To describe the numerous alterations would require a large volume, and the reader is therefore referred to an interesting account of this curious old structure called 'The Chronicles of Old London Bridge.'

Numerous speculations were made by scientific men, engineers and others, on the effect the removal of the old bridge would have upon the river. My father pointed out the probable results in a very simple manner; he said that the river was in an artificial state in consequence of the old bridge acting as a dam to the free passage of the waters upwards and downwards, both tidal and fresh; and the consequence was, that the river above had to a certain extent accommodated itself to circumstances. By the removal of this obstruction the river would soon be reduced to its natural level; the fall of 4 to 5 feet through the bridge would be removed, consequently the tide would rise so much higher and fall so much lower above bridge, and so much more tidal water would be admitted above the old bridge throughout the whole length of the tidal flow as far as Teddington Lock; and this increased quantity of tidal water passing up and down twice each way during the twenty-four hours would scour the bed of the river, and thus remove the great quantity of mud deposited along the shores. And further, that the drainage of the metropolis, and in fact the whole valley of the Thames, at least as far as Teddington, would be greatly improved; and the water, being constantly changed, would be clearer and fresher. He further said, that the actual level of high water would scarcely be materially affected, perhaps not exceeding 5 or 6 inches; and lastly, that the process would be gradual, and that it would take several years before the river would attain its final and natural state. Such has proved to be the result.

As the works of new London Bridge proceeded attention was drawn to the irregular outline of the wharves, which were not only unsightly but extremely injurious to the regular passage of the waters. At this time also, people having visited Paris and other great continental cities, were struck with the architectural beauties which they had seen, and became much disgusted with the mean, shabby-looking appearance of London; and well they might, for there were no great leading thoroughfares worthy of the name. Cockspur Street, leading from the Strand to Pall Mall, was scarcely 20 feet wide; the Royal Mews occupied Trafalgar Square; the Haymarket was encumbered by haycarts; Cross's Menagerie and Exeter Change blocked up the Strand near Waterloo Bridge; the connection between Holborn and Oxford Street was round by old St. Giles', and Farringdon Street was filled with a market, and surrounded by undertakers. Regent Street had been commenced under the direction of that clever architect Nash, which, from his novel mode of grouping shops into distinct masses of different styles, excited considerable attention, and was totally different to anything we had hitherto seen in the metropolis. This great and really magnificent street was, I am told, entirely his own idea, and according to the opinion of the late Sir Robert Smirke, was a Herculean task, on account of the great variety of interests to be dealt with, and he told me that nobody but the indefatigable Nash could have carried it through. He built a house there for himself, now the Gallery of Illustration, opposite to the Club Chambers near Waterloo Place, where George IV. honoured him with a visit. In fact, just about this time there was a perfect mania for architectural improvements.

A committee, called the Committee of Taste, was accordingly appointed, in order to design such improvements as were imperatively required in the neighbourhood of Charing Cross, the Strand, and Holborn and Oxford Street. This Committee consisted of the late Lord Farnborough, John Wilson Croker, Sir John Soane, Sir Robert Smirke, Nash, and two or three others, and certainly no committee ever discharged its duties better. To its labours we are indebted for Trafalgar Square and the improvements in the Strand, Cockspur Street, the Haymarket, the old Opera House, and those between Oxford Street and Holborn, which are really very good, and the architecture, although not altogether faultless, is nevertheless, taken as a whole, very effective; in fact, nothing like these improvements has been effected since. The new street from Waterloo Bridge to Oxford Street, undertaken soon after, has been a miserable failure; instead of taking a direct line, they availed themselves as far as they could of the old miserable intervening streets, so that this thoroughfare, which ought to have been one of the best in London, is now one of the worst, and the increase in the value of the property on each side has been very little. But if this street had been made in a straight line, and of ample width, the shops and buildings on both sides would have been of a superior character, and would have yielded far higher rents, which would have gone a long way towards paying part of the expenses, if not the whole.

About this time Sir F. Trench, who moved in the most fashionable circles and was a great amateur in architecture and fine arts, was seized and enraptured with the idea of constructing quays along the banks of the Thames, between Whitehall and Blackfriars Bridge, and converting the space so recovered from the shore of the Thames into a handsome carriage-drive and promenade ornamented with gardens and fountains. He applied to the late Mr. Philip Wyatt and myself to assist him in preparing the designs and in obtaining an Act of Parliament to carry it into effect. At the request of the London Bridge Committee I had previously, in company with the late Mr. W. Mylne, prepared a general plan for this object, but it went no farther on account of the difficulty of raising the funds. Trench, however, overlooked this, and said he had no doubt that sufficient money would be obtained. He accordingly, with his great influence and indefatigable activity, formed a committee of the highest class; neither were the ladies excluded; amongst others, the beautiful Duchess of Rutland took the greatest interest in the undertaking, and at the first meeting, which took place at Her Grace's house, she was unanimously voted to the chair, and conducted everything in the most business-like manner. Lord Palmerston, then Secretary for War, took a leading part, and it is singular that many years later his Lordship, then Premier, should have proposed a similar measure, and the continuation of the coal duties for carrying it into effect, which was adopted; but when we proposed the undertaking and the mode of raising the funds, notwithstanding our powerful committee, the idea was considered as chimerical. Trench, however, was so confident that the means would be found, that he went to considerable expense in preparing a book ornamented with numerous beautiful engravings showing the effect which would be produced by the undertaking, to which Wyatt and myself contributed our share. A solicitor, named Leech, was appointed, notices for going to Parliament were duly given, and the necessary plans and estimates were deposited; but when the question as to the means of raising the funds came before the Managing Committee, everybody was at a loss. To form a company appeared impossible, as it did not appear that sufficient revenue would be derived from the undertaking; and as to raising funds by increasing and extending the coal tax, the Government was decidedly opposed to it; they said they had done as much in this way as possible for London Bridge, and that the public would not submit to any further tax of the kind. Thus, after a considerable deal of useless trouble and expense, Trench, Wyatt, and myself were obliged to abandon this great undertaking, which has since been carried into effect nearly upon the same principles as we recommended.

As to the architecture of the approaches to London Bridge, I referred the subject to my brother-in-law, Cockerell, a very accomplished and competent authority, and I exhibited his designs to the Committee as well as some of my own. They, however, considered them to be too ornamental and costly, although they were as plain and simple as these important approaches rendered necessary. The Committee, having rejected them, referred the subject to the late Sir Robert Smirke, then one of the Crown architects, and he designed the present buildings on both sides of the bridge, as far as King William Street on the north, and the old townhall of Southwark on the south; and certainly, with all due respect to my late friend Sir Robert Smirke, a more unworthy set of buildings was never designed. Thus not only has a rare opportunity of making handsome and appropriate buildings to one of the greatest thoroughfares in the world been lost, but the buildings are so low and badly built, that the advantages of the ground, which it must have been foreseen were capable of almost unlimited development as regards rental, have been in a great measure thrown away.

Whilst building London Bridge I had also numerous other works to attend to, namely, the Admiralty works, the harbours of Ramsgate, Sunderland, Donaghadee, Port Patrick, Kingstown, and Port Rush, Staines Bridge, the bridge across the Serpentine in Hyde Park, finishing the Eau Brink Works, the Nene and Witham outfalls, the Ancholme Drainage, together with a good deal of miscellaneous business. As the harbours and Admiralty works are fully described in my book on 'British and Foreign Harbours,' I will proceed to the drainage: first, the Eau Brink. I have already said that the Great Cut was opened in 1821, just at the period of my father's death; according to the latest Act, the engineers of drainage and navigation were obliged to report within twelve months after the opening of the cut; they were obliged to examine the whole, and report how far they had been completed, and what further was necessary in order to render them efficient. Mr. Telford and myself accordingly devoted several days to this; on examining the Eau Brink Cut we found that it had been made in exact conformity with Huddart's design, as specified by the Act; but, notwithstanding, the upper end was too small, and the scour there was so great that it threatened to break through the bank across the upper end of the old channel, and thus revert to its old course. We therefore recommended that it should be increased one-third in area, the greatest part of the increase being at the upper end, and that the money destined in the estimate for clearing away the shoals in the river between the cut and Denver Sluice should be used for the purpose of widening the cut, as the bed of the river did not require clearing. This report took the Commissioners by surprise. They said they had been deceived, and did not believe that it was required, and would have the whole subject investigated by other engineers, which was accordingly done. However, it ended in proving that Mr. Telford and myself were right, and the enlargement of the Eau Brink Cut was made under my direction; this had the effect of lowering the low-water mark at the upper end 2 feet more, making 7 feet altogether. The scour of the cut was so much more than estimated, that the banks between Denver Sluice and the cut were in many places undermined, the channel was diverted from the old quays in Lynn, and several buildings on the opposite shore were washed down, and as the Eau Brink Commissioners were bound under the Act to compensate for any damages done to any interest or party, they, the Commissioners, were compelled to pay for all these damages—50,000*l.* to the bank owners, 28,000*l.*, and 700*l.* a year to Lynn Harbour, 10,000*l.* to the Marshland Drainage, and other minor sums. Having settled this, they obtained an Act to relieve themselves from all further liability.

My father had been employed by the Duke of Bedford, and other great land-owners in the north level of the fens, to consider the best plan of improving the Nene, so as to render it a good outfall for the drainage of the extensive low fen lands bordering it, which, on account of their bad drainage, were frequently subject to floods, and comparatively valueless. My father wrote a very able report on the subject, and recommended that the Nene should be deepened, enlarged, and lowered throughout its course from Peterborough to the sea; that a new channel should be made from a place called Rummery Mill above, to the Horse-shoe Bend, below Wisbeach (for the course followed by the river through the town was so crooked and confined that it could not be sufficiently improved without incurring great unnecessary expense) and that the navigation to and from the town should still be preserved by locks connecting the old with the new channel; and also to make a new outfall for the river from Kinderly's Cut to a place called Crab Hole, in the Great Wash, where there was ample depth of water. He said that the new outfall might be made partly within the estuary and partly in the marshes without, to Skates Corner, but that then it would neither be so direct nor so effectual as if made entirely within the banks of the Nene estuary. Mr. Rennie's report was approved of, but the necessary means for carrying it into effect were wanting, so the matter for the time lay dormant. Subsequently Mr. Telford and myself were appointed the engineers, and reconsidered the whole matter. Finding the people of Wisbeach were violently opposed to the main channel passing by their town, we were obliged to give up this part, as well as the upper portion of the channel to Peterborough, and confine ourselves to the improvement of the outfall below Wisbeach, and to commence the new outfall near the lower end of Kinderly's Cut. I strongly advocated my father's plan of making the new outfall direct to Crab Hole, within the old banks, but as the first expense would have been a little more, although far less in the end, as has been proved, it was resolved to make the new outfall to Skates Corner, partly within, partly without the old banks.

An Act of Parliament was accordingly obtained, and Messrs. Jolliffe and Banks became the contractors. The works commenced, and the outfall was opened in 1831. Whilst it was in progress Mr. Telford and myself frequently visited the works together, and in June we went down the old estuary of the Nene in a boat at low water, for the purpose of examining more minutely the state of the channels near Crab Hole and Skates, where we proposed that the new outfall should enter the estuary. It was a very stormy day, accompanied by lightning, thunder, rain, and a strong south-west wind. We got as far as Crab Hole at low water, when the weather beat us completely, and we were obliged to walk over the muddy shore half-way up to our knees, and drenched to the skin. We had sent some refreshment to an old house, called King John's House, near the bank, said to have been erected during his reign, and to have afforded His Majesty shelter after his retreat from Lynn. The rain now came down heavier than ever, so that we had no alternative but to retrace our steps back to the dirty old "public" at the Ferry, called Cross Keys, about 3½ miles distant. We got back, thoroughly soaked, about three in the afternoon. I immediately stripped and went to bed. Old Telford, being a strong, hearty man, of about seventy, instead of following my example, ordered a large fire to be made in the only sitting-room there was, called for the newspaper, and sat himself down to dry. After two hours' nap I was thoroughly refreshed, and went down to the sitting-room. When I entered there was such a steam that I could hardly see anything; but, approaching the fire, found Telford had nearly dried himself, and he abused me thoroughly for being so effeminate as to go to bed. He suffered, however, severely afterwards for his imprudence; for he was taken with a violent diarrhœa at Cambridge on his return, and was confined there for a fortnight, and escaped with difficulty with his life; but the diarrhœa haunted him more or less ever after, until his death. He was a most agreeable, facetious companion, and I passed many happy days with him. Previous to the diversion of the old channel through the new outfall, Mr. Telford and myself ordered the contractors to assemble as many men, horses, carts, and materials as possible, in order that the old channel should be stopped up during the neap tides.

When everything was ready we went down and met the contractors, Messrs. Jolliffe and Banks, and immediately gave them orders to commence filling up the old channel; they had about thirteen hundred men, and horses, carts, and materials, and appliances of all kinds, and set to work in right good earnest. The Corporation of Wisbeach, who had always opposed the measure, although they were compelled by the Act to contribute 30,000*l.* towards it, which was perhaps the cause, offered every obstruction in their power, and said that the new outfall was not excavated deep enough according to the Act, and came down in their barge with their law officers, giving us official written notices to stop all proceedings. At this critical moment the contractors were rather taken aback; Mr. Telford and I, however, nothing daunted, ordered the men to proceed stopping the channel, and to take no notice of the Corporation. We further told them, that if they did not go away, their barge and all in it would be swamped, and that the responsibility would rest entirely with them. Seeing that we were in earnest they turned tail, and, leaving their protest, returned to Wisbeach. The third day afterwards the old channel was completely closed, and the Nene diverted to its new outfall. It should be observed here, that Mr. Telford and myself, calculating upon the loose nature of the soil, which was silt, and which we felt confident would scour when fairly acted upon by the current, only made the contract for the excavation to the level of low water of spring tides; and therefore it would have wasted money to have excavated that which we knew the current would do for nothing. The current at first appeared to have very little effect; and the Duke of Bedford's manager, the late excellent and talented Tycho Wing, a schoolfellow of mine at Dr. Burney's, became much alarmed, and was sadly afraid that the outfall would be a failure. Telford and I knew better, and assured him that our only doubt was whether the current would not be too strong, and render it necessary to protect the sides with stone. This we considered to be no disadvantage—on the contrary a great benefit; for making the cut small in the first instance, we should always be able to regulate the scour whenever it might have a tendency to enlarge the cut beyond the size necessary to discharge the drainage water effectually, at the same time preserving a sufficient depth for navigation; but if it had been too large in the first instance, it could not have been properly adjusted afterwards. Mr. Wing was comforted by our assurances; still he had his doubts, and two months elapsed before any sensible scour appeared to take place. The fact was, the fall in the bottom was so little, that the current had to remove the obstacles to its progress, which it could only do by degrees, when it had accumulated sufficient fall or head; having done this, its progress was most rapid, and increased daily, so that within six months after it had been opened it had scoured out the bottom to 9 feet below low water of spring tides; the sides also had been regularly scoured away, and the area of the cut was increased to three times its original size. Spring tides, which had scarcely exceeded a few feet at Wisbeach, and not much more at Cross Keys, rose remarkably at both places, so that vessels of considerable tonnage could reach Wisbeach even at neaps, whereas before they could only get up there at spring tides. The trade of the port increased so rapidly, that they were soon enabled to pay off the 30,000*l.* which they had been previously obliged to borrow to contribute to the cost of the outfall.

The outfall by the scour had now attained its proper dimensions, and we recommended that the banks should be paved with stone, in order to prevent them from being enlarged, which was accordingly done. The outfall went on improving until the year 1837, when I examined it, and found that the low-water mark had fallen 10 feet 3 inches at Cross Keys Bridge, that there was a rise of tide of 20 feet at springs, and depth of 9 feet at low water, and a rise proportionate at Wisbeach at both springs and neaps; so that vessels drawing 16 feet could go up to the town at springs, and 12 feet at neaps, and the whole of the surrounding lowland country was completely drained and the property nearly doubled in value. Mr. John Young was appointed engineer under me for paving the new outfall with stone, and afterwards entered business upon his own account as merchant and shipowner at Wisbeach; and by his talents, energy, and industry has since realized an ample fortune, has been elected mayor several times, and has become member for the county of Cambridge.

It had long been a favourite idea with the late Lord William Bentinck and his friend Mr. Thomas Hoseason, of Banklands, to make a bridge across the Nene estuary, at Cross Keys, in order to shorten the distance between the south of Lincolnshire and Norfolk. The bridge over the lower end of the Eau Brink had been completed, and another had been made at the Fossdyke Wash by my father, for the Welland; so that it was only necessary to make another across the Nene estuary, at Cross Keys, to complete this desirable line of communication. A company was accordingly formed for this purpose, of which Lord William Bentinck was the head. An Act was obtained at the same time as the Nene Outfall Act, and I was appointed the engineer. The Nene Outfall Commissioners obtained a clause in the Bridge Act compelling the Company to build the bridge over the Nene Outfall Cut at the same time; this I told them was very unwise, for as the bridge was to be built of wood, with a drawbridge opening in the centre to allow vessels to pass, it would be impossible to drive the great piles forming the piers of the bridge sufficiently deep to be below the scour in the outfall; the better plan would be to wait until the outfall had been scoured to its full depth, and then build the bridge. My opinion was overruled; the bridge was built; and it was impossible, as I expected, to drive the piles to the requisite depth. Where the full effect of the scour had taken place it was found necessary to secure the piles of the bridge by throwing a great mass of stone round them. This materially obstructed the current through the bridge, until at length there was a fall through it of from two to three feet, which greatly injured the drainage, so that the Nene Outfall Commissioners ultimately got an Act to make a new bridge for the Company at the Commissioners' expense. All this might have been avoided if the bridge had been built as I originally recommended. The spot where this bridge and line of embankment is made is the same place where King John's army was lost, and where my father was nearly drowned some years before, crossing in his carriage, being overtaken by the tide. Six thousand acres of this Wash have been reclaimed from the sea by myself; and where once the tides used to ebb and flow, are now fields under culture producing the finest crops.

Notwithstanding all the attempts to improve the river Nene above Wisbeach, nothing had been done, and I was again requested by the Duke of Bedford's advisers to examine the subject and make a comprehensive plan by means of which Whittlesea Mere and all the low fenny country around it might be drained, to the extent of 55,000 acres, it being then little better than a marsh. I accordingly surveyed and levelled the whole country, and made my report in 1837. I showed that by improving the Nene from Peterborough to the outfall, and making a main drain to Whittlesea Mere to connect it with the Nene, and by making a catchwater drain round the base of the surrounding hills, so as to discharge the highland water into the Nene at Peterborough and the Ouse at Hermitage Sluice, the whole country would be thoroughly well drained, the navigation would be greatly improved, and there would be an ample supply of fresh water at moderate cost. This plan was approved of by the late Mr. Robert Stephenson, but was not adopted. The Middle Level Corporation, in whose district was the greatest part of the lowlands to be drained, would not listen to it, but insisted on draining them by the Ouse, 10 miles farther distant. This measure was carried out at double the cost of my plan, and a minor plan substituted for the improvement of the Nene, which is said to have cost a great deal more than any benefits derived from it, although the Eau Brink Cut had lowered the low-water mark on the Ouse nearly 6 feet. Still the sands below Lynn, at the mouth of the Ouse, accumulated to such an extent, that the navigation up to that town was so seriously obstructed that moderate-sized vessels could only come up to the town at spring tides, and they frequently got ashore upon the numerous shoals, lost their tide, and were detained for days together, besides suffering considerable injury. The drainage interests, moreover, complained that the water in the Ouse did not fall low enough to enable the middle and south level lands to be properly drained. In fact, the good effects produced by the Eau Brink Cut were decreasing, in consequence of the waters not being able to get off below Lynn, so that they were held up to the extent of 2 feet at the lower end of the Eau Brink Cut, thus reducing the original fall gained by that cut from 7 feet to 5, whilst the fall gained by the Nene Outfall had been fully 10 feet 6 inches to 11 feet, being a difference of 6 feet in favour of the Nene. A committee, consisting of Lord W. Bentinck, Sir William Foulkes, and others, leading proprietors and parties interested, requested me to examine into the whole subject and report as to what was best to be done. I accordingly employed nearly twelve months in surveying and levelling the Great Wash and the mouths of the Ouse, Nene, Welland, and Witham, which are the principal rivers discharging their waters into the Great Wash, and which drain all the adjacent fen lands, amounting to nearly a million of acres, besides the high lands. I found that by improving all the mouths of these rivers an additional fall of 7 feet might be gained for the Ouse, 2 feet for the Nene, and a similar amount for the Welland and the Witham, and recommended that all these rivers should be united and made to discharge their waters into one great main channel in the centre of the Great Wash, and that the main and minor channels should be properly embanked. By this means not only would all these rivers be much improved, and the drainage and navigation rendered as perfect as they could be made, but, in addition to this, from 150,000 to 200,000 acres of land would be gained from the Wash, or, in other words, a new county, of most valuable land, would be added to the kingdom. This project was so vast and important that it took the world by surprise. It was impossible to deny the soundness of the principles or data upon which it was founded, or the vast importance of it in a national point of view, if means could be found to carry it into effect; but here was one of the great difficulties, and another still greater presented itself, namely that of uniting together the vast number of conflicting interests concerned, so that they might combine together as one whole body for the completion of the undertaking.

After the plan had been published and promulgated for some time, all sorts of objections were raised and attacks made upon it. I replied coolly and steadily to them all, and the more it was investigated, the more the world became convinced of its practicability. Still it was impossible to combine the various conflicting interests, and equally difficult to form a company for such a vast and novel undertaking. The public naturally said, if the drainage and navigation interests and the landed proprietors, who were so much interested, did not see their way, how could it be expected that a company should? Thus the affair, although frequently agitated, lay in abeyance for several years. Its chief supporters, Lord William Bentinck and Mr. Hoseason, having gone to India, there remained none of sufficient energy and influence to push it forward. At length the late talented and indefatigable Lord George Bentinck became member for Lynn, in succession to his uncle, Lord William, when he became Governor-General of India. Lord George was unanimously chosen chairman, and examined most minutely, with his usual sagacity, every detail of the measure, and was perfectly satisfied of its practicability and value, but thought that it was too great to be undertaken as a whole, and that it would be better to divide it into two parts, one comprising the Ouse and Nene, the other the Welland and Witham. I must not omit to mention that the late Prince Consort was much pleased with the plan, and expressed his approbation to me of it. It was accordingly decided to form a company for the Norfolk half, including the Ouse and Nene, and to reclaim 35,000 acres of land from the Great Wash; and at the end of 1845 the requisite plans were prepared and notices given for a Bill to be applied for in the ensuing session, under the title of the Norfolk Estuary Act. The Company originally calculated that the land gained from the estuary would have indemnified them for making the new cut for the Ouse; and so it would, if they had been allowed all the land below high-water mark, without having any other burdens entailed upon them. But unfortunately this was not the case; they were saddled with the maintenance of the Great Cut, although, strictly speaking, it ought to have been maintained by the navigation and the drainage interests, which alone derived the benefit from it. The land frontagers claimed all the land, or green marsh, to be embanked at the expense of the Company, who were only to receive a certain portion of its improved value; the Ouse bank owners were to be indemnified to some extent also. The Crown was to have a commission of five per cent. upon the expenditure; and the Church was to have a certain portion, or tithe, upon the land gained: indeed, so many restrictions were placed on the Company that their calculated profits were materially reduced; nevertheless, as a number of the shareholders were otherwise greatly interested in the improvement of the drainage, frontage, &c., they calculated that if the shares were worth nothing, they would still be the gainers. I protested as much as I could, without avail, against all these restrictions, and I doubted much whether the Act could or would be carried into effect with any benefit to those shareholders who simply looked to their profit from the shares. It turned out as I expected. The Act passed during the session of 1846. Still there was no prospect of the Norfolk Estuary Act being carried into effect, on account of the restrictions above mentioned; and the Company wisely determined not to proceed unless the drainage and navigation interests, which were so materially concerned, came forward with a handsome contribution towards it. The Middle Level proprietors had obtained a Bill, in the year 1846, for the improvement of their drainage, which was much opposed, and in which I took a leading part. However, they carried their Bill, and the works were designed and commenced under Mr. James Walker; but they soon found that the work would be comparatively valueless. The Norfolk Estuary Act had now been carried, and they therefore entered into negotiations with the Norfolk Estuary Company, and these negotiations finally ended in the Middle Level Commissioners and the Lynn Corporation, who represented the drainage and navigation interests, agreeing to contribute each the sum of 60,000*l.* towards the completion of the new channel for the Ouse, contemplated by the Norfolk Estuary Company, upon condition that the late Mr. Robert Stephenson should be joint engineer to the Norfolk Estuary Company with myself, to which, of course, I had no objection; and it was agreed that an Act of Parliament should be obtained in order to ratify this agreement, which passed in the year 1850.

At that time a former Act rendered it necessary that before any improvement was made in any port or harbour, a Commission should be appointed by the Admiralty to investigate the plan, the Commission to hold its inquiry in public; Captain Washington and Captain Veitch were appointed by the Admiralty for this purpose. They held their court at the townhall, Lynn, and, singular to relate, the gallant gentlemen advocated a curved instead of a straight channel, which, under the circumstances, was so contrary to the practice of every good hydraulic engineer, that their report was simply ridiculous, and when produced by the Admiralty before the Committee of the House of Commons on the Bill, was proved by the best engineers to be wrong, and was therefore ignored by the Committee, and the plan proposed by Mr. Stephenson and myself was unanimously adopted. The Bill accordingly passed, and the agreement between the Middle Level Commissioners, the Corporation of Lynn, and the Company, the two former contributing 60,000*l.* each, was ratified; still there were several restrictive clauses, such as entailing the maintenance of the works upon the Company, giving up the whole of the green marsh to the several frontages, which materially abridged the profits of the Company, and increased their risks, and which I in vain protested against. The main cut commenced on the lower side of Lynn, and was continued in a straight line 2½ miles. We calculated that certain dimensions were ample in the first instance, because it was not necessary to excavate the cut artificially to the full depth, as we knew that as soon as the current began to act upon it, it would in a short time be adjusted to its proper capacity for the admission and discharge of the tidal waters, in the same manner as had taken place in the Nene estuary outfall already mentioned. From the lower end of the above cut the channel for a distance of two miles farther to deep water was to be trained through the sands, formed by guide walls of rough stone raised to about the level of half tide, with beacons upon it at certain distances to indicate the channel; and when once the channel had been thus trained, the remainder of the banks to their full height above high water would be raised naturally by the silting up on each side, combined with the gradual process of embanking the land from each of the shores of the estuary, if properly managed.

Messrs. Peto undertook the contract of the main cut. The land having been bought, and everything arranged, a day was fixed for turning the first sod, at the upper end near Lynn. The ceremony was performed by the late worthy Sir William Foulkes, the chairman of the Company, on the 1st of November, 1850. On that day a grand procession, consisting of the Company, the Corporation and trades of Lynn, the Earl of Leicester, Lord Lieutenant, together with numerous gentry, and other spectators, attended the ceremony, which went off with great *éclat*, and the whole, as usual, terminated with a grand dinner to about two hundred persons at the townhall. Unfortunately my friend Mr. Stephenson was abroad and could not attend. The excavation of the great cut proceeded rapidly for some time, until Mr. Stephenson and myself, judging that enough had been done, and that the current would do the rest more effectually, ordered the dams to be removed at both ends and the water admitted, which was accordingly done. As soon as the fen proprietors heard of this, they were greatly alarmed; they said that the Company had violated their agreement with them, and had gone contrary to the Act of Parliament, in not excavating the cut to the full depth required before letting the water in. We endeavoured in vain to persuade them to the contrary. They immediately applied for a mandamus to stop the works and restore the dams until the cut had been excavated to the full dimensions required by the Act, and obtained an injunction, so that we were compelled to restore the dams and stop out the water. We determined, however, not to give up the point, and argued the question before Vice-Chancellor Turner, and were beaten. We then applied to the Admiralty, who would not assist. We appealed from the Vice-Chancellor's decision to the Lords Justices of Appeal, and were again beaten; still we would not give up, and at last we found that there was no alternative but to get a new Act of Parliament allowing us to make the cut in any manner we pleased, provided that we made it of the dimensions originally agreed upon. The Eau Brink Commissioners opposed us by every means in their power, but our evidence, which was given by the first engineers of the day, so completely satisfied Parliament that we carried our new Bill as we wished it.

Thus, after a severe struggle for two years, we carried our point; but this was so much valuable time lost, besides a great deal of money spent in litigation. Having obtained our new Act, we set to work immediately to remove the dams of the cut, and to let the water in, and at the same time to commence the dam for stopping up the old channel of the Ouse at the lower end of the town of Lynn, near the upper end of the new cut. As the dams and bottom of the river near them were composed of strong clay, it took some time to remove them so as to admit the waters freely into the cut; but as soon as this was done, the scour began to have a sensible effect; this was increased by the closing of the dam across the old channel, so that within a few months afterwards not only had the new cut been scoured out to the full depth required by the old Act, but considerably beyond it. Thus Stephenson and myself were proved in the right, and the opposition entirely in the wrong.

The effect on the port of Lynn was very remarkable. The depth of water at spring tides was 18 to 20 feet, and neaps 14 to 16 feet; and there was a regular depth at low water of spring tides of 9 to 12 feet in the cut, so that the largest vessels could always come up and depart with their full cargoes either at spring or neap tides, and the channel was so direct and easy of navigation, that pilots, of whom there was a large establishment, became in a great measure unnecessary, and their numbers were considerably reduced as well as their charges, and the increase of trade soon enabled the town of Lynn to pay off the 60,000*l.* contribution to the estuary works.

The drainage interests also derived a similar benefit by the lowering of the low-water mark 6 feet, which, together with that obtained by the Eau Brink Cut, was altogether 11 feet. The Middle Level Drainage, upon which a large sum had been expended, obtained an increased fall of between 3 and 4 feet, which enabled the Commissioners to drain the greatest portion of their lands naturally instead of artificially. In fact, the Port of Lynn by means of these works has become one of the best on the east coast of England, at the least expense, and with the most moderate dues; indeed, if the Drainage and Navigation had paid double the money which they did to this great work, and which in justice they ought to have done for the benefits it has conferred upon them, they would have been more than amply indemnified.

The Estuary Company having now completed the cut, turned their attention to the best means of reclaiming from the Wash the 35,000 acres which had been allotted to them by the Crown, or at least as much of it as they could; but unfortunately there were so many different opinions among them and their advisers, that they lost a great deal of time and money in pursuing improper measures. I, who originally designed the undertaking, and had acquired great experience in this department from having carried on successfully similar works in the Great Wash and elsewhere, always adhered to one system, namely, to work with nature, and never to go against her; if we did, I invariably found that we were beaten.

We knew, from a variety of experiments that had been made, that the alluvial matter held in suspension by the waters in the Great Wash was an ascertainable quantity; that this alluvial matter was transferred from one place to another, according to the prevailing winds and currents; and that it was only deposited where circumstances were favourable, such as eddies and sheltered stagnant places. Now my great object (after having confined the fresh and tidal waters of the river Ouse to one adequate channel, so as to preserve the drainage and navigation in an efficient state) was to cause to be deposited the alluvial matter that was held in suspension in the waters spreading over the other parts of the estuary. This could only be done effectually by arresting the progress of the flowing and ebbing waters in such a manner, that as much of the alluvial matter as possible might be deposited in the places most convenient, that is, in those places where it was desired to raise the soil above the level of high water of neap tides. When this level is reached grass may be expected to grow, and in a very short time the whole is converted into a green marsh. The process simply this: when the deposit has reached a certain number of feet above low water of spring tides, a species of light green vegetation first covers the surface in patches, then by degrees extends over the whole; the reclaimed land still continues rising, and at a higher level above low water samphire makes its appearance; the accumulation of soil still goes on increasing, when the samphire disappears, and grass succeeds it, and in a short time afterwards the place assumes the appearance of a level green marsh well adapted for grazing cattle. After it has arrived at this stage, which it does at about the level of high water of neap tides, the accumulation is very slow; on our coasts it seldom attains a much higher elevation, except where the sand is blown up by strong winds from the sea, which forms dunes or banks, that, as in the case of Holland and other places, sometimes attain an elevation of 30 feet and upwards above the level of the highest tides. This, however, is not the case around the shores of the Great Wash; there the marshes are simply produced by the gradual deposit of alluvial matter in the manner above stated.

The great object, therefore, as I have said, is to facilitate this deposit or accretion as much as possible by artificial means. Nature, if left to herself, though sure, is very slow. We must therefore assist nature, by following and working in unison with her laws. We must go to work gradually, and not by great and expensive operations check the currents violently and at once, which would only produce an equally strong current elsewhere, so that while we gained in one place we should lose in another. By a series of light works composed of bushy fagots or other similar materials, raised about 12 or 15 inches above the level of the sands, and disposed in a series of lines at certain distances from each other, not continuous, but in lengths, so that the ends may overlap each other, the currents will be gently checked without being wholly obstructed, the water between them will be rendered stagnant, and the alluvial matter with which it is charged will be deposited. When the deposit has reached to the top of these works, the works themselves should be raised higher in the same place, or in an intermediate position, as circumstances shall render advisable. Where a certain space is intended to be raised or warped up, it is generally better to commence at the upper end and work downwards; the works themselves are less expensive, the height to be raised is less, and the water impelled by the mass of the tide behind brings up a greater quantity of alluvial matter, and in proportion as the space above is warped up it accumulates in a greater degree below. In carrying on works of this kind, wherever we find that there is a tendency to make a channel, and with it a strong current, it must be checked gradually from the upper end, so that the quantity of water passing through it may be decreased, and this channel will soon fill up. When a certain space has been naturally or artificially raised to the level of a fine green marsh, provided that there be a sufficient quantity to pay the expense, it should be embanked entirely from the sea. As to the expense incurred in embanking it, and the value of the land when embanked, generally speaking, we shall seldom err if we take as a rule that the land should be worth double the cost of embanking. When we have an estuary to deal with from which we may expect to reclaim several thousand acres of land, it resolves itself into a serious question both of time and money as to the best mode of accomplishing it. One object should be as far as practicable to reduce the extent of main or barrier banks, and if the situation be well adapted for the purpose, a barrier bank may be commenced at the lower end, and gradually pushed forward in proportion as the space above it exhibits a tendency to silt up; in combination with this the minor operations should be carried on, so that the one may assist the other. When a sufficient quantity of marsh above the barrier bank has been formed, it may be embanked from the sea, and by keeping the barrier bank always sufficiently in advance the interior banks will become less costly, as they will not be exposed to the main force of the waves during storms; otherwise each separate bank must be made a barrier bank, and the whole cost will be materially increased. The propriety of adopting either the one system or the other will greatly depend upon the peculiar local circumstances.

With regard to the quantity of land, when fit for the purpose, to be embanked or enclosed at one time, this also will, like the other, depend upon the local circumstances. Generally speaking it is safer to confine the operation to about 400 or 500 acres; and the proper time for closing the embankment is during neap tides, when the work will be much facilitated.

It is true that much greater quantities may be taken in at one time, according to the Dutch system, but then several closing spaces must be left open, and these must all be specially prepared for the purpose by lining the bottom and sides with wicker-work and fascines to prevent the scour, and then filling them up with clay, stone, fascines, and earth, as the case may require; and if by chance a breach should take place, which occasionally occurs under the best of management, the internal space to be filled with water is so great that the violence and strength of the inpouring current is increased in the same proportion, so that it carries all before it, scoops out great channels in the interior space, and damages the land by the great quantity of sand brought in, and it becomes difficult to clear the internal space of water. And although by this plan a less quantity of embankment is required, nevertheless the expense per lineal foot becomes greater; but where the length of the embankment is small compared to the depth of the land to be enclosed, this system, if properly managed, may be adopted with advantage. The most advisable plan must be left a good deal to the judgment and skill of the engineer-in-chief taking advantage of the local circumstances.

The above principles for warping and enclosing land I recommended in several reports to the Norfolk Estuary Company, but they were not adopted to the extent I recommended; the consequence has been that several years' time and a good deal of money have been unnecessarily expended. But I believe that they have been finally convinced that my system was the proper one, and it has since been adopted to some extent, with considerable success, and about 1000 acres (in 1867) have been enclosed, the property of the Company, at the cost of about 15*l.* per acre, the land being worth 40*l.* They have, moreover, enclosed 600 acres more for the Prince of Wales, upon which they will be paid one-third of its improved value. As several thousand acres in the estuary belonging to the Company are rapidly approaching that state when they may be profitably embanked, I confidently believe, by proper management, that the shareholders will ultimately recover a good deal if not the whole of their capital, and perhaps a good portion of their interest also; but I shall always contend that if my recommendations had been followed from the beginning, a great deal of money would have been saved. The Company ought not to have submitted to the restrictions imposed by the Acts. The contribution from the drainage and navigation interests ought to have been double, for which they would have been amply repaid, and the land gained would have been far greater, so that even by this time it might have been a very profitable concern.

Whilst the Nene Outfall was in progress, I was employed by the Corporation of Boston to improve the outfall of the river between the Grand Sluice at the upper end of the town, and Hobhole near the mouth of the river, a distance of nearly four miles. This had previously been proposed by my father, in the year 1845, to the Corporation of Boston, previous to commencing the great drainage of the East, West, and Wildmere fens bordering upon the Witham between Lincoln and Boston, amounting to about 125,000 acres of lowland, which for the most part was little better than a marsh. It was here he proposed his grand system for the drainage of lowlands by means of catchwater drains for the waters from the highlands, and main and minor drains for the waters from the lowlands, both systems of drains being distinct and separate from each other, and which he afterwards most successfully carried into effect. In order to obtain the requisite fall, it was necessary that the outfalls of the two main drains should be carried into the Witham below the town of Boston; but in order that this might be effectual, it was also necessary that the Witham should be greatly improved between Boston and the mouth at Hobhole; and as the navigation between these two points was very defective, he thought that the Corporation should bear the expense, as they would derive the greatest benefit. For this purpose my father proposed two plans: one was to improve the old channel, partly by cutting off the bends and confining and straightening its course, and from Hobhole making a new cut to Clayhole, where there was ample depth of water at all times of tide; and the other plan was to make a direct cut within land from the lower end of Boston to Clayhole. He proposed also, for the drainage, to make an outfall for the highland water at Maudfoster, just below the town of Boston, as he said that by bringing the highland water there, it would effectually serve to keep the river open between that point and its mouth at Hobhole, and to make an outfall for the lowland waters at the latter place; and he recommended the drainage interests to contribute a certain portion of the expense towards the improvement. The Corporation of Boston declined doing anything, and the drainage was left to take care of itself. Mr. Rennie, however, foresaw that if the Witham was not improved by either one or the other of the plans that he recommended, the highland water would not be effectively discharged by the sluice at Maudfoster, and therefore he made a communication between the Maudfoster drain and the Hobhole drain at a place called Cowbridge, a few miles above Boston, where there was a gauge, so that whenever the water in Maudfoster drain exceeded the gauge it passed into the lowland main drain, and from thence into the Witham at Hobhole. What Mr. Rennie foresaw came true; the Witham, not having been improved, became worse, and the river in front of the Maudfoster Sluice was silted up, so that it could not discharge its water, and therefore the whole of the water, highland and lowland, was obliged to go by Hobhole, which drain and sluice, foreseeing what would take place, he had enlarged for the purpose.

In the year 1827, the outfall of the Witham between Boston and Hobhole had become so much silted up that at high water of neap tides there was scarcely water enough for fishing boats to come up to the town during the summer months and dry seasons, and at spring tides only sloops of very small draught could get up to the town; in fact, it might be said that at that time Boston as a seaport was lost, and the trade and navigation of the port ruined. At this the Corporation became greatly alarmed, and sent for me. I directed the late Mr. Francis Giles to make a complete survey of the river, which he did in his usual able and correct manner, and no person could do it better. Being provided with this survey, I made my report, and saw clearly that there was no remedy but to carry into effect either the one or the other of my father's plans of 1805. As the Corporation funds were very limited, I recommended the plan for improving the old channel, partly by cutting off the bends and confining the river as far as Hobhole, and from thence making a new cut to Clayhole, as recommended by my father. There were three parties who were to contribute towards the improvement of the outfall of the Witham, namely, the Corporation of Boston, and the Witham and the Black Sluice Commissioners. The Black Sluice Commissioners demurred to the plan proposed by me, which was my father's, on the ground that the cut ought to be made from the Black Sluice inland direct to Hobhole, otherwise they would not derive the advantages they had a right to expect from their contributions; so the whole matter was referred to Mr. Telford, who said that no improvement in the river below Boston would be of any use, unless the Grand Sluice above Boston was removed, so as to admit the tide to flow farther up the river. Now, although Telford was to a certain extent right, but by no means wholly so, because by carrying into effect either of my father's plans there could be no doubt that considerable improvement would be made on the outfall of the Witham below Boston, as the sequel will show, nevertheless, if the Grand Sluice had been removed the tide would have flowed farther up the river and the increased quantity of tidal water passing upwards and downwards would have improved the outfall still further; but there were insuperable objections to the removal of the Grand Sluice, which neither of those parties above mentioned who were to make the improvement in the outfall could control, for the river above the Grand Sluice was under a different body. The banks must have been raised to admit the tide; compensation must have been made for the loss of fresh water, and various other interests must have been consulted that were hostile. The Black Sluice Commissioners stuck to Telford's report, and withdrew from the contribution, so that the whole matter fell to the ground.

The outfall of the Witham became worse, and the Corporation of Boston, being left single-handed, and having determined to do what they could to improve the river, again requested my advice. I said that the best thing they could do under the circumstances would be to carry into effect by degrees, as far as their means would allow, the plan of my father already referred to for improving the old channel of the river, and to begin by cutting off the bend between Hobhole and the upper end of Burton's Marsh (this cut would be about half a mile long), and blocking up the old channel immediately above it, which was about half a mile wide. These two works would shorten the navigation quite half a mile, and admit and discharge the tidal and fresh waters more readily, and thus produce a corresponding scour and lowering of the low-water line and bed of the river all the way up to the Grand Sluice above Boston.

The Corporation adopted my recommendation, and entered into a contract with Jolliffe and Banks for that purpose. Although a small work, it was attended with considerable difficulty, particularly in closing the old channel, on account of its great width and the great body of tidal water which passed through it. The effect of this work exceeded my most sanguine expectations; in a short time it improved the channel upwards to Boston to such an extent that spring tides rose at Boston Bridge 14 feet, and neaps 10 feet, and the bed of the river was deepened from 3 to 4 feet below low water of springs, so that vessels drawing 15 feet and 16 feet could come up to the town at springs, and vessels drawing 12 to 13 feet could come up at neaps; moreover, all the silt was scoured away from the front of the Maudfoster Sluice, so that it discharged the highland water from the fens, which it had not done for years before, and improved also the discharge of the waters from the Grand and Black sluices. I must confess that I was not a little elated at this successful result, as it most completely established the correctness of my father's opinion as well as my own, and demonstrated the fallacy of my friend Telford's judgment. The cost of the above works was 33,000*l.*, which was very small compared to the advantage obtained. The Corporation of Boston were so much pleased with the success that they determined to carry into effect the remainder of the improvement in the old channel to Boston, which was afterwards done by confining the channel by degrees to a proper width by means of fascines and loose stone and clay properly combined together up to the level of half tide, so that the flood and ebb always acted to the greatest advantage in one and the same channel without materially diminishing the quantity of tidal water.

The effect of these additional works was to still further deepen the bed of the river and increase the flow of the tide by lowering the low-water mark, which improved the navigation and drainage still further, so that the trade of Boston revived and increased in prosperity, and all this was effected by the resources of Boston alone; and it is only to be regretted that the drainage interests, who derived so much benefit, were not compelled to contribute their just proportion.

In 1852 a Bill was obtained to carry into effect my plan for improving the mouths of the Witham below Hobhole, and the Welland below Fosdike Bridge, and reclaiming 35,000 acres of land from the Great Wash, as formerly described, but the Bill was clogged with so many restrictions, and neither the drainage or navigation interests would contribute anything towards it, although they would have been greatly benefited, that it was found impossible to carry it into effect so as to remunerate the shareholders, and therefore it was abandoned.

I about this time finished some minor works which had been designed and partly executed for the improvement of the Witham near Lincoln. This was part of a great plan of my father for improving the river Witham, so as to make it navigable for the Yorkshire coasting vessels, drawing 6 feet and carrying about 70 tons. From Lincoln they proceeded to the Trent by the Old Foss navigation, which entered the Trent at Torksey, and from thence to the Humber and the adjacent coasts. This improved navigation of the Witham answered very well as a commercial speculation, and in the year 1847 was sold to the Great Northern Railway Company, who established a railway on its banks, which now forms part of their loop line between Peterborough and Lincoln, which has in a great measure superseded the navigation.

During this period I was requested by the Commissioners of the Ancholme Level, consisting of about 50,000 acres of low fen lands, bordering upon the Ancholme in North Lincolnshire, to give them my advice as to the best means of improving their drainage, and at the same time the navigation of that river.

The Ancholme takes its rise near Market Rasen, in the highlands of the north of Lincolnshire, and after a course of some miles it enters the lowland district of the valley, and proceeds through it in a northerly direction for a distance of about 18 miles, when it joins the Humber at right angles to its course. The valley varies from one mile in width at the upper end to three miles at the lower end, where it joins the Humber. It is bounded on the south by a ridge of chalk hills of considerable elevation, and on the north by a similar ridge of sandstone hills.

The Ancholme having very little fall where it enters the lowlands, and being prevented from discharging its waters with facility into the Humber on account of the great mass of water and high tides in the latter river, was forced back upon the lowlands, and frequently inundated them, so that they became little better than marshes, and the river itself was necessarily extremely circuitous. In 1806 the Commissioners applied to my father for advice, and he recommended, in the first place, that the river should be straightened as far as practicable, in order to utilize the fall of the current to the fullest extent; also that a sluice should be constructed to exclude the tides at Terreby; and a catchwater drain made on the south side, with separate sluice to discharge the highland waters into the Humber. This advice was only partly followed: the river was straightened; the catchwater drain only extended as far as Brigg; a lock was made for the navigation at Hortestow Green, where the river entered the lowlands; and a sluice with a lock was made at Terreby. In principle, these works were correctly designed and well adapted for the purpose as far as they went; but I am not exactly aware whether they were carried out according to his plan, or under his immediate direction; and twenty years had elapsed before I was invited to give my opinion.

When I visited them the level was very badly drained; the river was full of shoals; the navigation, which was intended for the Yorkshire coasting vessels up to Bishop's Bridge, was only practicable a few miles beyond Brigg, and that in a very imperfect manner; and the works generally were in a very bad state.

After having inspected the works, I recommended, first, that the main river should be deepened, widened, and enlarged throughout the entire length of the level, so as to accommodate the full-sized Yorkshire coasting vessels drawing 6 ft. 6 in.; and that the river should be of ample capacity to contain the floods fully 2 to 3 feet below the level of the lowlands, so that they could not be overflowed, and might always be able to drain into it.

Secondly, that a lock should be made with a lift of 6 feet, so that coasting vessels might be enabled to get up to Bishop's Bridge; and that as a great deal of sand was brought down by the upper part of the river, which continually produced shoals and filled it up, I recommended that at the upper end of the lock an overfall should be made, together with a capacious reservoir, into which the sand might be discharged, in order to prevent it from getting into the river; then it could be removed whenever necessary.

Thirdly, that a new sluice and lock should be made at Terreby, where the Ancholme joined the Humber; that all the new sluices should be laid 6 feet below the old one; and that new bridges should be made across the main river to replace the old ones, which were not of sufficient capacity.

Fourthly, that the south catchwater drain should be drained out and enlarged, and extended to the upper end of the level.

Fifthly, that a catchwater drain of the requisite dimensions should be made on the north side of the level, from the new sluice to the farther extremity of the level, and that the water should be discharged through a separate opening in the new sluice.

Sixthly, that wherever a brook entered the catchwater drain on either side, there should be corresponding weirs and reservoirs for receiving the sand and other deposits, so as to prevent them from getting into the drains.

These several works were executed, under my directions, by Messrs. Jolliffe and Banks, as contractors; and Mr. Adam Smith, as resident engineer, to whom the real credit is due for the very able, honest, and zealous manner in which he discharged his duties, particularly for the execution of the new sluice at Terreby, which was done without a contractor, and is one of the cheapest and best works of the kind.

These works have answered their object completely; and the Ancholme district is as well drained as any level in the kingdom, and the navigation is complete of the kind.

Whilst carrying on these works I was the frequent guest of the Earl of Yarborough, the Commodore of the Yacht Club, and received the greatest kindness and attention from his lordship; without his support, ability, and firmness, these works would never have been attempted, nor carried to a successful conclusion.

CHAPTER IV.

Sketch of the Risk and Progress of the Railway System—The Manchester and Liverpool, London and Birmingham, and other early Lines.

I will now proceed to a very important epoch in my life, namely, my first introduction to railways upon the locomotive system. Railways of wood were first introduced on the Tyne, for the purpose of bringing down coal from the adjacent collieries, to be shipped from Newcastle and the vicinity for exportation to London and other places. These were 4 feet 6½ inches wide from centre to centre, and the coal waggons were specially adapted to them. The wooden rails wore out rapidly, and were subsequently improved by having cast iron edge rails fixed upon the wooden ones, and the wheels of the waggons were made of cast iron also, having a flange on the inside to keep them in their places. It reduced materially the friction of the road, enabled the horses by which the waggons were drawn to take a greater load, and the expense of maintaining the ways was greatly reduced. Matters proceeded in this manner until towards the end of the last century. At that time, the celebrated James Watt began his experiments upon steam, and then turned his attention to the improvement of the steam engine, which had previously been so far perfected by Savery, Newcomen, and Smeaton, as to be used for pumping water from mines with considerable effect, by employing steam and atmospheric pressure alternately, for raising and lowering the piston in the cylinder to which the pumping apparatus was attached. The waste of fuel and the expense of working these engines were very considerable, and they were inapplicable to any other purpose but pumping. Watt saw these defects, and immediately set about devising means for remedying them. He first enclosed the steam cylinder at top and bottom, and elevated and depressed the piston by means of steam only; and instead of allowing the spent steam to escape into the atmosphere, it was discharged into a separate vessel, into which a jet of cold water was constantly playing, so that the steam was condensed there, and hence this vessel was called a "condenser." In this condenser there was an air pump constantly working, so that to some extent a vacuum was produced, which facilitated the discharge of the steam from above and below the piston, relieved the pressure upon it both ways, and added considerably to the effect of the whole machine, as well as economizing the fuel required to work it. He also added a crank to a connecting rod at the opposite end of the beam to which the piston was attached, and by means of this crank communicated rotary motion to any machinery connected with it, adding still further to the value of the steam engine, and rendering it universally applicable. The boiler also he greatly improved, so as to produce a larger quantity of steam with less fuel. The ingenious idea of the crank was pirated from him before he could patent it, and he resorted to another invention to produce rotary motion, which he called the sun and planet wheel; this consisted of a toothed wheel attached to the lower end of the connecting rod fixed to the end of the beam, which wheel worked into another attached to the end of a horizontal shaft, upon which were fixed other wheels to give motion to any system of machinery which might be required. It should be observed that in this improved engine the connecting rod of the piston was attached to one end of the beam, the connecting rod attached to the crank, or sun and planet, was fixed to the other, and the air and cold water pumps were attached to rods connected with the intermediate part of the main beam, so that they were all worked together by the rising and falling of the piston, and thus formed one whole compact machine. He also added several minor contrivances, which it is unnecessary to mention, and which rendered the steam engine still more complete. His improvements did not end here, for he made numerous experiments upon the expansibility or elasticity and effects of steam at various temperatures, constructed a high pressure engine, and subsequently one with a condenser mounted upon a carriage supported by wheels, which was nothing more than the locomotive engine, a model of which still exists. Watt, however, as I have already observed, did not like high-pressure steam: he was fully aware of its importance; but at that time, from the backward state of the iron manufacture, he did not see his way to controlling it with safety, and he considered that his low-pressure condensing system was the best and most economical, and he therefore gave up all idea of pursuing the investigation of the locomotive engine and its applicability. Fortunately, however, everybody was not of the same opinion. Watt had clearly invented the locomotive engine, and his able and faithful assistant, William Murdock, afterwards made another working model of it on the same plan, with which he used to amuse himself by setting it in motion to run about his room. But the idea and its practicability once established, the locomotive was not to be thus abandoned. Amongst those who heard of it, and who appreciated its value, were two Cornish engineers, Trevithick and Vivian, who had been accustomed to work high-pressure steam, which was largely used in the Cornish mining engines about this time. Trevithick and Vivian soon saw that Watt's waggon boiler was too cumbrous, and not capable of producing steam fast enough or in sufficient quantity for a locomotive engine. They therefore invented a new kind of boiler with a tube in the centre, around which the heat from the furnace circulated, so that a greater surface was exposed to its action, and, consequently, steam was generated much more quickly and in greater volume and elasticity. This boiler was, moreover, more compact, lighter, and more portable than Watt's; the engine also was more simple; and the cylinder and piston being vertical, the latter was attached to a crank, which again was applied to the axles of the wheels, which made them revolve at every double stroke of the piston. The whole of this engine was mounted on a carriage; and this may be termed an improvement upon Watt's engine, and the second stage of this great invention.

Trevithick and Vivian being convinced of success, took out a patent for the tubular boiler and the engine, for its application to common roads and railways; and foreseeing that if the wheels were applied to the smooth surface of a railway, the adhesion of their surface, combined with the weight of the whole machine, would be sufficient to impel it forward when worked by the engine, proposed that the peripheries of the wheels should be smooth; but in other cases, where more grip or action might be necessary, they proposed to add bosses or cogs to the peripheries of the wheels. What they now required was to apply the invention to practice. They accordingly were either invited or went to the Merthyr Tydvil Iron Works, where iron tramways were in extensive use, and there proposed an engine for drawing the waggons, instead of manual and horse labour. The invention was greatly approved of, and the proprietors of the mines determined to give it a fair trial. A locomotive engine was then made by Trevithick and Vivian and placed on the railway. Waggons laden with ten tons of iron and coal were attached to the engine, and, to the surprise of all, it drew them at the rate of six miles an hour. This was in the year 1802.

One would have thought that the principle, practice, and value of the invention having been thus fairly demonstrated, the wealthy and enterprising ironmasters would at once have adopted it and have brought it into general use. Indeed, it is difficult to conceive why this was not so. It is true that the machine was still cumbrous and difficult to manage, and the cast-iron tramways were probably too light and badly laid, so that they were frequently broken; still sufficient must have been proved to show that these were defects that might easily be remedied, and that continued practice would enable the inventors to render both the locomotive and the railway more perfect. In almost all new inventions, nothing is made perfect at first, and it requires constant trials and much perseverance to remove obstacles which cannot be foreseen, and the existence of which can only be proved by experience, when the proper remedy can be applied. The attempt to introduce locomotives on railways was not, however, for the time prosecuted further at Merthyr Tydvil.

It may be asked, why did not Trevithick and Vivian pursue the matter further elsewhere? I believe the answer to this is simply that they had not the means; they had already expended so much in prosecuting the invention that their resources were exhausted, and unless fresh pecuniary aid came in they must necessarily abandon it, at least for the time. Nothing is more difficult than to introduce a new invention, however plausible it may appear at first sight. We become by habit so wedded to our old ways that we are apt to regard anything new with indifference. When it has succeeded, the more simple it is the more we are astonished that we should not have perceived its value before; hence, unfortunately, we so frequently find that many able men, who have made brilliant discoveries and inventions which have conferred so much benefit upon mankind, have exhausted their all, and have died of starvation, just at the very moment when they have succeeded. Then some lucky one steps in and derives all the advantage. Such was the fate of poor Trevithick, who may be said to have been the inventor of the modern railway system. Even Watt himself would have probably shared the same fate, if he had not met with the great Matthew Boulton, who appreciated his inventions, and furnished the means for carrying them into effect. The same may be said of Cort, who introduced puddling, the simplest mode of converting cast into wrought iron. Bessemer himself told me that he was nearly ruined before his great discovery succeeded. In fact, numberless examples might be adduced of this melancholy truth; but to my story.

The next example we find of the employment of the locomotive engine was that of Blenkinsop, of Leeds, which was similar to, but more compact and lighter than that of Trevithick and Vivian, and was applied to draw the waggons laden with coal from Middleton Colliery, near Leeds, to that town; I saw it at work in the year 1814. It then drew 20 tons at the rate of seven miles an hour, at which I was much astonished. Although quite a lad, I thought to myself, "Something more will come out of this hereafter." But the most singular thing connected with this was, that the principle of adhesion to the rails by the smooth tired wheels, and the fact of the power of a locomotive being in its weight, as laid down by Trevithick, was completely forgotten, and the locomotive was propelled forward by means of a toothed wheel driven by the engine, acting on a corresponding toothed rack laid alongside the rails; this was nothing more than Trevithick's idea of bosses or teeth attached to the wheels, but intended to be applied in a different way. Still the invention was making its way. Blenkinsop's engine excited great interest in the north, especially in the neighbourhood of the Tyne and Wear, and numbers of engineers, scientific men, and others went to see it, and being convinced of its value, determined to introduce it into that district. Amongst others, Mr. Blackett, of Wylam Colliery, who had seen Blenkinsop's engine and railway, perceived the error of attempting to propel the locomotive by means of the toothed wheel and ratchet, and determined to revert to Trevithick's original design of the smooth tired wheels. Whether he had heard of Trevithick's invention or not, does not appear, although, as the subject had become generally known, and numerous intelligent minds had been directed towards it, it is very probable that he had. However, be that as it may, he has the credit of reintroducing Trevithick's invention, and a locomotive, with improvements, leaving out the toothed wheel and ratchet, was applied to a railway at Wylam with complete success.

Amongst others who visited Blenkinsop's railway and locomotive engine was Mr. James, a general land agent and surveyor, at Newcastle, who was in large practice, and had a respectable fortune. He was a man of enlarged mind and great intelligence, and although not a practical mechanician, he was so much struck with the effect of Blenkinsop's engine and railway, that he at once said it was a new mode of conveying passengers and goods which must supersede all others, and become universal. In his enthusiasm he wrote a long letter, addressed to the Prince Regent, on the subject, in 1815, pointing out the value of this new mode of transport, the saving which it would effect in manual and horse labour, the ease and expedition with which goods and passengers would be conveyed, and the vast benefits which would be conferred upon the country by the general introduction of the railway system.

Immediately he set to work and devoted his whole energies and time, regardless of his own business, to the promotion of this new mode of transit. The celebrated George Stephenson, who up to this time had been a working collier, began to emerge from his obscurity and exhibit his genius to the world. James got acquainted with him, and being greatly taken by his talent, imparted his views about the introduction of railways, and, it is said, supplied money towards aiding him in prosecuting the work. Be this as it may, Stephenson, who had seen Blackett's engine, was fully convinced of the soundness of the system, and that it must become universal; and therefore, with his usual indomitable energy and talent, he applied himself to promote and develop the new scheme by every means in his power. He became acquainted, and ultimately entered into partnership, with Mr. Lord, and established a manufactory for locomotive engines at Newcastle, which maintains its celebrity at the present day. He first made a locomotive of a new and improved construction, which worked with considerable success upon the Hetton Colliery railways. Next he was employed on the Stockton and Darlington, for which he made several locomotives, all of which, be it observed, were employed solely for carrying goods at the rate of about eight miles an hour; and it was by no means anticipated that they would be able to exceed that, or that they could calculate much upon general passenger traffic. However, the success that had already attended their efforts, made Stephenson confident that the railway system could be equally well applied wherever there was a large traffic. James and Stephenson then entered into a kind of partnership, and first proposed to make a railway, in 1817, from Manchester to Liverpool, for between those towns the traffic was so enormous that the Duke of Bridgewater's Canal and the Mersey could not accommodate it properly, and great dissatisfaction was expressed at the delays and high charges. Nothing, however, was done at the time, and it was not until the year 1825 that the project was so far matured as to enable a company to be formed to carry it into effect. This company was principally composed of gentlemen from Manchester, Liverpool, and Newcastle; and George Stephenson was their chief engineer. They went to Parliament in the year 1825, but were so ill prepared to encounter the violent and powerful interests by which they were opposed, particularly that of the Bridgewater Canal, which enjoyed a very large portion of the traffic, that they lost their Bill.

The late Earl of Lonsdale, then Lord Lowther, one of the Lords of the Treasury, a very able and intelligent young man, knowing me from my connection with London Bridge, and with the Whitehaven Harbour, of which I was the engineer (where his father was the chief owner of all the great collieries round the town), asked my opinion about the proposed new system, and whether I thought that it was likely to succeed. I told him frankly that I thought it would. His lordship replied, "I think so, too;" and he offered me the post of engineer to the Manchester and Liverpool Railway, adding, "Although it will be greatly opposed, I think we shall carry it." I replied that my brother and myself would be happy to undertake it, provided that we did not interfere with Mr. Stephenson or any other engineer who had been previously employed. Lord Lonsdale said that he would arrange all that with the Company, and my brother and myself were accordingly appointed engineers-in-chief.

As we were left entirely to our own discretion to adopt the old or to choose an entirely new line, we selected the present energetic and talented engineer, Mr. Charles Vignolles, to make the necessary surveys for Parliament. After examining the old line and the surrounding country, we finally decided upon adopting the present one, which passes over Chatmoss. Whilst we were proceeding with the survey, my brother George and Mr. Vignolles met Mr. Bradshaw, who was the sole and independent manager of the Duke of Bridgewater's Canal, on Chatmoss, not far from the line, and near his own residence. He went up to my brother and introduced himself as the manager of the Duke's Canal, and my brother at once mentioned his name. Mr. Bradshaw shook him by the hand very cordially, and said that he was glad to make the acquaintance of the son of his old friend, Mr. Rennie, for whom he had the greatest respect and friendship; and knowing what we were about, said that he, on the part of the Duke's Canal and the other water carriers, would oppose the railway by every means in their power, and he felt pretty confident that they could throw out the Bill a second and even a third time, if it were attempted. At the same time he expressed the most friendly feeling towards my brother, and invited him and Mr. Vignolles to his house, where he received them most hospitably, and conversed, amongst other subjects, about the Bridgewater Canal, and the great difficulties they had to overcome in completing it. Pointing to a little whitewashed house, near the Moss, about half a mile distant, he said to my brother: "Do you see that house? Many a time did the late Duke of Bridgewater, Brindley, and myself spend our evenings there during the construction of the canal, after the day's labours were over; and one evening in particular we had a very doleful meeting. The Duke had spent all his money, had exhausted his credit, and did not know where to get more, and the canal was not finished. We were all three in a very melancholy mood, smoking our pipes and drinking ale, for we had not the means to do more, and were very silent. At last the Duke said: 'Well, Mr. Brindley, what is to be done now?' Brindley said: 'Well, Duke, I don't know; but of this I feel as confident as ever: if we could only finish the canal, it would pay well, and soon bring back all your Grace's money.'" After remaining a little longer, the party broke up in melancholy silence, and each went his way. It happened shortly afterwards that the Duke managed to get money enough to complete the canal, and Brindley was a true prophet, for the canal has paid well, and has been mainly the making of the great houses of Sutherland and Ellesmere.

The surveys for the line over Chatmoss were completed and deposited in due time, and the usual notices were given. The Bill was read a first time in the Commons, and after the second reading was referred to a committee. Certainly Bradshaw had not exaggerated the opposition, for the Bill was most energetically contested, the leading counsel against the Bill being the late able and amiable Baron Alderson. The Bill, however, passed the Commons, and ultimately the Lords, after an equally strong contest; but Lord Lowther and his friends were indefatigable, and to his lordship's great exertions the success may be mainly attributed. At the time I was so completely prostrated by the effects of the fall which I had met with at London Bridge, as previously mentioned, that I was unable to take any very active part, which therefore chiefly devolved upon my brother and Mr. Vignolles. After the passing of the Bill, my brother and myself prepared working drawings and estimates for carrying the work into effect, and we naturally expected to be appointed the executive engineers, after having with so much labour and anxiety carried the Bill through Parliament. The Executive Committee of the Company behaved extremely ill to us. Stephenson, although he had failed in carrying the first

Bill, still possessed considerable influence with the Executive Committee, who proposed that Mr. Stephenson should be united with us. We said that we had no objection to Mr. Stephenson taking the locomotive department, which should be distinct from the other works. This, however, the Committee declined, and at once appointed Mr. Stephenson the chief engineer, and Mr. Vignolles the resident.

My brother and myself designed that the width of gauge should be 5 feet 6 inches from centre to centre of the rail, and if this had been adopted we should never have heard of any other. This was the proper gauge for which an engine could have been made of the most powerful description, without being too long. Moreover, the centre of gravity being lower, it would have been more steady, being better adapted to going round sharp curves. The same advantages would have been gained by the passenger and goods carriages. Unfortunately, Mr. Stephenson was of a different opinion: he thought that the old colliery waggon gauge of 4 feet 8½ inches from centre to centre of the rail was ample, and it was adopted. Hence all the enormous litigation and expense which afterwards ensued. Brunel subsequently, on the Great Western Railway, feeling confident that Stephenson's gauge was too narrow, proposed and ultimately adopted 7 feet as the proper gauge. Now this was as much too wide as Stephenson's was too narrow. Stephenson's party have ultimately prevailed, and the narrow gauge has been generally adopted, much to the efficiency and economy of railways.

Before leaving this part of the subject, it may be right to remark that the opposition endeavoured to make a strong case against our line crossing Chatmoss, which they said was utterly impracticable at any reasonable expense. This we knew from our own experience to be simply ridiculous, and so it was found to be during the execution; it has now proved to be the best part of the line, and the easiest to keep in repair. All that was required was to drain the surface by moderate-sized drains, so as to get rid of the superfluous water, then the foundation for the rails would be sufficiently solid to bear anything; moreover, it possesses a certain degree of elasticity which prevents the concussion or jolting that is usually found on a hard unyielding road. In 1828 the line was so far advanced that the Company determined to advertise for locomotives, and to give a premium of 1000*l.* for the best one that was produced. Amongst the competitors were Messrs. Stephenson and Lord, Messrs. Braithwaite and Ericsson, and Mr. T. Hackworth.

The competition took place at Rainhill. The Company restricted the weight of the engines to seven tons, which shows how little the subject was at that time understood, for the power of a locomotive engine is in proportion to its weight and the steam generated to work it; hence it was quite clear that in restricting the weight they restricted its power. Notwithstanding this, the engines performed wonders, and attained the speed of 28 to 29 miles an hour. This was so contrary to the general expectation, for even the makers did not expect above 8 or 10 miles an hour, that everybody was astonished, and from that time forward the glorious career of railways was established, and the old road system for goods and passengers was condemned as a thing of the past. The shares, which had been at a discount, now rose to a premium, and numerous new lines were in contemplation. Stephenson's engine, the 'Rocket,' gained the premium, as it complied with all the required conditions, although that of Messrs. Braithwaite and Ericsson was in other respects considered the best.

The rails used upon this line were what is technically called the edge rail. These were of cast iron, weighing 30 lb. per yard, and they were cast on lengths 3 feet long, resting upon blocks of stone about 12 inches thick and 2 feet square, like the old tramways. Here is another example of the little that was known upon the subject; for it was forgotten that the old tramways were worked by horses, which seldom exceeded the speed of 2½ miles, and, consequently, the vibration upon a tolerably well made road was scarcely sensible, although it prevailed to a certain extent; but when the velocity was increased to above 20 miles an hour, the case was totally different, as should have been anticipated; yet the line was opened for traffic on the old principle, until it was found that the concussion and vibration produced by the rigidity of the road were so serious as to cause considerable trouble. On account of its elasticity, the Chatmoss section was found to be the easiest and best part of the road, yet it is singular that this did not occur to the Company. They continued to use stone blocks more or less up to the year 1837, when they discovered their error, and replaced the blocks with wooden sleepers, and large hillocks of these stone blocks may still be seen lying along the Manchester and Liverpool, and Birmingham lines.

But to return to the edge rail, which was certainly a great improvement upon the old flange rail. This edge rail was, I believe, first invented by William Jessop, a celebrated man of his time, and tried under Smeaton, who proposed it for the Leicester and Loughborough Tram Railway. The flange was transferred from the rail to the inner edge of the waggon wheel; and the edge rail having no flange occasioned less friction to the shaft. As the rails were manufactured of cast iron, they could not be made longer than 3 feet without materially increasing their liability to break; but there were so many joints that it was very difficult to make a smooth road and keep it in order, particularly when the speed of travelling increased. Hence the use of stone blocks had not been abandoned. This difficulty was at length overcome by the substitution of wrought-iron rails, which were first proposed by Buckenshaw. These were laid on wooden sleepers, and this was another great improvement; but the numerous joints were still a great difficulty in the way of making a smooth road and keeping it in order. This was at last overcome by making the rails in one single piece on rollers patented for the purpose.

Rails have now been increased from 30 to 80 lb. per foot, and have been fished at every joint; that is, a plate has been fixed on each side with sufficient room in the connecting bolt-holes to allow for expansion and contraction.

With regard to locomotives, numerous improvements have been made. We have first Watt's idea of making a steam engine, with its boiler complete, upon an independent carriage, mounted upon wheels, so that it could move in any direction and propel itself; as well as a weight attached to it, mounted upon wheels also. As the tramways of cast iron were then established, this locomotive machine could have readily been applied to it, so that, in fact, to Watt may be attributed the first practical idea of the locomotive engine, although there are some doubts about this, for a M. Cugnot is said to have made a working model of a locomotive engine upon a considerable scale, at Paris, in the year 1783. Be this as it may, they were both made so nearly about the same time that it is difficult to decide which had the priority. At all events, Watt is justly entitled to the merit of having been the first to invent it in England.

Hackworth's (of Darlington) engine was made about the same time, and was similar to Stephenson's.

The next improvement consisted in placing the cylinders of the engine horizontally instead of vertically, so that the piston acted directly upon the axle upon which the driving wheels were placed, instead of by the intervention of a double crank; this made the engine more compact.

The next was the multitubular boiler, by means of which the generation of steam was greatly accelerated, in consequence of the increased surface exposed to the action of the heat.[4]

I will now revert to 1826, the time when I was asked my opinion as to the value of railways, and I said, in the most decided terms, to Lord Lowther, that I thought very highly of them, that they must succeed and eventually supersede every other mode of transport for passengers and goods. Being quite convinced of this, with which opinion my brother George cordially agreed, I set about projecting lines to those places where I thought they were most applicable; that is, where there was actually a large and constant traffic, and where a more facile means of locomotion would be attended with an increased trade. Next to the Liverpool and Manchester, I selected the London and Birmingham line, whilst my brother, in company with the late Jonas Jessop and William Chapman, chose a direct line, from the termination of mine at Birmingham, to Liverpool, so that the two lines together would have considerably reduced the distance between London, Birmingham, and Liverpool. My line proceeded by Aylesbury, Banbury, Bicester, Kenilworth, and War-

[4] The blast-pipe, also, was one of the most important improvements. Previous to this invention it was necessary to employ bellows to keep up the fire in the boiler, and these were worked by the engine, so that a good deal of power was wasted in order to keep the furnace going, and the greater the speed, the greater the power necessary to work the bellows; moreover, the waste steam ejected from the cylinders was constantly puffing out in the faces of the drivers, so that they could not see clearly before them. This was a great annoyance, which they were most anxious to get rid of, but nobody seemed to know how. At last, either Stephenson, Hackworth, or Booth, or somebody else, whilst driving an engine, and being much annoyed by the blowing in their faces, said, "Confound the steam; let us send it up the chimney." A pipe was accordingly made, connecting the cylinder with the chimney, so that the discharged steam might be ejected through it into the chimney. When this alteration was made, the engine was again tried, and to their astonishment they not only got rid of the annoyance of the steam, but the bellows were of no use, and the faster the engine went the more vigorous became the fire. Upon thinking over this coolly, the mystery was soon explained, for the hot steam being discharged into the comparatively cold chimney, a vacuum was produced, and the air rushed through the furnace to fill up the vacuum; the faster the engine went the greater the amount of steam sent up the chimney, and the more active the draught through the furnace. The bellows became wholly unnecessary, so much power was saved, and the nuisance of the escaping steam was entirely got rid of. This was really a great step towards in rendering the locomotive more effective; and it is very difficult to ascertain to whom the improvement is justly due.

wick, to the higher part of Birmingham; and my brother's from thence by Stafford and Runcorn—with a bridge across the Mersey at that place—to Liverpool. Thus London, Birmingham, and Liverpool, the three largest and most important commercial towns in the kingdom, would have been connected together in the shortest possible distance and with the least expensive works practicable. The project, upon the whole, was well received, but the public were not quite prepared for such an undertaking; in fact, it was in advance of the time, and for a while fell to the ground. Canals had not fallen into disrepute, and Mr. Telford, who succeeded my father in carrying into effect the great improvements proposed by him for the Birmingham Canal, suggested a continuation of this (through the very district which my brother, Jessop, and Chapman had proposed to carry their railway), to connect it with the Bridgewater and Mersey canal to Liverpool; this was accordingly executed, and, as regards a commercial speculation, failed entirely, as it was soon superseded by railways. The fact was, that Telford, having been bred in the old school, and having seen the triumph of canals, could not, or would not, believe in the efficacy of railways, or that they would ever succeed; and, indeed, he laughed heartily when he had succeeded in supplanting my brother's line of railway from Birmingham to Liverpool by a canal. He had a strong prejudice against railways, which he maintained until his death, in 1835. He had just finished his great work the Holyhead Road, with the great connecting suspension bridges of Conway and Bangor, and of which, with good reason, he was justly proud, and could not bear the idea of their being superseded by any other system of locomotion. Although an able engineer in many respects, he was not much of a practical mechanic, and very likely conscientiously thought that railways would not succeed. He had done his work well in his own department, and was too old to learn anything new. He died at a good old age, much respected and beloved by those who knew him, and leaving numerous monuments behind him of his engineering talents.

Another important line of railway which I proposed at this time was one between London and Brighton, and I employed two very experienced and competent surveyors, Messrs. Edward Grantham and Jago, who had frequently been employed by my father to make surveys in various parts of England, particularly in the region between London and Portsmouth, and the Weald of Kent, so that they had a thorough knowledge of the district. After examining the country myself, I directed that the line should commence at Kennington Common, and proceed from thence by Clapham and Streatham to the lower end of Croydon; from thence up Smitham Bottom valley to the hill at Merstham, which was to be pierced by a tunnel; from thence to Redhill, Horley, and the valley to the Cinder Banks at the base of the ridge of hills called Tilgate Forest, where there was to be another tunnel; from thence, skirting the left side of the valley near Balcombe, to the valley of the Ouse, which was to be crossed by a viaduct, and thence to Hayward's Heath, between Lindfield and Cuckfield, where it passed through another ridge of hills by a deep cutting; it then proceeded direct to the Southdown hills, near Clayton Hill, which was to be pierced by a tunnel; thence down the valley to Brighton, where it terminated at the upper end of the town on the right side of the valley. This line extended from Kennington Common to Brighton. The country is very rugged, having three lofty ridges of hills running east and west, which it was necessary to pass through, as there are no leading valleys or gaps to facilitate the passage without going a long way round, which would have defeated my object. This line might have been shortened nearly a mile by going direct from Tilgate Forest to Cuckfield, but the works would have been much heavier, and could scarcely have been justified at the time, as the public were not prepared for such expensive operations. I also employed Mr. Vignolles to survey another line from Nine Elms, Vauxhall, by Dorking, Horsham, and Shoreham; from thence along the coast to the west end of Brighton. This line, upon the whole, was easier of execution than the other, but it was five miles longer, which I considered objectionable, as my object was to lay down the shortest possible line between the two termini, so as to render all future competition out of the question. By this time I was so fully convinced of the ultimate success of railways, both for speed and economy, that I announced in the prospectus, that when the railway system had been introduced into France, the journey from London to Paris might be made by this route in twelve hours; however, at this period no passenger railway had been completed, and therefore my statement was considered only as a rough guess, which might never come to pass, and therefore a company could not be formed. At the same time I employed competent persons to make the survey for a coast line from Brighton to Worthing, Arundel, Chichester, Havant, Portsmouth, Southampton, Salisbury, and thence to Warminster, with the intention of extending it hereafter to Bristol.

The Manchester and Liverpool Railway was opened with great ceremony, when the Duke of Wellington, then Premier, and Mr. Huskisson, the President of the Board of Trade, attended. Unfortunately, the latter most able Minister was killed, by being run over by one of the locomotives. Poor Huskisson was standing between the two lines of rails, with the Duke and several others, when the engine came up unexpectedly; he lost his presence of mind, and took a wrong step, which ended in a death universally deplored by the nation, as well as by his numerous friends.

The success of the Manchester and Liverpool railway having been established, the next thing was to extend the line to Birmingham, and a company was immediately formed for the purpose, with Mr. George Stephenson as engineer. About the same time another line was projected by his son Robert, from London to Birmingham, pursuing pretty nearly the line of the old Grand Junction or Paddington Canal; this line was longer than mine, led over much more difficult country, and did not pass through the same number of towns and population, although it touched the important city of Coventry.

My line commenced at Blackwall, and pursued the route of the Regent's Canal, nearly similar to the line of railway which my father had formerly laid down.

There were two strong parties, one of which supported Stephenson's line, and the other mine; but ultimately Stephenson's party was the strongest, he carried his line by one or two votes, and the present London and Birmingham railway was executed.

The route to Liverpool was unnecessarily long, and therefore a new line was started to shorten the distance, called the Trent Valley Line, which commenced at

Rugby, and joined the Birmingham line at Stafford, thus cutting off the angle at Birmingham, and saving a considerable distance. This line was strongly opposed, but was carried, it is said, by the influence of the late Sir R. Peel, and was executed, I believe, at the cost of three millions. If my own and my brother's line had been carried out, this would have been unnecessary, as a branch to Stafford would have sufficed for that and other towns, and Birmingham would have been upon the main line. It is singular, also, that Telford's canal, which supplanted our line, has been absorbed by the adjacent railways.

Having now, in a rather rambling manner, brought my professional diary so far, I must revert again to some incidents connected with my private life. On the death of Sir Joseph Banks, Sir Humphry Davy was unanimously elected President of the Royal Society in the year 1820, and I had the honour of being elected Fellow, 1823.

I was elected member of the Travellers' Club, 1822. This club was formed in the year 1818, for the express purpose of associating travellers together, and for promoting travelling. It was enacted that no person was eligible to become a member unless he had travelled on the Continent, in a direct line, 500 miles from London, and great things were expected from it; but although it was the first travelling club established in London, and contained amongst its members the most distinguished travellers, it shortly degenerated into an ordinary club, and nothing has ever emanated from it towards extending our knowledge of the globe, or in publishing the travels of the numerous able men who belonged to it. Sir Arthur de Capel Brooke, who was a member of it, and who travelled a good deal himself, particularly in the north of Europe, going as far as the North Cape, and who published an interesting account of his journey, spoke to my brothers, myself, and numerous other travellers on the subject, proposing to establish a new club, composed of none but distinguished travellers; he suggested that this, in the first instance, should be nothing more than a dining club, to meet once every month, saying that travellers meeting together in this social manner would communicate to each other their various voyages, and would stimulate each other to further discoveries; that although a mere social club in the first instance, in time some good would result, and that in the end a regular scientific society might be established for the promotion of geography. He accordingly collected together all the distinguished travellers of the day, naval, military, and civil, and a most delightful society it was; the result clearly proved Sir Arthur Brooke's sagacity, for from the Raleigh Club was originated the Royal Geographical Society. I am not quite certain with whom the idea first originated, but I recollect that at one meeting, when, amongst others, the late distinguished traveller and Secretary to the Admiralty, Sir John Barrow, was present, an animated conversation took place, to the effect that the Raleigh Club had been in existence many years, and a very agreeable club it was; but, except amusing each other with our adventures, we had done nothing towards promoting our original intention, which was to stimulate discovery in foreign lands, and to extend the knowledge of geography throughout the world. I think Sir John Barrow then said, "Why cannot we establish a real geographical society, and read papers, and publish transactions, like other scientific societies?"

The idea caught at once: the whole of the members then present applauded it, and resolved to carry it into effect. A committee was appointed, and every member, I think, of the club joined. A general meeting was then called, members soon joined, and Sir John Barrow was appointed first president; a council and vice-presidents were chosen, a house was taken in Waterloo Place, Colonel Jackson was chosen secretary, a royal charter was applied for and obtained, and the society was duly instituted, under the name of the Royal Geographical Society. This society, like all others in their infancy, had a good deal to contend with; it went on well for a time, but at length it began to languish. It required some man of weight and influence to devote his time to it, to enlist the Government heartily in the cause, and to make it understand that it was its interest, on the part of the public, to promote the Society by every means in its power. Fortunately, at that time the Society had in Sir Roderick Murchison the very man most competent to undertake this arduous office. He worked hard; he convinced the Government of the utility of the Society, and the many advantages that would be afforded by its existence, as it could collect information upon all geographical subjects, at far less expense than could be obtained by any Government establishment; it would, moreover, stimulate travellers to increased exertions, and accumulate a fund of geographical information—knowledge that is so important to a commercial country like our own.

The then head of the Government was so much struck by the representations made to him of the value of the Society and of the advantage it would be to the Government, that he resolved in the House of Commons to grant 500*l.* a year in aid of its funds. This at once revived the members' spirits; they started, as it were afresh; the Society became exceedingly popular, and there was no end of applications to be enrolled as members, both from ladies and gentlemen, and ever since then it has been one of the most, if not the most, popular societies of the day.

The old Raleigh Club, from which the Geographical Society originated, having done its duty, and most of its original members having succumbed to time, there was no longer any necessity for its existence; but as all scientific societies have their dining clubs, which meet on the days of the societies' meetings, it was resolved to merge the Raleigh into the Geographical Club.

In the year 1824, John Wilson Croker, Esq., originated the Athenæum Club, for men distinguished in science, literature, and art, and asked me to become a member. I was only too happy to be associated with such a company as he collected together.

Sir Humphry Davy, in the year 1825, originated the Zoological Society, and asked me to join, which I did most willingly, and perhaps it has been the most popular and successful of any modern society of that kind. It commenced operations by purchasing the well-known Cross collection of Exeter 'Change, in which in my early days I took an especial delight; for, considering all things, it was a very wonderful collection, and it is difficult to understand how, in such a confined and unhealthy spot, it could have been maintained in such good condition. The only other exhibition of the kind in London was at the Tower; the collection of animals there consisted of presents from the sovereigns of different countries. These were afterwards lent to the Zoological Society, who established their museum in the Regent's Park, and, taking it altogether, it is probably the finest and best maintained in the world.

CHAPTER V.

Travels in the North of Europe and Spain.

Having been very hard worked, and being in bad health, owing to my still suffering from the effects of my fall into the cofferdam of London Bridge, I felt the necessity for some relaxation. I therefore made my arrangements for a short continental tour, resolving to visit the north of Europe. Adhering to my original plan, I shall only mention those places and occurrences which, for particular reasons, I think, may possess some little interest, and shall pass over the ordinary descriptions of places and things which are now so well known as to be hackneyed. My journey, principally by sea, from London to St. Petersburg passed without incident, except that off the Island of Bornholm we met the Russian fleet, consisting of seven sail of the line. It was a fine sight, and the ships seemed well handled; but one line-of-battle ship in going about missed stays, and got into what the sailors term irons, that is to say, she would move neither way. Our sailors laughed heartily, and we could see by the signals made from the flag-ship that the admiral was very much annoyed.

On board the steamer I made acquaintance with Lieutenant Conolly, who intended to take the route from St. Petersburg overland to India—a particularly difficult and dangerous journey; but Conolly was an intelligent, ardent, and courageous person, anxious for information, and ambitious to explore a route which had hardly ever been undertaken before; but he unfortunately fell a victim to his ambition. He was a little man, determined and energetic, capable of undergoing great fatigue, and a very agreeable and sociable companion. He and I took to each other and became great friends, and entertained the pleasing idea of meeting again after his return from India; but this never took place, for he was murdered with his companions.

Conolly and I got our luggage together as soon as possible after landing, and drove to the Grand Hotel, which was then a sort of barrack or caravanserai, near the Newski Prospect. Before seeing the city, I presented the various letters of introduction with which I had been provided; amongst others were several from my old friend, Chevalier Benkausen, the Russian Consul-General in London.

I first called upon Lord Heytesbury, our ambassador, whom I had previously known as Mr. Abbott, at Naples, by whom I was most cordially welcomed. I also waited upon all the authorities, and was well received by them; but the person to whom I was most indebted was my friend, General Wilson, a sensible Scotchman, who had entered the Civil Service of the Government, as engineer to the cotton, gun, and general iron manufactures, for which he had erected a large establishment at Kolpnau, about 13 or 14 miles from St. Petersburg. This place I visited with him; it was very complete, for the Emperor Nicholas, who was at times very fond of Englishmen and everything English, had taken Wilson and his establishment under his especial protection, and had given him authority to incur any reasonable expenditure to make it perfect. Wilson, therefore, imported the newest and most improved machines and tools of every kind from England, and obtained also the best of English workmen. The cotton manufactories were upon a most extensive and imposing scale; in fine buildings, with hosts of employés, they were conducted in the true imperial style, that is, with great show, but little profit. Still, they were not without their advantages; they were the means of training a set of native workmen, who, although not inventive, are excellent imitators, consequently after a time their services must have been valuable. The same may be said of the iron manufactories, particularly as Russia has some of the richest iron mines, and they only require skilled workmen to develop them to their utmost; Wilson contributed materially towards this; and it was impossible to have selected a better man for this purpose. He possessed a calm even temper, firm, but just, and conciliating, with a competent knowledge of what he undertook to perform, without possessing any considerable amount of invention. He spoke the Russian language like a native, besides French and German. He therefore most justly possessed a good deal of influence, and was thoroughly liked and respected, from the humblest workman under his orders up to the Emperor, who was very fond of him. Lastly, he was thoroughly honest, a rare thing in Russia in those days, where peculation was rife from the highest to the lowest; and if Wilson had followed the universal example, which he might have done with impunity, he would have made a large fortune; but after many years' service he died comparatively poor. Although anything but a military man, he was a general in the Russian service, and was decorated with several stars and orders. All persons officially employed like Wilson had military rank; he concealed this as much as he could; but in St. Petersburg he was always obliged to wear uniform, and as there were guardhouses in almost every street, whenever he passed the guard turned out to salute him, which annoyed him much.

There was another remarkable Scotchman, of the name of Baird, with whom I made acquaintance. He was, however, a totally different character—a shrewd, intelligent, clever, active, indefatigable person, wholly devoted to making money. He was in constant communication with England; and as soon as a patent was taken out there for any new invention, if it was applicable in Russia he at once imported and patented it, and thus obtained a monopoly. He cultivated, with the greatest tact and assiduity, all the officials, from the highest to the lowest, as well as all persons of any influence, and had a thorough understanding with them, particularly with the police and officers of customs; thus he contrived to gain the greatest influence, and secured almost a monopoly of everything worth having. If ever the Government was desirous of contracting for any large work, Baird was almost sure of obtaining it, at the best price. Whether his numerous friends were interested in the profits resulting from these multifarious undertakings is best known to himself; but according to the ordinary practice of human nature, particularly in Russia at that time, everybody endeavoured to make the most he could, and it is probable that there was no exception in this case; but suffice to say that Baird made a very large fortune, with which he retired to his native country; and we must not omit to mention that Baird, whilst benefiting himself, was of the greatest service to Russia, and tended materially to advance her prosperity by the numerous valuable inventions he introduced, and by training the natives, and inciting them and urging them to make all sorts of improvements, which, without such a man, would never have been undertaken, so that nobody grudged his wealth, and he left the country to which he had been a real benefactor universally liked and esteemed. As to myself, personally, I feel much indebted to him; through his kindness I had the opportunity of knowing many persons and seeing many things which otherwise would have been out of my power.

I frequently dined at the cafés and restaurants when not otherwise engaged, and at some of the best of them met persons of the first class, who were always very well bred and polite. One day I dined at one of the best table d'hôtes, when several persons of high rank were present, and amongst them a young naval officer of good family. Amongst other subjects of conversation, the recent taking of Varna came forward, and the naval officer seemed to speak rather disparagingly of it. I observed one of the superior waiters looking at him and listening attentively for some time, then he quietly went up to him and whispered a few words in his ear, which my neighbour told me was an order to hold his tongue, for such conversation would not be allowed. The fact was that all these waiters were employed by, or in the pay of, the police, and all conversations were reported. The young naval officer held his tongue immediately, and the party broke up very shortly afterwards, and I was told that the matter would not end there. I took the hint also, for nothing was more dangerous than to talk politics, and I avoided them ever after.

In considering the position of St. Petersburg, it is, perhaps, difficult to find a more inconvenient and unhealthy spot for a great seaport town than that chosen for the Russian capital, at the mouth of the Neva, at the head of the Gulf of Finland, where there is no tide, and where the greatest depth over the bar is only seven feet, so that none but vessels of a small class can ascend the river. Hence, though within the bar there is ample depth of water at all times, large mercantile vessels are obliged to stop at Cronstadt, seven miles distant, and there discharge their cargoes, which are transferred to St. Petersburg in lighters; in like manner they receive their cargoes from the capital; this, of course, is attended with great delay and considerable expense. Moreover, during the prevalence of strong westerly winds, the waters of the Gulf of Finland are heaped up at the upper end, and those of the Neva are driven back, so that it frequently happens that a large portion of the city is inundated; and in November, in the year 1827, a terrible example of this occurred. In many parts of the city the waters rose more than fourteen feet, many thousand persons were drowned, and a vast amount of valuable property was destroyed. These inundations might be avoided, and the port improved to a considerable extent, although it would necessarily require a very great expenditure. Still the object to be gained is of such importance that every reasonable means should be adopted to effect it, and compared with the enormous sums which have been expended in establishing this capital, the cost of improving the port would appear trifling.

Cronstadt is, properly speaking, the port of St. Petersburg. Here all vessels, whether of war or merchandise, must stop; and the southern side of the island, where there is the deepest water, from 24 to 27 feet, has been chosen as the site for the naval arsenal, as well as the port for merchant vessels, where docks had been made when I was there by enclosing a portion of the water space from the gulf; the two basins or harbours were close together, separated only by a partition wall. The accommodation for the mercantile vessels, although not small, nevertheless was in a rude state, devoid of the usual mechanical appliances which we have for many years been so accustomed to in England.

The naval arsenal, which was made in the time of Peter the Great, and was considered perfect at the time it was made, consists of a long canal leading from the outer to a small circular basin. Connected with this there were four dry docks for the line-of-battle ships of that period. From this the circular basin on the east side was connected with two other dry docks; and around these canals, basin and docks, there were several storehouses and magazines; but upon the whole they were badly arranged.

I may here as well mention that my brother and myself afterwards, at the request of the Emperor, designed a complete naval establishment for this place, utilizing as much as we could of the old construction, and this design, I believe, was to some extent adopted.

We afterwards built four iron steamboats, with their engines, for the Caspian Sea, which were the first ever afloat there. These vessels were built in England, then taken to pieces, and sent with the requisite number of workmen to Odessa. Thence by land they were transported to the Caspian, where they were again put together, with their engines, and answered their purpose well. We afterwards made the iron gates for the docks of Sebastopol, a pair of which were subsequently brought to England and France as war trophies. We also constructed several vessels of war, worked by the screw, for the Baltic and Black Seas; amongst others a yacht for the Emperor, to review his fleets in the Gulf of Finland, as well as for pleasure excursions. This was a small vessel, about 260 tons, with a pair of oscillating engines of the nominal power of 120 horses, although capable of working up to three times that power, and making fully 14 knots an hour. She was fitted up plainly but very neatly. From circumstances over which we had no control, the completion was prolonged to a later period of the year than we anticipated; and it was not before the middle of October, 1850, that we were ready to leave England, when my brother and his son George determined to go with her and deliver her in person. She was well insured, and was navigated by an English captain—who, I observed at the time, was not a very sharp fellow—and an English crew. As she was a small vessel she had to take a considerable quantity of coals on deck, which brought her down beyond her usual line of floatation. I went with her as far as Gravesend, and saw them fairly on their voyage, but before she reached the Baltic she encountered a severe storm, when she behaved admirably. At Copenhagen she took in more coals, and started again, and in the Baltic encountered another severe storm, which she got through equally as well as before. The weather then became fine, and they thought that all their troubles were over. The last storm had driven her considerably out of her course, which the captain had not taken a correct account of; and one fine starlight night, about ten o'clock, they were steaming away with a smooth sea, at the rate of about 10 to 12 knots an hour, when the engineer, putting his head out of the engine-house to enjoy the fine evening, suddenly called out, "I think I see land," and went at once to the captain. The captain said, "It is impossible. We cannot be nearer land than 30 or 40 miles." The captain and the Russian officer and my brother were at the time in the cabin taking their grog comfortably before going to bed. But he quickly went upon deck and soon discovered his error, and ordered the vessel to be put about; but before this could be done she struck upon a rock, and in spite of all their endeavours she could not be got off. The Russian officer declined taking any of the responsibility, saying it lay with the captain, and he would have nothing to do with it. At daylight they found themselves hard and fast upon the Island of Dago. My brother, finding that nothing could be done to get the vessel off, resolved with his son to make his way at once to St. Petersburg, and report the loss of the vessel. As the winter had begun, the journey was attended with considerable difficulty. However, they reached St. Petersburg, and had an audience of the Emperor Nicholas, who, when he heard the story, laughed heartily, and said, "Now, if this had occurred under the command of Russian officers and sailors, what would the English have said? why, that no wonder an accident had occurred, when the vessel was confided to those stupid fellows, the Russians. Now, you see, it has happened with the English; and they, with all their pretended knowledge, don't know the Baltic yet, and are more stupid than the Russians." Again he laughed heartily, and said, "I am delighted to find that my sailors are, after all, as clever as the English, and under the circumstances I am not sorry that the yacht is lost. It will teach the English in future not to be so proud of their knowledge, and to give the Russians credit for knowing the Baltic as well as they do."

Although the yacht was wrecked, we lost nothing, neither did the Emperor, for he was insured to the full, and we received an order to make another of precisely the same size. The underwriters, as soon as they heard of the accident, sent out a clever fellow, Captain Farr, who, upon arriving at the spot where the vessel was stranded, found that she had not received much damage; he, therefore, finding that the season was too far advanced to take her back to England, or even to get her into any Baltic port, at once weighed her and then sank her in deeper water in order that she might be protected during the winter. In the spring he returned, weighed her again, and took her to Revel, where he repaired the trifling damage she had received, then returned with her to England, calling at Hamburg on his way, and from thence she made a quicker voyage that had ever been made before. I saw her in the East India Dock, and she looked quite new; and unless I had known I should never have supposed that she had been stranded, and been a whole winter under water in the Baltic. She afterwards sold, I think, for 9000*l.*, having originally cost 14,000*l.*; and as the insurers had been paid nearly 4000*l.* for the original insurance, the underwriters scarcely lost anything. Thus we gained considerably by the accident, having to make two yachts instead of one, and neither the Emperor nor the underwriters lost anything.

Leaving St. Petersburg I started for Moscow, traversing 400 miles of most uninteresting country in an open droschky, drawn at the rate of 10 miles an hour—as I paid liberally—by four horses abreast. The only noticeable place I passed through was Novogorod, a considerable town, with a good deal of activity, and apparently an extensive trade. Here the two great water-carriage systems met, connecting the Baltic with the Black Sea, and also with the Caspian. The canals were crowded with vessels laden with the products of the East and West; natives from the East clad in their flowing garments, Tartars, with their bows and arrows, Cossacks from the Don, Armenian, Greek, and Turkish merchants, and the never-failing Jew, English, French, and Germans, all mixed together, and carrying on their particular business, formed a very amusing and busy scene. I contrived to get a tolerable dinner there, the only one deserving of the name since I left Petersburg; I devoted two or three hours to looking over the canal-works, which, for that time, were not badly executed, but the town contained nothing remarkable.

On reaching Moscow, what struck me most was the Eastern appearance of the inhabitants, particularly when compared with the extremely modern look of the town itself. The Kremlin, of course, I need not describe. I visited the celebrated Riding House, which is 1200 feet long, covered by a wooden roof, of the single span of 240 feet; it is without doubt the finest and largest shed in the world, and a splendid piece of carpentry, well worth going some distance to see. It is frequently used for reviewing troops in unfavourable weather, and it is said that ten thousand men can go through their exercises under its shelter.

On my way from Moscow to Warsaw I had no idea that this great road—one of the principal in Russia, and, upon the whole, not through an unproductive country—I should find so totally unprovided with anything for the accommodation of travellers. In those days I certainly did not expect much, and having travelled in more barbarous countries, I could submit cheerfully to a good deal; but I certainly did not expect that it would be so bad as I found it, and therefore I carried little with me, contrary to my usual practice. This I certainly repented of, for our fare was most miserable; if we got eggs, butter, cheese, and bread, we thought ourselves lucky, though sometimes we got better provisions; however, perhaps it was well, for although ill when I started, I began to get better, and slept soundly. My man Weiss consoled himself with plenty of vodki, and considering the rough fare we got I could not altogether blame him. We passed Smolenski, and a very poor place it was; there was nothing like an hotel, nor any accommodation for travellers. Although seventeen years had elapsed since the invasion of the French in 1812, the remains of the fire and battle which took place when the French captured it were still very visible. When we entered Poland cultivation appeared to be carried on more extensively and with greater skill, and the people seemed more intelligent; we passed several large proprietors' houses.

At Warsaw I attended a review of the garrison, in the Great Square, in honour of the recent victories of the Russians over the Turks. It consisted of about twelve thousand men of all arms, commanded by the Grand Duke Constantine, the Viceroy. I never saw finer troops in my life, nor any that manœuvred better. Before the review mass was said in their midst, and when *Te Deum* was sung by the whole of the men—and they sang it with great skill—the effect was very fine, and was rendered still more so by a salvo of one hundred and one guns, fired from the forts.

I returned rapidly through Germany, and reached home after an absence of sixty-five days, during which period I had travelled nearly thirty nights.

I immediately visited London Bridge, and found everything going on well. In fact, during my absence my brother George had diligently looked after my business as well as his own.

My brother George married the only daughter of Sir John Jackson, Bart., in June 1828. I had then taken a house for myself, No. 15, Whitehall Place, where my two younger brothers, Matthew and James, lived with me for about a couple of years.

At this time, as a bachelor, I saw a good deal of society, and made acquaintance with most of the celebrated scientific men of the day, also the leading literati, artists, painters, and sculptors. In fact, when I could tear myself away from my business I passed my time most agreeably and profitably in that distinguished society. I numbered amongst my acquaintance Sir H. Davy, Dr. Young, Wollaston, Dawes, Gilbert, Sir A. Cooper, Sir D. Home, Laurence, Greene, Playfair, Leslie, Whewell, Peacock, Hopkins, Liston, Barlow, Irving, Bailey, Colby, Parley, Sedgwick, Greenough, Delabeche, Meecham, Lyell, Brande, Faraday, Christie, Allen, Pepys, Sir James M'Intosh, W. W. Scott, J. W. Croker, J. Barrow, Sir S. Raffles, Marsden, Sir F. Chantrey, Sir T. Lawrence, Turner, Calcott, Stansfield, Behnes, Chalon, Sir A. M. Shee, Eastlake, Varley, Martin, Philips, Theodore Hook, Samuel Rogers, Southey, Robert Brown, Hallam, Sir A. Alison, Sir J. Parry, Sir J. Franklin, Sir John Ross, the late Sir J. Lubbock, the late Admiral Fitzroy, Professor Owen, and many others, and last, not least, Mrs. Somerville. Amongst foreigners—Humboldt, Laplace, Cuvier, Arago, Pring, Gerard, Cardinal Mezzofauité, Mailenette, Wree-Viking, and Von Breek.

In 1833, having now completed London, Hyde Park, Staines, and Crammond bridges, the great naval works at Sheerness, Woolwich, and Chatham, the Victualling Department, or Royal William Yard, and a large portion of the breakwater at Plymouth, Sunderland, Port Patrick, Donaghadee, Port Rush, and a large portion of Kingstown Harbours, the Eau Brink Cut, the Nene Outfall, the Witham Outfall, the Ancholme Drainage, and several other minor works, I was almost knocked up, and was recommended to take another continental journey for recreation. As I had never seen Spain I determined to go there, and accordingly started in the mail for Falmouth, and reached Cadiz on the fifth day.

Whilst at Gibraltar there was a grand military and civil fête, given by the governor of the fort Algesiras on the opposite side of the bay, to celebrate the establishment of the constitution, when he invited the governor of Gibraltar, Sir W. Houston, one of King William IV.'s most attached followers, and who had received me most kindly, and all the officers of the garrison. The governor of course could not go, nor all the officers; but a considerable number did, and I went in a boat in company with a number of the officers of the Rifles, while a great many rode round by land. We were received in the most courteous manner by the governor and the Spanish officers, and were most hospitably entertained.

On leaving Gibraltar I bargained with a respectable Spanish muleteer named Manuel, well known and recommended by my landlord, for the hire of four good mules, one each for myself and servant, and two for the baggage—which Manuel and his man occasionally mounted also. My idea was to go by Ronda to Malaga, but I was advised not to do it, as it was reported to be greatly infested by brigands; moreover, Spain at the time of my visit was in a very disturbed state on account of the Carlist war, and to add to this the cholera spread dismay and terror in most quarters, so that the time was very unfavourable for travel; still I determined to go on, and trusted to Providence for a happy deliverance, only instead of going by Ronda I determined to take the coast road. We proceeded through a wild, uncultivated country, and after two days' travelling reached Malaga about sunset, and took up our quarters at a comfortable little hotel situated in a narrow street near the Alameda. Here our consul, Mr. March, warned me against going any farther without waiting for some companions; but as my time was valuable, I thanked him for his advice, and determined to proceed, and after two days' stay started again, sleeping the first night at Velez, or Old Malaga. After supper Manuel came in and said that there were several suspicious characters about the village, and that, moreover, our arrival had caused some commotion; he had therefore told the landlord that we should start at daybreak, but strongly recommended our setting out two hours earlier. This we accordingly did, and at midday reached the old Moorish town of Alhama, perched upon the summit of the pass which separates Malaga from the vale of Granada; from here we descended into another rich vale, which, well irrigated and cultivated, teemed with wine, oil, corn, and fruits, and was filled with villages and chateaux, all indicating wealth and prosperity; yet withal there was a certain degree of wildness intermixed with it, which made it appear as if they were only half-civilized. It was long after dark before we reached Granada, then we had to go to the custom-house, where the officers were very much inclined to be troublesome; but I made friends in the usual comfortable manner, and got to a posada near the centre of the town—not a bad place, and which I was very glad to get into. Manuel came in whilst I was at supper and congratulated me upon our safe arrival; he said that for the last two hours he expected that we should be attacked every minute, and he therefore had urged us on as fast as possible; in fact, we came latterly at the rate of about eight miles an hour, which, he said, had saved us. The city of Granada, although extensive, appeared to be, like most of the towns I there saw, in a state of decadence, little trade, and consequently no prosperity. The lords of the soil seldom visited their estates, but left the whole to their factors or managers; and when the proprietor did come, he seemed to take no interest in his tenants or labourers, but lived in a half-ruined château in a miserable manner, reserving all for show and extravagance. Manuel said, that as our route lay through the mountains, and as it was very unsafe, it was absolutely necessary to take one or two escopoteros or armed police, to escort us at least as far as Andujar: this I at once consented to do, and we were now no contemptible party, consisting as we did of six well-armed men.

After halting at midday at Jaen, we proceeded through an open country, which presented anything but a thriving and prosperous appearance; the peasants were returning from their labour armed with guns, and they had a savage and discontented look. Manuel, who did not like the look of things, went up and spoke to one of them. When he came back to me he said that we must not go to Andujar that night, as the peasant had told him that there was a strong band of brigands in the neighbourhood, who had plundered the country right and left, so that all the country people were obliged to go armed and keep together. On hearing this news I resolved to stop for the night at a small village about two miles in front of us, which Manuel said was decidedly the best plan. We halted there a little before sunset, and a wretched place it was, without even a venta or public inn of any kind; I therefore hired one of the most respectable of the cottages, which was more like a stable or cowhouse than anything else, although it had a kitchen, and one or two rooms abovestairs; the floors were of broken brick, there were no windows, and only some planks on tressels for beds, with one or two broken chairs. However, we were able to purchase materials for supper, and with cloaks, saddles, and bags very soon made beds. I confess I did not like the place at all. As we might be attacked during the night, we barricaded the house as well as we could, and slept in our clothes, with our arms ready, and one man keeping watch. Having done this we went to sleep, but were awoke soon after midnight by loud cries and screams, and a man began knocking violently at our door asking admittance, crying out that the robbers were come. We were up in an instant, prepared to give the rascals a warm reception. We had scarcely made our dispositions for defence when the robbers, to the number of at least a dozen, made their appearance, well armed, and demanded our money, horses, and baggage, on pain of death. These I determined not to yield, and defied them. They, seeing that we were well armed and prepared for a stout resistance, hesitated for a minute; and I, not wishing to push matters to extremities, called to Manuel to tell them, that as for yielding to their demands it was ridiculous, and if they did not go away at once, I should be joined by some troops, and then every rascal of them would be shot; but if they chose to send four or five of their men to escort me towards Andujar I would pay them liberally. They then consulted, and agreed to the proposal, when I told them that I should be ready to start at five in the morning. Accordingly, at the appointed time, our friends, armed to the teeth and well mounted — as rascally-looking a set as ever one saw — made their appearance. I gave them a cigar and glass of brandy each, which put them in good humour. I could get on tolerably with Spanish, and entered into conversation with them; they were very agreeable, and told some curious stories. Manuel came riding up to me and told me to be upon my guard, for that one never could be sure of them for a moment; however, I thought the best plan was to show no fear. At nine o'clock we got to our destination, at a miserable village between Andujar and Cordova, where we went to a wretched venta; I gave them a breakfast of such as we could get, with cigars. I paid them handsomely, and so we parted, apparently the best of friends. My guide and worthy friend Manuel, as we left, said, "You seem to be very well pleased, but you don't know these ratteros. I hope that we have done with them, but I very much doubt it; you have paid them too well not to make them wish for further acquaintance with you, and depend upon it we shall meet them again, when we shall perhaps not be so well prepared." "Well," I said, "we have got rid of them for the present, and if we happen to make their acquaintance again, depend upon it we shall have the best of it." We jogged on all the day through a wild but not uncultivated country; the land was rich—plenty of vines, olives, corn, maize, and fruits, and everything, if properly cultivated, well calculated to make the people comfortable and prosperous; yet everything around denoted misery, poverty, and wretchedness. When I talked to the people they seemed reasonable enough; they said that they toiled from morning to night, but never got paid, or at least so little that they could not live upon it; and then, what with the government taxes and the priests, it was impossible to live, so that there was no use in working. I certainly could not help sympathizing with them, for they are really a fine generous people, and if they were properly treated, there is not a finer race anywhere. Unfortunately there is no middle class, and the nobility are completely worn out, so that the unfortunate peasants are ground down to the lowest misery; yet with all this there is a nobleness, independence, and enduring fortitude about a Spanish peasant which causes you to admire them the more you know of them. I soon recognized their character, and appreciated it accordingly. Whenever I entered a venta or posada I always made it a point to treat the host and hostess with frankness and courtesy as if we were equals, also to show myself ready to oblige and to assist in any preparations that might be going on. Thus I secured the utmost attention, and they readily produced their best at the cheapest rate, a result which no amount of money would have obtained.

But to return to my story. When within seven miles of Cordova, while passing near the small fortress of Ercaloro, at about half-past five in the evening, i.e. not long before sunset, I met a priest, who told me that it would be impossible to proceed, as he had observed five mounted robbers prowling in the olive woods between the fortress and Cordova, who would be certain to fall upon us; and that, moreover, they were in all probability only the scouts of a much larger body. Manuel exclaimed, "Did I not tell you that we should meet these rascals again?" I, however, replied that I was determined to reach Cordova that night, and asked the priest to introduce me to the governor of the fort. To this he willingly consented, and the governor having heard my story, was so obliging as to say that he intended sending fifty men to Cordova next day, but that they might accompany me now. We accordingly started and soon came in sight of the five mounted men, who sure enough proved to be our five old friends, who very soon turned and galloped off as hard as they could. We fired one or two shots at them; but as night was coming on we thought it best not to pursue, and continued on our way to Cordova, which we reached at eight o'clock, very glad to get off so well.

From Cordova we reached Seville, where Mr. Wetherall, the Consul, strongly advocated the introduction into England of Manzanilla, a wine then scarcely known. I requested him to send me a hogshead, which was universally approved of, and henceforth the taste for pale dry sherry has entirely superseded that for the old golden and brown, and there is no doubt that as a tonic it is far superior.

The road across the Sierra Morena was kept clear by the singular expedient of intrusting its defence to a body of German colonists, who held considerable lands and dues on condition of keeping the roads free from brigands. This plan answered admirably; in a very short time the brigands were exterminated, and after that, though no patrols were ever seen, yet if any fresh bands ventured to appear, the Germans were instantly under arms and never relaxed their pursuit until the brigands were either destroyed or driven out of the country.

We proceeded by diligence from Seville to Madrid; and when we reached Ocã-na, after having travelled three days and nights, our majoral or conductor had compassion on us, and said we should halt for the night; that is, we arrived about nine o'clock and he said we must be off again at five in the morning; this, however, was a great release, and we all thanked him; but I believe we had no great reason for being so very grateful for his kindness, as it was rumoured that if we had proceeded we might have been attacked by the Carlists. However, be that as it may, we got a comfortable bed, to my great surprise. The cholera had been flying about the neighbourhood, and I felt a slight attack of it, which I got rid of by a few drops of sal volatile and camphorated spirits in a wineglass of cold water. At five in the morning we started from Ocãna, after getting a biscuit, a cup of chocolate, and glass of cold water, which one finds almost everywhere in Spain; indeed, rough as the travelling was in those days—and it could not be worse—we always got most excellent bread, eggs, and sometimes milk and wine, although the latter was generally new, and as thick as porridge and almost undrinkable; still with bread, eggs, and milk one could always get on.

When I arrived at Madrid, the city was in the greatest state of excitement; the Carlists were making war in the most vigorous and successful manner, headed by the celebrated Zumalacarragui; and it being dangerous to talk politics, I particularly avoided them, and went about seeing everything I could as a stranger. I also called upon our minister, Sir George Villiers, whom I had known in England, and was most kindly received by him. I here met a Colonel Downie, who had served under Wellington and afterwards settled in Spain. He spoke Spanish perfectly, and was much respected by the natives, which was saying a good deal for him; for the Spaniards are a most peculiar people, and especially the upper classes, extremely reserved and exclusive towards strangers. Personally, however, I have no reason to complain, for they were very civil to me. I was introduced by Downie and other friends, and had an opportunity of visiting some of the first families of the place, amongst others the Veraguas, the descendants of Columbus, of Cortez, Viluma, Frias, and others. I went to their tertulias or conversaziones, which were the only kind of society to be had when I was there; in fact, parties were so divided on account of the civil war that many of the great houses were shut up. At the tertulia there was nothing but conversation, so that with the exception of the lights there was no expense of entertainment. As you were leaving the house, in the hall the servant presented you with a glass of cold water, with a biscuit of flour and sugar, which, when taken with the water, was not unpalatable. At one of these tertulias the servant, in the midst of the conversation, brought to the lady of the house her supper, which she set to work on, without making the least remark; in fact, it was usual, and no person thought anything of it.

Having now viewed everything worth seeing in the city, I determined to see something of the environs, and accordingly asked Downie to accompany me to

Toledo; and as politics were very uncertain, I thought it better to be doubly armed with passports. I therefore got our minister, Sir G. Villiers, to verify mine, and also the minister of police, a very gentlemanly man, the Marquis of Viluma, who was unusually civil, and gave me a capital dinner and a special passport for Toledo.

The next morning at daybreak we started, with four horses, and were just leaving the town when we were stopped by a messenger running after us, who proved to be the servant of a lady whom I had met the day before at the Marquis de Viluma's, and who was a niece of the Archbishop of Toledo. She had sent her servant with a letter of introduction for us to her uncle the Archbishop, for which we were thankful, as it is rather difficult to obtain permission to see all the different objects, such as the treasury. Accordingly we started off afresh over a very wild country, with a very indifferent road, or rather none, and reached Toledo about noon. We immediately went to the cathedral, which is a very rich and imposing Gothic edifice, and although we did not see the archbishop, who was not there, yet the dean most kindly showed us everything, particularly the treasury, which was replete with a most gorgeous collection of vestments, mitres adorned with precious stones, silver croziers, gold and silver chalices, cups and basins and priestly utensils without number. We then went and paid our respects to civil and military governors, and got an order to see the Royal Sword Manufactory, which formerly was celebrated throughout Europe for its excellent blades, which were said to be equal to those of Damascus, but had for some time past been rapidly on the decline; there were some two or three hundred men employed where I was, but they were making only ordinary blades for the army. Having seen everything, and perambulated this curious old city, which was in anything but a thriving state, we returned to the fonda or hotel, miserable as it was, to our dinner, previous to starting for Aranjuez.

We had just done dinner when an officer of police made his appearance, and said very politely that the civil governor wished to see me. I said that I had already seen him, and paid my respects, and shown my passport; I asked if anything was wrong with it. He said no, then went away, and again returned, saying the civil governor must see me. I again asked if the passport was right. He said perfectly. Then I said that I thought the governor's conduct was extraordinary, and I declined going. Two more police officers then came, and said that the gates of the city were closed to me, and that I should not leave without first going to the civil governor. My friend Downie then got alarmed, and said that he would go with the officers, see the civil governor and explain matters. He accordingly went, and asked the civil governor what he meant. The governor replied that he was surrounded by Carlists, and he was obliged to be constantly on his guard, for fear of an insurrection; but he had been informed that I had brought a private letter to the archbishop, who was the greatest Carlist in the place, and he must know what that letter contained. Fortunately it was open, and both Downie and I had read it before delivering it, and it was nothing more than a request that we might be shown everything in the cathedral. With this explanation the governor was perfectly satisfied, and dispensed with my personal attendance, which I was glad of, and determined to carry no more introductory letters, for at that time it was most dangerous to both parties. Off we started, but were stopped at the city gates, at which we were both much annoyed, and being determined that we would stand this annoyance no longer, were just going to force our way through, when a messenger from the governor told them to let us go, and off we galloped as hard as possible.

Before leaving Madrid I visited some of the convents, though with great difficulty, for a short time before my arrival the mob, in a fit of revolutionary excitement, had attacked several of them, and murdered many of the unfortunate inmates, whilst the rest were obliged to fly for their lives. When I applied for admission it was refused, until learning that I was an Englishman, they opened the outer gates, which had been well secured, and admitted me, and a melancholy spectacle I beheld. Very few of the monks remained, and those that I saw had their heads and arms bandaged up on account of the wounds they had received. A great deal of property had been destroyed, and a still greater quantity had been stolen; in fact everything had a most wretched and desolate appearance. Shortly afterwards the whole of the convents and religious establishments for monks and nuns were shut up by order of the government, and all the property was seized for the benefit (?) of the nation.

Leaving Madrid with several others, we proceeded to France *viâ* Saragossa and Barcelona, as the direct route through Bayonne was of course quite impracticable. Barcelona was then, and is now, taking it altogether, the finest and most thriving town in the peninsula, and may be called the Manchester of Spain. It is situated on the shore of the Mediterranean, in a rich fertile plain, backed by a lofty range of mountains about three or four miles distant. The central streets are very narrow, but the Marina and artificial harbour, with the fine spacious quays by which they are surrounded, have a noble effect, although the mole, a fine work as it undoubtedly is, has been badly designed with respect to the currents and the great quantity of alluvial matter held in suspension by the waters, and carried along the shore from the deltas and mouths of the Hobugal and Ebro. The consequence is, that the space covered by the mole is constantly filling up, and requires incessant dredging at great expense; notwithstanding which, the trade is so great that the harbour is always more or less full of shipping, and besides its manufactures of cotton and silk, it exports largely wine, oil, bark, fruits, and timber.

I made an excursion with my servant to the celebrated convent of Monte Serrata, situated upon the mountain of that name, about 30 miles to the west of Barcelona; starting early in the morning, and stopping at the bridge of Mastoul across the Hobugal, about 15 miles from Barcelona. Here I enjoyed, from below the bridge, one of the most interesting and beautiful views, I think, that I ever beheld. Facing me was the bridge, consisting of two Gothic arches; the south one was the largest I ever saw, being about 140 feet span. On the north side of the bridge was a Roman arch, in tolerable preservation, except the cornice; and on the south side of the bridge were the remains of a Moorish fort. Thus I had before me, at one *coup d'œil*, the ruined works of three great nations; in the distance was the convent of Montserrat, perched upon the mountain side, with its numerous pinnacles rising above it, and these overtopped by the numerous lofty peaks of the Pyrenees behind.

Whilst examining the bridge, I perceived on a sudden a large body of troops, with several mounted officers, rapidly approaching; upon inquiring the cause, I was told by an officer that I had better get out of the way as soon as possible, as there were a great many Carlists about, and they expected to be attacked every minute, as they had to defend the pass to prevent the Carlists from getting to Barcelona. I therefore went off at once to the convent. I left the carriage at the bottom of the mountain, at a small inn, and got a man to carry up our things; and having a letter to the worthy superior, was most hospitably received. They gave us a comfortable dinner. We then set out to examine the numerous hermitages which were perched upon different peaks of the mountain, which resembled the teeth of a saw, from which the hill takes its name. These hermitages consisted of a small hut, just large enough for the hermit's bed, and table, and chair; here they remained winter and summer, and only occasionally descended to the convent. All the hermitages were empty, the hermits had fled, and there were scarcely a dozen monks in the convent. Whilst climbing about the mountain I observed several Carlist scouts, well armed, lurking about, and I saw others at a distance—no doubt they were part of the attacking force expected at Barcelona; they, however, did not molest me, and I was too happy to leave them alone. The view from the mountain all round is very fine; all the leading valleys are filled with manufacturing towns and villages, amongst the principal of which was Manresa, in the valley immediately below us. We got back to the convent soon after dark, where the prior had ordered us a good supper, and afterwards invited me to his apartments, where he gave an excellent concert, which was very well executed by the choristers and monks attached to the convent. I soon found out that the prior and all around him were most devoted partisans of Don Carlos, and wished the Christinos and all revolutionists to perdition, to which place they said they would be most assuredly consigned. Knowing that I was an Englishman, and being recommended to him, he felt himself perfectly at home with me, and seemed to be tolerably well informed about English politics: he understood perfectly well the difference between Whigs, Tories, and Radicals, and had not a doubt but that the good sense and talent of the Tories would soon upset the Whigs and Radicals; and so far he proved right; but he went on to say that they would assist the Spanish Carlists, and send all the Christinos and their wicked associates to the devil, where they ought to go; for the Catholic religion could never thrive and Spain never could prosper so long as the Christinos were triumphant. By this time I could get on tolerably well with Spanish. We conversed on various topics, and passed the evening very agreeably, as the prior was a very superior person, and really, considering that he was a Spanish priest, he was an enlightened man, and by no means bigoted. He said that no nation could prosper without religion; and according to his belief he considered that the Roman Catholic was the true faith; at the same time he did not mean to say that a person professing any other religion could not be saved, but that God in His great mercy would pardon their ignorance. We bade the prior good-night, thanking him much for his kindness, and retired to our dormitory, which was very cold and solitary, near the chapel: we had plenty of cloaks, which were indispensable; and I told my man to make a stiff glass of hot punch, not only for myself, but also for the worthy monks who attended us, and I took care that they should be offered supper; but although they had a glass of punch and a cigar, they would take no more. The night was very cold, the moon shone bright, and the stillness was remarkable. I awoke long before light, and heard the monks saying their matins; feeling it was cold, I got more covering, and again fell asleep. Awaking soon after eight o'clock, we got up, and had chocolate; then, making a handsome present to the poor-box, I retraced my steps downwards, and I must say that I never passed a more agreeable day. The scenery was magnificent; that alone was sufficient to recompense anyone for the journey; and in addition to this there was the visit to this great convent, at one time one of the most powerful in Spain, but now in its decadence, and its once powerful inmates degraded so far as to hope for deliverance from the formerly hated and persecuted heretics of England! It was a most singular sight; and it is still more extraordinary, that after a lapse of thirty-four years, these very Christinos, with the queen at their head, should now be persecuting all liberal Catholics, and again threatening the establishment of the Inquisition.

Leaving Barcelona, I traversed France, passing through Toulouse, Bordeaux, and Paris, and reached London in the beginning of 1834.

Although not strictly in chronological order, I will here mention one or two incidents which occurred to me shortly before, and which may be of interest. In the early part of 1831 the cholera visited England for the first time. Everybody was alarmed at its approach: it was a mysterious disease, nobody seemed to know much about it. They knew that it came from India, and that it was most fatal and capricious, sometimes attacking those on the mountains, and sometimes those on the plains; sometimes, in passing through towns and villages, carrying off in its strange and deadly course all those on one side of a street, leaving the other side untouched. It was making its way regularly from the East to the West, deviating rather northwards in its course, and hence it approached Europe by Russia, thence to the north of Germany. Its regular and gradual approach struck everyone with awe: we saw the enemy coming, but how to avoid or how to attack him nobody knew. The medical profession were completely at fault, all suggesting different remedies, each proposing what he considered his own specific, yet doubtful of the result. At last the dreaded mystery leaped across the German Ocean, and made its appearance in Sunderland, where it spread alarm and dismay far and wide. It then turned southward, and made its appearance in London, in the month of March, and numbers left the city, flying in all directions. The French, dreading its invasion, closed their ports, and placed England in quarantine, but in vain attempted to shut out the terrible malady, for, in two or three months apparently glutted with death, it jumped over the Channel into France, and became more virulent than ever. The French physicians, who had visited England during its prevalence, and thought they had acquired good knowledge of the disease and its treatment, were, if anything, more at fault than our own medical men; for the disease held them as it were in derision, and in Paris seven thousand fell victims in one day. I was attacked with it one morning in June, about four o'clock, with a sudden shock, and felt as if death had clutched me in his jaws. I had been some time previously thinking what I should do in the event of an attack, and consulted my medical advisers, but could elicit nothing satisfactory. I then made up my mind that, as the attacks of the disease were sudden and violent, the remedy must be something of the kind; and the most likely remedy, if any, appeared to me some strong stimulant, such as camphor, sal volatile, and opium, which, having previously been subject to diarrhœa, I had frequently tried with effect. I therefore always kept a mixture of this kind ready. Feeling the attack, I jumped out of bed and staggered to the table, took a strong dose, rang the bell violently, sent for my doctor directly, and went back to bed, and for an hour suffered terribly. I then became calmer, but excessively exhausted, and lay almost motionless. The doctor came about seven, when I was much better: he asked me what I had taken, and I told him: he then said, "I do not know that I should have prescribed exactly what you have taken, but the principle is correct," and he gave me something of the same kind, but in a milder form. In a couple of days I was quite well, and I have ever since carried camphor, sal volatile, and opium with me; and subsequently, when travelling in Spain, Portugal, and Sweden during the prevalence of cholera, I frequently took these remedies myself, and administered them with success to others.

The same year I took a trip to Austria and the south of Germany, and attended the meetings of the scientific society, the Natur Geforsches, then being held at Vienna. I was well received, and made a member of the society; and there I made acquaintance with some of the most distinguished professors of Germany, and a most sociable set of gentlemen they were. The assemblage consisted of about three hundred; we dined every day together, and received much gratification and instruction. I attended the sections regularly, and there I saw the Austrian archdukes, distinguished for their scientific acquirements, who took part, and no mean one, in the discussions, without the least pretension, precisely upon the same footing as the other members; foremost amongst the rest was the celebrated Prince Metternich, then in the height of his power and consequence, as Prime Minister of Austria. He certainly was a very remarkable man; of the middle size, extremely good-looking, with an aquiline nose, sharp, intelligent eyes, a firmly compressed lip, a thoroughly gentlemanlike manner, a dignified appearance, complete self-command, and altogether impressing you with the idea that he was the great grandee and sovereign minister of the ancient and then all-powerful empire of Austria. Notwithstanding his high position, he attended sections in the most unassuming manner, like the most insignificant member present. He attended the different sections daily, and always sat amongst the crowd, not on the bench near the president, and took his share in the discussions as if he was a simple citizen, and any point that he happened to take up he well maintained. Having heard so much of the all-powerful Prince Metternich, I was rather astonished to see him act so amiable and distinguished a part in such a society. I was introduced to him, and was very kindly received. He opened his palace, and he and his distinguished and handsome consort, the Princess, received the whole of the members and their ladies with the greatest cordiality. I also was invited amongst the rest, and remarked that, although princes of the imperial family and the most distinguished nobles were present, I did not observe many of their ladies; and I was told by a friend, that as the ladies of the German professors could not afford to dress in the splendid style of the great Austrian ladies, these latter were excluded for fear of their eclipsing in dress the German professors' wives. The Emperor Francis gave the

Association a most magnificent entertainment at one of his palaces, Laxembourg, about 12 miles from Vienna. His Imperial Majesty sent sixty-five royal carriages, with outriders in uniform, to conduct the association to Laxembourg; and when we arrived there, we found sixty-five other imperial carriages, with servants in the imperial livery, to take us to the palace, and drive us about the park and environs to show us everything worth seeing. At three o'clock dinner was announced in the palace, and a most imperial dinner it was. Several of the archdukes, and Prince and Princess Metternich were there, with all the members and their ladies; and before sitting down to dinner the Prince said that His Imperial Majesty the Emperor Francis intended to have been present to receive the Association, but unfortunately His Majesty was very unwell, and was unable to attend; he therefore requested the Prince to apologise to them for his absence; His Majesty hoped that we would excuse his absence, and make ourselves as comfortable as if we were at home, which we certainly did, and a splendid affair it was. A magnificent band played during dinner, Tokay champagne flowed in abundance, and at five o'clock we started on our return to Vienna, conveyed in the same royal carriages and accompanied by the same escort of imperial servants that brought us to Laxembourg in the morning; in fact, it was impossible that anything could have been better done, or that royalty could have shown more respect or deference to science than was done to us.

On the following Sunday, the municipality of Baden, a place some miles from Vienna, invited us to another banquet, and sent comfortable carriages to take us there and back. The burgomaster and councillors received us on our arrival, and gave us a most excellent entertainment, accompanied by the greatest cordiality and kindness. When we arrived it wanted about two hours to dinner; and a friend of mine, high in office, asked me if I should like to be introduced to the celebrated Archduke Charles, the Commander-in-Chief of the Austrian armies, and who was the first general who defeated the great Emperor Buonaparte, viz. at Aspern. The proposition was quite unexpected, and I eagerly accepted it. We accordingly went to the Archduke's palace, where, on account of ill-health, His Imperial Highness was living very quietly, taking the baths. Upon arriving, we were shown into an ante-room furnished in the most simple manner. After waiting a few minutes, a chamberlain made his appearance, and ushered us into the Archduke's presence, when we were formally introduced to His Imperial Highness — a most simple, unaffected, dignified gentleman, characterized strongly by the features of the imperial family, at the same time possessing all the dignity and command of a great soldier. He received us with great courtesy, and after bidding us be seated, entered into conversation with us in the most easy and familiar manner. He addressed me very kindly in French, asked after the Duke of Wellington in particular (whom fortunately I had the honour of knowing), and expressed in the highest terms his admiration of him as a soldier and statesman, and said that England owed much to him. He then entered into a general conversation about England, her great importance and power, saying that she was the saviour of Europe, and expressed an ardent wish that she might long retain her present influence. He said that he should have been most happy to have dined with the municipality, to meet us there, but unfortunately his health would not permit. After an excellent dinner the Association returned in the evening to Vienna, much gratified with their entertainment.

CHAPTER VI.

Ship Canal from Portsmouth to London—Machinery and Engine
Making—Screw Steam Ships—Hartlepool and Coquet Harbours—Railways round London—Railway Mania—South-Eastern Railway—London, Chatham, and Dover Railway.

Railways had by this time made rapid progress, and had been completely
established as the future means of conveyance for goods and passengers. The
Manchester and Liverpool and the Stockton and Darlington had been completed
with the most successful results. The Grand Junction between Liverpool and Birmingham, the London and Birmingham, and Great Western, were making rapid
progress towards completion, and numerous other lines were either projected or
about being carried into effect. Still the canals were not altogether supplanted;
and it was proposed to make a ship canal from London to Portsmouth, by means
of which the dangerous, tedious, and expensive navigation between those places
would be avoided. The late Mr. Horace Twiss, M.P. for Wootton-Bassett, and afterwards Under Secretary of State for Foreign Affairs, took the greatest interest in
this undertaking; and from calculations which he had carefully made from official
returns, he stated that a clear revenue of 1,000,000*l.* per annum might be derived
from it. The first Lord Ashburton, then Alexander Baring, with whom I had a long
conversation on the subject, said that, if practicable at reasonable cost, it would be
a valuable national work.

During the height of the great revolutionary war with France, long before the
public had been accustomed to the vast sums which have been raised and expended upon great works in modern times, a canal between London and Portsmouth
had been considered as a very desirable and profitable work. My father made a
plan for it in 1807, and the then Earl of Egremont offered to subscribe largely to
it: a money crisis however occurred, and it was abandoned for the time. A canal,
however, upon a much smaller scale was afterwards made by Mr. Josias Jessop,
between the Wey and Arun, and from thence through Chichester and Longston
harbours to Portsmouth. It was upon too small and imperfect a scale, and therefore did not answer.

A ship canal, however, capable of transporting a 74-gun ship and Indiamen
of the largest class, was afterwards contemplated. A very influential committee
requested me to investigate the subject thoroughly; firstly, as to its practicability;
secondly, what was the best time for such a canal; and thirdly, what would be the
cost. I accordingly, with the assistance of the late Mr. Francis Giles, who took the
levels and surveys, explored and examined the three lines which were most prac-

ticable—the first by the Merstham, the second by the Dorking, and the third by the Guildford valley. The last was decidedly the best line, having the least height, the easiest route, and the best supply of water for the lockage. It commenced at the Thames, and continued up the valley of the Wey to Guildford, where it crossed the summit, descended into the vale of the Arun, which it crossed by an aqueduct, and thence along the base of the hills to Portsmouth Harbour.

The canal was to have been 100 yards wide at the top, and 24 feet deep. At the summit there were to have been ample reservoirs, and capacious basins or docks at each end. The voyage from London to Portsmouth would have been made in two days—that is to say, by common haulage—but steam tugs would have reduced it to twenty-four hours. The estimate was 7,000,000*l.*, which was considered so large at the time, that all idea of prosecuting the undertaking further was at once abandoned. The world had not then been accustomed to the enormous sums since spent upon railways, and then they would never have believed that 16,000,000*l.* would be spent upon the London, Chatham, and Dover Railway, only the same length as the proposed canal, or that a similar amount would be spent in the same county upon the South-Eastern Railway.

That the canal is perfectly practicable there can be no doubt, and it would have been of great public advantage; but whether, after all, it would have yielded a reasonable profit for the capital expended, is a question which I will not undertake to determine.

One part of my father's business was the making of machinery, of which he was very fond, being a first-rate theoretical and practical mechanic; but the machinery department formed only a very small part of his extensive business, although he constructed several important works, such as the Albion Flour Mills near Blackfriars Bridge, afterwards destroyed by fire, and where he subsequently had his own works, which still exist. This is admitted to have been one of the best pieces of that class of engineering ever constructed, either before or since, and performed a quantity of work in proportion to the power employed, such as has never been surpassed. He also designed and constructed the rolling mills in the Royal Mint, which have been in full work for more than half a century, and are still in as efficient a state as ever. The diving bell may almost be said to have been an invention of his, for he effected such great improvements in it that he was enabled to apply it to building under water at Ramsgate Harbour for the first time in 1813. It was worked from a scaffold above water, to which were attached movable trucks with windlasses, working upon a rack-and-pinion railway, so that the diving bell and the apparatus for raising and lowering the bell, together with the stones, could be worked, and the building carried on with the same certainty and nearly the same expedition as above water. He also erected a similar apparatus applied for raising heavy blocks of mahogany at the West India Docks: this is the Gantry crane, which has been since very generally employed in almost all building operations. My object in mentioning the subject here is merely to say that when my father died he left this machinery department to my brother George and myself, though I believe that if he had lived a few years longer he would have given it up altogether. The site of this manufactory was formerly one of the most fashionable suburbs of the metropolis, and here the celebrated Nell Gwynne had her country house. My brother and self continued the business, rebuilt the place entirely, with considerable improvements, and did a large amount of business here. We constructed the rolling mills for the Calcutta and Bombay mints; the great flour mills and baking machines of Deptford, Portsmouth, and Cremill Point yards; numerous locomotive engines for different railways, amongst others the 'Satellite' for the Brighton Railway, which was one of the first that attained the speed of 60 miles an hour upon the narrow gauge. We made the steam engines and machinery for Her Majesty's ship 'Bulldog,' the yacht 'Dwarf,' and others; also for the famous Russian steamer 'Wladimir,' which did so much mischief at Sebastopol; two yachts for the Emperor of Russia, and other vessels for the Russian navy, together with the whole of the iron gates for the dockyard at Sebastopol, two pair of which were brought back as trophies by the British and French armies. We built four iron steam vessels and their machinery for the Russian Government, for the Caspian Sea, which were the first that floated on its waters; they were first built in London, then taken to pieces, sent to St. Petersburg, and thence down the Volga to the Caspian; men were sent with them by us, who put them together there, and launched them successfully. We made and erected the small-arms manufactory at Constantinople, for making five hundred muskets per week. We constructed the engines and machinery for the 'Archimedes' screw-vessel, which was the first screw used in this country; and again, the iron vessel, engines, and screw for the 'Dwarf,' which was the first screw-vessel ever introduced into the British Navy, in the year 1839, for which I take the credit myself; for after we had succeeded so well with the 'Archimedes,' I waited upon Sir George Cockburn, then Senior Naval Lord of the Admiralty, and proposed to him to make a small iron vessel worked by a screw. I engaged that the vessel should make the speed of ten knots an hour by the measured mile; and that if after she was completed and tried she did not come up to the required conditions, of which their own officers were to be sole judges, I would take back the vessel and machinery, without any compensation; but if they were satisfied, they themselves were to fix the price to be paid for the vessel and machinery. Sir George said the offer was so fair, that if I would put it in writing in the form of an official tender, he would recommend the Admiralty to accept it; this I accordingly did. The vessel, engines, and screw were completed to the satisfaction of the Admiralty officers, the price settled by them was at once paid, and so the 'Dwarf' was the first screw-vessel introduced into the British Navy. It was certainly no small gratification to myself to have introduced the first vessel propelled by the screw into the Royal Navy, as I felt convinced that it was the only proper method of propelling vessels of war: it was the more gratifying, because my father was the first who, in 1819, introduced the paddle-wheel system into the Navy; and thus our family have had the honour—and a great one it is—of introducing into the Navy the two greatest improvements of modern times. My father, who was always consulted by the Admiralty, proposed machinery in every department where it could be applied with any advantage; such as railways; the Gantry crane, and others, worked by machinery; heating anchors in furnaces, by means of which only could they be properly made; employing convicts to do the labour, with a moderate gratuity to stimulate their exertions, and thus reduce the expense of their keep; and employing private establishments wherever they could do the work cheaper than in the dockyards. The state must and ought to have such establishments as should be able to do their own work when occasion requires; but in a country like England, where the arts and manufactures are carried to the highest possible extent by individual competition, and where the field of exertion is so vast and the prizes of success are so great, no government establishment can compete with them. It cannot hold out sufficient inducement for exertion, and hence we find that no great invention has ever emanated from a public establishment. Certain officers the government must have, and these must be at fixed salaries, for which they have to do a certain quantity of work, and for this the hours are fixed; they have no inducement to go beyond this. Yet this perhaps is the wisest course for a government like ours; it can always command the talent of the day, and it is far more economical for a government to pay the market price, whatever it may be, than to take persons, however well qualified, wholly into its employment; the moment this is done the inducement to extra exertion ceases, and the government must go again to the market for the next best talent, and so on. Hence it is my opinion that a government should have the fewest possible establishments it can get on with, so as not to leave itself wholly dependent upon private firms; and that it should go liberally to the public, specifying in general terms what is required, then it will obtain the best workmen in the wisest manner, without being taxed by extra pensions or any other drawback; by this means a government would command all private establishments, and make the most of its own.

In 1832 I was requested by the authorities of the Isle of Man to examine the whole of the coast of the Island, and to give my opinion as to the best plan for improving the harbours. I accordingly sent over my assistant, Mr. Coombe; and having carefully surveyed Douglas, Derby Haven, Castleton, Peel, Ramsey, and Laxey, made complete hydrographical surveys of the whole, and detailed plans for the best way of improving them. At Douglas I proposed to make an extensive low-water asylum harbour, and also at Derby Haven, which were the most important places, and possessed the greatest capability of making good refuge harbours at the most economical rates for the numerous vessels trading between Ireland and England, and also for foreign vessels bound for Liverpool. The other ports susceptible of improvement were chiefly local, and therefore only a moderate sum was proposed to be expended upon them. A great harbour might, indeed, have been made near the Calf of Man; but this would have involved an expense which the revenues of the Island had no means of paying, though a harbour there would have been of importance to the vast number of vessels of all classes trading between Liverpool and America; and therefore, if anything was to be done there, Liverpool ought to have contributed largely towards it. Liverpool, however, thought differently. They had no idea of encouraging their vessels to stop so near home. So that all idea of making a refuge harbour near the Calf of Man was abandoned. Neither could the Islanders obtain foreign aid for Douglas, or Castleton, or Derby Haven. They were therefore left to their own resources, and were obliged to confine their operations to making a small addition to Douglas Pier, which I designed for them, and which was carried into effect by Mr. James Brown—a most excellent practical engineer—who had been employed many years by my father at

Holyhead Harbour and elsewhere.

About the same time I was asked to make a plan for the improvement of the ancient port of Hartlepool (I think one of the oldest in England), for shipping coal from the coal-fields of South Durham, which were then being developed to an extraordinary extent. I made a plan, which was afterwards carried into effect under my direction, the late Mr. James Brown, above mentioned, being the resident engineer. When I visited the place, it was the most secluded, primitive fishing village I ever saw. It has now become one of the most thriving and populous towns of Durham.

At this time I was also requested by a society of gentlemen—amongst whom were Messrs. Ladbrooke, Mills, Smith, Webb, &c.—to examine the mouth of the Coquet, near Warkworth, in order to make a harbour there for the shipping of coal from some collieries which were about to be opened in that district. This harbour, which consisted of a north pier and south pier, was made under my direction; Mr. George Remington being the resident engineer. It was merely intended as a tidal harbour, with floating docks attached to it; which latter, for want of funds, were never made. By extending the northern pier farther seaward, they could easily obtain 12 feet or more at low water; and the Coquet Island outside forms an excellent and safe roadstead for vessels drawing 20 feet at low water—an advantage that no other port on the east coast possesses. There is plenty of coal in the vicinity, that has never been developed for want of capital; but no doubt the day will come when this port will be of considerable consequence. Whilst superintending these works, the late Earl Grey, then Prime Minister, who lived at Howick, about nine miles to the north of the Coquet, invited me to spend a day there. I accordingly went over, and was most kindly received. I there met his son, the present Earl, and the present Sir George Grey, and passed my time very agreeably and instructively. The late Earl, when Commissioner of Portsmouth Dockyard, knew my father well, and had great respect for his talents.

Whilst employed in constructing the piers of Sunderland Harbour, I made acquaintance with Mr. Lambton, afterwards the Earl of Durham, who married one of Earl Grey's daughters. He was one of the commissioners of the harbour, and a great coalowner. He was a slight-made person, of the middle size, with an olive complexion, dark, piercing eyes, and a profusion of jet-black hair. He possessed considerable talent, great firmness, and a stern, haughty, proud bearing, with great impetuosity of temper. Being heir to a large fortune, he was spoiled in his youth, although not without kindly, generous, and noble feelings, and where he took a liking, was a firm friend. He was always very kind and friendly to me, and took a leading part in promoting the success of Sunderland Harbour. His violent temper, and inability to control it, was the cause of his failure as a public man. When Governor-General of Canada he, contrary to all rules of order and subordination, threw up his command in a pet on account of some trifling provocation, and returned to England without authority. The obloquy which this foolish and imprudent step entailed upon him for ever after rendered him unfit to take a leading part in public affairs, for which otherwise he was well calculated, and he died soon after, it is believed of a broken heart.

Railways having been now fairly established, and having to a considerable extent superseded roads and canals, the Brighton Railway scheme was started again. The history of this work I have already described. During the years 1837-1838 it occupied a considerable portion of my time, and although I was then very ill, and totally unfit for business, I was obliged to struggle through it, and carried it at last. At times I was so nervous and unwell that I scarcely knew what I was about. I felt perfectly stupid, and thought that my life must end in a lunatic asylum; and many, even of my friends, considered that my career was over. When cross-examined before parliamentary committees, which examinations I was obliged to undergo at this time, after two or three hours my head got so confused that I could see nothing distinctly—everything appeared either double or upside down. However, I got through not only the Brighton, but also the Blackwall, Railway Bills. Apropos of the Blackwall Railway, I long had an idea that Blackwall, including the East and West India, the Regent's Canal, and the London Docks, should be connected with London by a railway, and that this line should form the grand trunk and terminus for all the railways which were to connect the eastern counties of Essex, Suffolk, Norfolk, and Cambridge with London. A Bill for making the Eastern Counties Railway through Essex and Suffolk had then passed, having their terminus in Shoreditch, which was quite out of the way, and left out all the docks; whereas the line which I proposed would not only have connected them with London at Fenchurch Street, which is within half a mile of the Royal Exchange, but would have brought the great traffic of the eastern counties there also. Mr. Stephenson, seeing the importance of my Blackwall line, started another in opposition, which was defeated in Parliament, and my line was carried; but my party was not strong enough to carry it into effect; Mr. Stephenson's was, therefore they took up my line, and he was appointed the chief engineer. He wholly ignored my principle of making the Blackwall line the main trunk for the eastern counties' traffic, but declared that the Blackwall line should be considered distinct; and as he thought that so short a line was not adapted for locomotives, he said that it would be far better to work it by means of ropes attached to powerful engines fixed at each end of the line. I foresaw that this plan would not succeed, and told my friends so; however, it was of no use, the rope system was adopted, until it was found that the constant breaking of the ropes, their great friction, and the power required to work them, entailed so much expense and inconvenience, that the line would not pay a dividend. The company therefore resolved to abolish the rope system altogether, and adopt locomotive engines, according to my original plan; and the line was extended to join the Eastern Counties Railway at Stratford, and the Tilbury and Southend Railway, and it has been connected with the Victoria, East and West India, London, and St. Katharine Docks, and now pays a dividend of about 3 to 3½ per cent. In fact, all that I recommended has been done, and the result has been successful; but it would have been much more profitable if my plan had been adopted in the first instance, while the extra expense of the rope system would have been saved.

In the year 1844 came the great railway mania. Railways were considered as a mine of wealth to whoever would undertake them; and consequently new lines were projected in all directions, and I had my full share of them: amongst others the Great Northern, the Leeds and Carlisle, the Leeds and Bradford, the York and Scarborough, the Bristol and Monmouth, the Bishop's Stortford and Thetford (called the Direct Norwich), the Birmingham and Boston, the Newry and Enniskillen, the East Lincolnshire, the Lincoln and Hull, the Cannock Chase, and the North Wales, &c. The consequence was that the demand for engineering surveyors and assistants was very great. Engineering was considered to be the only profession where immense wealth and fame were to be acquired, and consequently everybody became engineers. It was not the question whether they were educated for it, or competent to undertake it, but simply whether any person chose to dub himself engineer; hence, lawyers' clerks, surgeons' apprentices, merchants, tradesmen, officers in the army and navy, private gentlemen, left their professions and became engineers; the consequence was that innumerable blunders were made, vast sums of money were recklessly expended, and the greater part of the lines were thrown out of Parliament in consequence of the innumerable errors committed in them.

The committee rooms of the Houses of Lords and Commons were thronged to such a degree with engineers, lawyers, and witnesses, that it was scarcely possible to find sufficient room for them. The barristers, solicitors, and parliamentary agents made enormous sums, and so did those engineers who were fortunate enough to get paid. My labours were most arduous. I had to work night and day for several weeks in preparing plans for Parliament, and if I had only got paid, I should have made a good fortune, for I employed in one department or another above three hundred assistants. It is true that I received a great deal of money; but the expenses were so great that the advances made to me were immediately absorbed, and before I could balance my accounts, most of the companies had vanished, remaining largely in my debt. As to recovering my debts in a court of law, it was impossible on account of the difficulty of proving whether there was a sufficient number of directors present when the order was given, and what were the names of the directors present; because, as they were not legally constituted companies, the different members of the board could only be sued in their individual capacities, whilst I, who employed the different parties to make the surveys and work out the details, was clearly liable to them; so that the claims against me were innumerable, and made without mercy, and I had the greatest difficulty in satisfying them without material loss.

To give an example of the difficulties I had to contend with in establishing my case against one of the companies that employed me—viz. the Cannock Chase Railway, an essentially good concern, which has since been carried into effect with considerable profit to another company who took it up afterwards: My company had paid me 2500*l.* on account of the expenses, and they owed me 2700*l.* more, which they never disputed. They had 12,000*l.* in hand, and could have easily paid my bill; but finding that they could not at that time carry their line, they united with another company without paying me, and handed over to them the 12,000*l.* subscribed for my company. I then got them to call a meeting, and said that as their original company had collapsed, I was prepared to meet them upon the most equitable terms, namely, that they should pay the balance of my disbursements actually out of pocket; this they declined to do. I then offered to deduct my share of the disbursements if they would pay the balance; this they declined also, and they would pay no more. Finding that I could do nothing with them, I sued them at law, and brought an action against the chairman; he, however, proved that he was not present when the order was given to me, I was therefore nonsuited, and had to pay my expenses and his, which cost me 500*l.* I then brought an action against two other parties, who were present when the order was given; but then it was proved against me that there was not a quorum, and as I could not get hold of the books, the secretary having absconded with them, I was nonsuited again at the cost of 500*l.* more. At last this secretary having got into difficulties, I got hold of the minutes of proceedings by an extraordinary combination of circumstances, went to the Court of Chancery, and eventually established my case, and recovered the whole of my claim, viz. 2700*l.*; but without the costs. So that after seven years' litigation I recovered the 2700*l.*, which was wholly absorbed in the expenses; and therefore I was where I began.

The principal parliamentary battle I had was in 1844-5; and in the following year I had another with the Great Northern Railway. The late Mr. Gravatt had the dual line from King's Cross by Barnet, Welwyn, Stilton, Stamford, Corby, Lincoln, Asking to York; this was in every respect the shortest and the most easy of execution; unfortunately, our company was not formed until the end of September, so that it was extremely difficult, in so short a time, to get the surveys and levels made correctly. The late Mr. Francis Giles undertook to have the whole completed in the most perfect manner by the 30th November, 1844, for depositing with the respective Clerks of the Peace, as required by the Standing Orders of Parliament. Mr. Giles's well-known reputation as a first-rate engineering surveyor appeared to Mr. Gravatt and myself a sufficient guarantee that the surveys would be well completed by the required time; in this, unfortunately, we were most grievously disappointed, for Mr. Giles, who had a good deal of other business in hand, could not devote his whole attention to it, which was absolutely necessary; in fact, he ought never to have undertaken it; but he always assured us that it would be properly done in time. However, finding that he did not go on so well as we expected, we endeavoured, as far as practicable, to remedy the evil by setting on additional surveyors ourselves, under the control of Mr. Giles; but in spite of all our exertions, Mr. Giles failed completely, and our line, in consequence of the numerous defects in the survey, was thrown out upon Standing Orders, and the present Great Northern line, which was our rival, but acknowledged to be not so good, notwithstanding a strong opposition, was carried through Parliament, very much to the annoyance of Mr. Gravatt and myself, and Mr. Gravatt never forgave Mr. Giles's neglect.

My rule on all these occasions was to endeavour to conciliate the landowners through whose estates we went, always asking and obtaining their permission before entering upon their lands, and by this means we made friends wherever we went; amongst other great proprietors we went through a considerable portion of the Marquis of Exeter's estate, near Stamford. Upon going down the line I found one of our surveyors drunk, and he had so completely departed from his instructions that I paid and discharged him at once. I called on the Marquis to explain this, but he was not visible, as he was busy with preparations for the reception of

Her Majesty and H.R.H. the Prince Consort, who were expected to arrive the next day. I then went down the line as far as York, and upon my return called again upon the Marquis at Burleigh, who received me very coldly, and said when he gave me permission to go through his estate, it was upon a particular line, which I faithfully promised to adhere to; but he was much surprised and sorry to find that I had broken my word, for that when H.R.H. the Prince Consort was shooting in his preserves he found one of my surveyors with several assistants breaking into and carrying the line through them, which he had strictly forbidden, and which I had as strictly promised to his Lordship that I would not touch; and it was most fortunate that the surveyor and his assistants were not shot, for it was never for a moment expected that they would be there. After having heard his Lordship quietly, and having asked the day, and the name of the surveyor, which his Lordship told me, I said that the man was not in my employment, for the very day on which I last called at Burleigh I found this same surveyor drunk, and carrying the line into the preserves, which I had strictly forbidden, and I immediately discharged him. And I added that as his Lordship found him in the preserves, he must have been sent there by somebody else. His Lordship was at once perfectly satisfied with this explanation, and became as friendly as ever, and pressed me to stop and dine with him; this invitation I courteously declined, for I was so much occupied that I had not an hour to spare.

One very important feature of our line, besides making it shorter, more direct, and easier of execution, was the position of the station at York, which we proposed to make on the main line, immediately outside the walls of the city; and in order to effect this we carried our line by a bridge across the railways then entering the York station, while our rivals proposed to carry their line into the station itself; which scheme having been adopted, compels them to back in and out, and not only occasions considerable loss of time, but materially increases the risk of collisions.

Another important line was the Bristol and Chepstow, which would materially have shortened the distance between Bristol, Birmingham, and Liverpool, instead of going round by Gloucester, and would have enabled the South Wales Railway to shorten materially their distance to London. In order to effect this I proposed to carry the line across the Severn at the old passage by an iron bridge, with a clear height of 100 feet above the high-water level of spring tides, so as to enable the largest ships to pass under. It happened that the rocks in the river afforded excellent foundations for the piers.

The late Mr. Cobden was chairman of the Committee in the House of Commons to whom this Bill was referred, and they were all astonished at the boldness and grandeur of the undertaking; although the late Mr. Brunel and others did not deny its practicability, yet the promoters of the undertaking could not see their way to find the means for carrying it into effect, and therefore the Bill was withdrawn.

I forgot to mention the Central Kent Railway line, 1838. It had long been considered a desirable object to connect Dover and London by a railway for the Continent, and the South-Eastern had already obtained an Act to make a line by Redhill, thence to Tunbridge, Ashford, and Folkestone, to Dover, the distance being 86 miles, whereas the old coach road was only 72. Moreover, the South-Eastern avoids all the principal towns and population in Kent; so much so, that it was considered to be very objectionable, and that it would not pay.

I was accordingly requested by a most influential committee to examine the county of Kent carefully, and endeavour to find out a better line. I was not long in discovering one, namely, to commence at London Bridge, thence by Lewisham, Eltham, the Crays, the Darent, 4 miles above Dartford, thence by Gravesend, through Gad's Hill, crossing the Medway a mile above Rochester, thence, within a mile of Maidstone, to Eastwell, where it was to separate into two branches, one to Ashford, and thence on to Folkestone and Dover; another to Canterbury, thence to Sandwich, where it was to terminate; while from the Darent another branch was intended to run up the valley of that river, with a tunnel at its head, and thence to Sevenoaks and Tunbridge. From this it will be seen that the main line connected all the principal towns in the county together; each was at the same time within the shortest distance from the metropolis, and nearly 14 miles nearer to Dover than the present South-Eastern line; and there was no inclination steeper than 1 in 264, or 20 feet to the mile, and the New Cross inclined plane of the Greenwich and Croydon line of 1 in 100 for three miles would have been avoided. This line was so obviously the best for Kent and the sea-coast, that when submitted to the South-Eastern Company, who had not commenced theirs, they admitted it, and told their solicitors, Messrs. Fearon, to tell our solicitors, Messrs. Freshfield, that they would make terms with us for carrying it into effect. How the negotiations fell through I never heard; whether it was from the opposition of Maidstone and Lord Winchelsea, who, as well as Sir Percival Dyke, violently opposed it—although since, I understand, they have sincerely repented—I never could learn; but the negotiation failed, although we were perfectly ready to give up the line to the South-Eastern upon a reasonable compensation, and they (the South-Eastern Railway) commenced and completed their line round by Tunbridge, and bitter cause they had to repent it. Two of my assistants—Crampton and Morris—after leaving me, considering that it would be a good speculation to get up a shorter line to Dover, persuaded a very worthy nobleman, Lord Harris, who has considerable property near, between Sittingbourne and Canterbury, and some other influential landowners on the line, to form a company to make a line between Rochester and Canterbury, and Mr. George Burge, the contractor of the St. Katharine Docks and Herne Bay Pier, under the late Mr. Telford, joined them.

Burge had invested a good deal of money at Herne Bay, and naturally expected that one day a line of railway would be made to it, and that the value of his property would be considerably increased thereby. Morris, a very honest, painstaking, and industrious man, who had been in my service many years, and afterwards became one of the contractors of the South-Eastern Railway, and made a good deal of money there, had the sagacity to purchase the old harbour of Folkestone, it is said for 10,000*l.*, and sold it to the South-Eastern Railway Company, profiting considerably by the transaction. Crampton had made some improvements in the locomotive engine, and afterwards became the principal executive engineer to Messrs.

Samuda. At the time Crampton came into my employment Messrs. Samuda had had a vessel constructed, and had made the engines for propelling her upon a new principle. The vessel was called the 'Gipsey Queen,' and a day was appointed for the trial. Whether Crampton had some misgivings about the success of the experiment, or whether he was tired of the employment, I do not know, but he was anxious to come under me. Knowing him to be a clever, hard-working person, I took him, and he continued serving me faithfully for four years. Crampton entered my service four days before the experimental trial of the 'Gipsey Queen,' which took place, I think, in the year 1840. The result of the trial was that the boiler blew up, Samuda's brother and four men were killed, and if Crampton had remained in their service, he would probably have been killed also. During the time Crampton was in my service, he made the acquaintance of my solicitors, the Messrs. Freshfield, who conceived a high opinion of his talents and energy.

Morris, Crampton, and Burge, then, commenced the London, Chatham, and Dover Railway with comparatively very little support for an undertaking of the kind, and experienced very great uphill work; so much so, that Burge got alarmed, and Morris and Crampton bought him out. Morris and Crampton still struggled on with it, and then Morris went out, and Crampton remained alone. At last he got Peto and Betts to join him, and then the concern went ahead. Lord Sondes, a large proprietor in Norfolk and in Kent, also joined them, and they completed the original line. They then went to Parliament to extend their railway by an independent line to London, and from Canterbury to Dover, and, subsequently, by lines to the City, so as ultimately to join the Metropolitan Railway, and from thence with the Great Northern Railway at King's Cross. How they raised the enormous amount of capital to execute these works was a miracle, but the tale has at last been unfolded, and the unfortunate subscribers have found it out, to their cost; the concern has become bankrupt, and the great contracting firm of Peto and Betts, as well as Crampton, have ended in the 'Gazette,' as a melancholy example of what energy and capital will come to when pushed beyond their just limits. The original shares of 100*l.* may now (at the time of writing) be bought at 18*l.*, and the South-Eastern Railway have been compelled to expend nearly 700,000*l.* to cut off the angle between London and Tunbridge; whereas, if they had only adopted my line of the Central Kent Railway, as they agreed to do in 1838, all this would have been avoided; the London, Chatham, and Dover would never have been made, the enormous losses would not have occurred, thirty-two millions would not have been spent in railways for a single county, and the South-Eastern shares would not (at the time of writing) be at 65.

CHAPTER VII.

In 1844 Count Adolphe Rosen obtained a concession for making railways in Sweden, and offered me half the concession, provided that I would go over to Sweden, lay out the lines, and bring the matter before the English public. I accordingly employed a Mr. Tottie, a Swede, who had been employed by Mr. Rastrick, to make the surveys of the lines laid out by me, which consisted of a main one from Gothenburg right through the kingdom to Oxhoe, as the central portion for the iron trade; from thence one branch went to the upper end of the Lake Wener, and the other to the Lake Malar, to communicate with Stockholm; the line then proceeded north by Westerâs to Upsala, and thence to Stockholm.

The same year I went from Hull with Count Rosen to Gothenburg by steam, and spent two or three days there. The country through which I passed, though not rich, was much more so and better cultivated than I expected to find it, and the people honest, simple, and industrious, and extremely civil. The general appearance rather picturesque, and in places wild, being covered with dense forests of firs, larch, beech, &c.; large spaces had been cleared of wood and brought under cultivation. The houses were for the most part built of wood, in the Russian fashion, the logs being laid close together, dovetailed at the ends, and the joints caulked with moss, the inside being closely planked, and in every room was a stone or porcelain stove; the windows were double, and in winter the outer and inner were both shut up, and all the joints pasted with paper so as to prevent the admission of the outer air. The houses were generally very comfortable. In some of the larger towns, such as Orebro, many of the houses were solidly built of stone, others had stone foundations and wooden superstructure; most of the country churches were of stone, with a detached building of wood for the bells, which were frequently of a large size, with a very fine full melodious sound. Gothenburg is a very well and regularly built town, chiefly of stone, in the Dutch style, with canals running through the streets. Some of the houses of the principal merchants, as well as public buildings, are spacious and handsome, although the town generally has a heavy, dull, yet substantial appearance.

I resolved to visit the celebrated iron mine of Daunemora, some miles farther northward. I accordingly started off with Count Rosen, and reached it the same evening, and visited the mine next day; it consisted of a mass of rock, cropping out to the surface, of almost solid magnetic iron, containing about 75 to 80 per cent. of the finest metal. There were extensive forests round, so that there was no want of fuel; the wood was converted into charcoal, and the finest iron was extracted, the best for making steel; it fetched the highest price in the English market, where there was a great demand for it. The machinery employed was very rude and simple, the bellows for the forges being in some cases driven by manual or horse labour, in others, where it was accessible, by water power. I was anxious to go to the great mining district of Dalecarlia, about 100 miles farther north, but the season was getting late, and therefore I was obliged to return to England.

When I had got all my surveys finished, I made a report upon the whole line; but I found that the Swedish Government was not sufficiently alive to the importance of railways at that time, or rather the Government did not see its way to giving them encouragement by subscribing or rather taking a pecuniary interest in them. Though by no means undervaluing the importance of railways, yet, being naturally cautious and economical, with only moderate funds at its command, the Government doubted much whether a reasonable profit would be derived from them, but at the same time wished us every success in obtaining the money in England. I represented to them that people in England, knowing nothing of Sweden, or her capabilities, would hesitate to subscribe their money without a certain guarantee by the Government of interest of 4 or 5 per cent. upon the amount of capital expended, and that as it was quite clear that the railways would pay that, there could be no risk; in fact, the guarantee would be merely nominal, but that it would have the effect of obtaining the money in England, and thus conferring upon Sweden a great national benefit, by saving their capital without running the least risk. I was recommended to give a grand dinner, which several of the ministers and all the other notables of Stockholm attended. Everybody was enthusiastic, and a great number of speeches were made as to the importance of railways, and the great national benefit they would confer, and my health was proposed by the Minister of Commerce, Skogman, and was received with the greatest enthusiasm, but the effect was nothing, for we neither got private subscriptions nor public guarantee. However, after all we had done something; we had introduced the subject of railways into Sweden, we had shown the importance of them, and we had in some measure opened their eyes, and we trusted that in time, when they had maturely reflected upon the advantages, they would view them more favourably, and contribute liberally towards them; but as there was no use in then pressing the subject further, we returned at once to England.

In all fevers there is a climax; the railway fever had its climax like the rest, and it was then upon the decline. The vast multitude who had expected to make their fortunes found at last to their sorrow that their money was gone; but that was not the worst, for they would have been very happy if they had not had to pay more. As for any new railway speculation, that was entirely out of the question, and therefore it turned out that the Swedish railways were too late for the market; nobody would entertain the subject; the very name of railway was sufficient to drive everyone away, so that there was no help for it but to abide better times. Accordingly the Swedish railways remained in abeyance until the year 1852; by this time the Swedish Government had considered the subject maturely, and felt that, as every other European nation had adopted them, Sweden, if she desired to keep pace with other countries, must either make the railways herself, for which at the time the Government had not the money, or she must encourage others to make them by guaranteeing a sufficient interest for the capital expended. Accordingly I went there again, and was as usual very kindly received by the King and his Ministers, and I saw that they were becoming more anxious than ever that the railways should be made. I had been there in October, 1848, and had the honour of being invited to dine at the palace in Stockholm, as I had previously the honour of dining with their Majesties at the summer palace of Hoga, near Stockholm. Upon arriving at the palace I was most courteously received by King Oscar, who did me the honour of presenting me to his handsome, graceful, and intelligent queen. When her Majesty heard that the railways were to be commenced, she said that she had heard so much talk about them and nothing had been done, that she feared they never would be made, "therefore talk no more about them, but set to work and make them." The dinner party consisted of about thirty. I had the honour of sitting next the Lord Chancellor, a very agreeable, intelligent person, who sat next to the Royal Family. We had an excellent dinner, without the least restraint, and the common topic of conversation seemed to be, who would be elected President of the French Republic, Cavaignac or Louis Napoleon, and everybody seemed in favour of Cavaignac as the proper person; they all spoke very disparagingly of Louis Napoleon. It seems curious to think how little the world knew of that extraordinary man, and how completely he disappointed all previous expectations.

When at Stockholm I was presented to his present Majesty, King Charles XV., then Prince Karl, a very handsome intelligent young man. Since his accession to the throne he has done me the honour of conferring upon me the order of Knight Commander of the Order of Wasa, for what his Majesty was pleased to term the great services which I had rendered Sweden. This was the more agreeable, as it was sent to me through my personal friend Count Platen, then the Swedish representative at the Court of London. Many years ago I made acquaintance with his father, the celebrated Count Platen, who was the chief instrument in changing the dynasty from the worn-out old race, and in placing Bernadotte on the throne. My introduction to the Count arose in this manner. The Count took the utmost interest in the completion of the great Gotha Canal, which unites the river Gotha below the falls of that river and the lower end of the great Lake Werner, thus completing the navigation between the whole of the towns bordering upon that lake and Gothenburg. A canal had formerly been made between the lower end of the lake and the river Gotha (which issues from it) below the falls, and at the time a very great work it was, but the locks were so unequally distributed, and the rise of some of them so great, that the navigation was very much impeded; the Count, therefore, came over to England, after the death of my father, to consult Mr. Telford, who was then considered our first engineer, as to what was the best means of improving the Gotha Canal, so as to avoid the inconvenience complained of. Mr. Telford went over, prepared a plan, and carried it into effect, much to the satisfaction of the Swedish Government. During Count Platen's visit to England, Mr. Telford brought him to Lynn, where we met as engineers of the Eau Brink Drainage and Navigation, and we explained to him the whole of the Eau Brink Works, with which he was much pleased. He was a very superior person, grave and dignified, with great intelligence, and of easy, affable manners. The Mayor of Lynn being informed by

Mr. Telford of his visit, called upon him, and being introduced, invited him, in the name of the Corporation, to one of their civic dinners, which he accepted. He had, however, brought no dress suit with him, and indeed was perfectly indifferent about it, and would have gone to the mayor's feast in his travelling costume, which, to say the least, was very rough. Telford consulted me about it, and we both agreed that the Count could not attend the dinner without the usual evening dress costume. Upon this being delicately explained by Mr. Telford, he took the hint and sent for a tailor, who in the course of a couple of days equipped him properly. The greatest attention was of course paid to him by all the company.

To return to my subject. As time went on, the money market got in such a bad state—no end of failures—that nothing could be done with the Swedish railway, and I began to think all my money and labour would be thrown away. The period for which Count Rosen's concession was granted had elapsed, and it was very doubtful whether it would be renewed at all, but it certainly would not in any case be to the extent of the original concession, which comprised the whole kingdom.

Fortunately, however, in 1852, confidence in the money market had been restored, and we received a renewal of the concession for so much of the line as extended between the Lakes Malari and Werner, with a branch to the iron mines of Nora from Orebro, which was as valuable as any part of the line; and the Government gave a guarantee of 4 per cent, on the amount of capital required, namely, 420,000*l.*, with power to raise 167,000*l.* more. The line was very easy, and Mr. Burge, the contractor, agreed to complete it for that sum. A company was accordingly formed, the capital was immediately subscribed, and the shares went to a premium. I went over to Sweden again; appointed Mr. Watson resident engineer, and the works began and proceeded very well for one year; unfortunately the chairman and leading man on the committee was the notorious John Sadlier, M.P., who afterwards made away with himself near Jack Straw's tavern, Hampstead Heath, when the whole of his proceedings were made public; and, amongst others, his mismanagement of the Swedish railway. What became of the money I never could make out, for I never could account for above 60,000*l.* as having been expended on the works and land combined; but it is certain that accounts far beyond that were presented to me by the directors for my certificate, which they were obliged to get before they could be passed by the Government, though I never would give the certificate, because they kept everything back from me; and finding that matters were going on in such a discreditable manner, I felt that with any regard for my character I could no longer remain their engineer. I therefore resigned at once, and fortunate it was that I did so, for I afterwards was informed that they had not only spent the whole 420,000*l.*, but also 167,000*l.* of debentures, and had issued 167,000*l.* more without authority; so that in round numbers they had expended about 700,000*l.*, and had not completed 50 miles of single line, which required nothing more than surface formation, and one or two short lengths of embankment and cutting scarcely exceeding 20 feet high; for which 6000*l.* per mile complete was ample. In fact, there never was a good affair so completely mismanaged, not to use much stronger terms. The consequence was that after Sadlier's death the whole came out, and everyone laid the blame upon him. The particulars

I do not know, and thank God I got clear of it in good time, finding that it was impossible to remain with honour. I never would certify the accounts the directors presented to me, because I believed that at the least they were in error, if not something worse. All I did was to certify to the contractor, Mr. Burge (who, as I knew, behaved very honestly), to the extent, I think, of 25,000*l.* or 30,000*l.* Thus this fine concern went to the dogs from sheer mismanagement, to speak mildly, and the shares, which had stood at a premium, were worth nothing; whereas in the hands of any sensible, honest body of directors, the line ought to have been completed for the estimate, and would have paid well. The Swedish Government could not with justice pay any guarantee when no part of the line had been completed and no satisfactory accounts rendered; and I have reason to believe that they were so disgusted with the way in which this railway had been mismanaged by the Board that they would have nothing further to do with English or any other companies, but determined to make all the rest of the lines themselves—which I understand they have done—and that the cost did not exceed my estimate of 5000*l.* to 6000*l.* per mile, including stations and rolling stock.

I had previously examined several other parts of Sweden, and in November, 1850, I had gone there, undertaking to deliver on my way a confidential letter from Count Reventlow, the Danish Ambassador in London, to the Minister of Foreign Affairs at Copenhagen. We went by Hamburg and Kiel, for at this time war was again expected to break out between Schleswig-Holstein and Denmark, and both parties were watching one another with the greatest anxiety.

Schleswig-Holstein is not a rich country by any means, although there are apparently some substantial farmers. The country is generally flat, with here and there some gentle eminences, with patches of fir trees; but there are some extensive mosses which might be reclaimed and converted into valuable land if properly drained. These mosses in many cases rest upon a bed of marl, which upon being dug up and mixed with the peat forms most valuable manure. We saw many examples of this in the fens of Lincolnshire, and in the north of Denmark. I have been informed upon credible authority that there are vast tracts of moss lands lying upon marl, and small lakes which might easily be drained and become most fertile; and it is to be hoped that the Danish Government will find the means of reclaiming them, as they will well repay the expense. Rendsburg, through which we passed, is a little town, well fortified, and may be made much stronger. Kiel, situated upon a splendid bay in the Baltic, is the university of Holstein, and a very neat little thriving town, with prettily wooded environs. From here we proceeded to Schleswig, the capital of the province, its inhabitants being partly Germans and partly Danes; in the northern part of Schleswig the inhabitants are wholly Danes. The town of Schleswig is about 12 miles from the Baltic; it is well built, surrounded with a very pretty fertile country, well cultivated, and diversified with wood. There is a good deal of agricultural traffic with the surrounding country, and the place appeared very thriving and prosperous. We arrived there about six o'clock in the evening, about two hours after dark, and it began to feel very cold. We got to a very comfortable inn, and were dining, or rather had finished our dinner, when the waiter entered with a quiet, mysterious air, and said, or rather whispered, that a gentleman wished to speak to me. Knowing no one in the place, I could not conceive what he meant; he, however, repeated the whisper in my ear in the same mysterious manner. When I told him to show the stranger in, there was no person in the room but my friend, Mr. William Sim, afterwards the solicitor to the Swedish Railway Company, who had accompanied me. Immediately after the waiter's departure the door opened, and in came a gentleman muffled up to the eyes in a blue military cloak. He cautiously shut the door, then uncovered his face, and looking stealthily round, and observing nobody but Sim and myself, advanced at once to the table, threw off his cloak, took off his cap, and saluted me as a naval officer. I motioned him to be seated, which he courteously declined, and addressing himself to me, said that he had been informed that I was the English gentleman who had brought despatches of importance from Count Reventlow to the Danish Government, and that he was a Danish naval officer, and had been sent by the commanding officer of Alsen to request that I would deliver up my despatches to him, as it was of the utmost consequence that they should be delivered to the Danish Government as soon as possible. I told him that I was perfectly unacquainted with him, and that I could not do so without a personal conference with the General commanding at Alsen, because these despatches had been confided to me by the Danish Minister in London, and as a point of honour I could not deliver them to any but an accredited officer of the Government. He then said that if I would be at Dussel, opposite to Alsen, the following morning, at eight o'clock, the General would have a boat ready to take me over to the island, and would have a war steamer in readiness to take me to Kosoa, where I could readily get to Copenhagen. I accordingly agreed to be at Dussel the following morning at eight o'clock. The officer, who was a perfect gentleman, then muffled himself up in his cloak, and requesting that we should say nothing about his arrival, quietly left the room. Sim and myself ordered a carriage with four horses to be ready the following morning at six o'clock, paid our bill, and went to bed. During the night there had been a very heavy fall of snow, and when we started there was some difficulty in moving forward. We were therefore obliged to take extra horses, and passing by the now celebrated heights of Dussel, reached the shore opposite the island of Alsen precisely at the time appointed, where we found a boat with four oars ready to take us across the Sound (which was about half a mile wide); here we were received by the General in full uniform. I then delivered him the despatches, for which he felt very much obliged, and we went on board a small war steamer, which got under way immediately, and landed us at Kosoa, not far from Copenhagen, where we slept. Next day we examined the cathedral at Roskeld, and reached Copenhagen in the evening.

When I first visited Sweden, a vast quantity of brandy distilled from potatoes was consumed; so much so, that the country was, to a certain extent, demoralized, and drunkenness was very common; and the farther north you went, the worse it was. The Government therefore determined to take every means in its power to suppress it. Heavy duties were imposed on potato brandy; and since that time, I am glad to hear that drunkenness has considerably abated. The Swedes are an excellent, open-hearted, gallant, and generous people, and most amiable and hospitable. I was treated with the greatest kindness by them, for which I shall always feel grateful. Christmas is a general holiday throughout the country for a month, commencing with Christmas Eve. During that time, scarcely any business is done. There is nothing but visiting and social parties from morning to night; and it requires a strong stomach and head to go through that festive ordeal without feeling the worse for it. When once entered the house, you cannot refuse to accept their kind and profuse hospitality. As a specimen, whenever a marriage takes place, the bride not only appears before her friends, but the house is thrown open to the public, and everybody is at liberty to enter and pay his respects to the bride and bridegroom, who receive all comers in full-dress bridal costume; and refreshments of all kinds are in great profusion. When I was there a marriage took place between two noble families, to which I was invited. The bride and bridegroom were both young, and bride very handsome and splendidly dressed. I was most kindly received and entertained, and did not get home until very late in the morning. I found that I had taken as much as I could decently carry, and if I had not escaped at the time I did, I must have remained all the next day, as many of the bridegroom's friends did.

In Sweden I made acquaintance, amongst many others, with our excellent Consul, Major Pringle, who during the last American war had the command of Washington for twenty-four hours. I was frequently a guest at his house, and I owe many thanks to him and his amiable wife and family for their great kindness and hospitality. Another excellent fellow was Mr., or Colonel, Elsworthy (as he called himself), the American Consul. He was a bachelor, and kept open house, and he was so exceedingly hospitable, and pressed his guests so strongly, that it was with the greatest difficulty you could escape sober enough to reach your quarters with safety.

The woods of Sweden were being fast cleared away, both to furnish fuel for the iron manufactory, in which a great quantity was consumed, and also that the land might be used for agriculture, so that this source of fuel for iron making is rapidly disappearing. In addition to these causes may be mentioned the great export of timber to every part of Europe for building and other purposes. It is true that large forests of fine timber still exist farther north, and also in Norway, but then the expense of transport to the iron districts will be very heavy; yet as Swedish iron is so very valuable, on account of its magnetic properties, for making steel, a large quantity must always be required for the southern markets of Europe, and with the greater cost of fuel the price of Swedish iron must increase.

The Bessemer process has considerably reduced the expense of producing good iron in England, and also for converting it into steel, still the Swedish iron is so much better that there will always be a certain demand for it.

It becomes a question, then, whether it would not be worth while to export the Swedish ore to England, where there is plenty of fuel, and where it could be converted into the best iron at the least expense. Would it not be cheaper to do this, than to manufacture the iron in Sweden, where fuel is so much dearer? and as the Swedish iron ore is very rich compared with ours, except the haematite, the extra freight would soon be recouped. I think it is by no means improbable that it will come to this at last.

About the time that the Railway guarantee was obtained from the Swedish Government, and the company was successfully started, another undertaking was proposed to me by a M. Von Alstein, a Belgian proprietor, and a man of some influence and property in that kingdom. The Dutch Government was anxious to get a more direct and constant navigation up the Scheldt, as that round by Bergen-op-Zoon was only practicable from half tide to high water, while at low water the whole channel was completely dry. It was considered, also, that the whole of the old channel might be filled up and converted into valuable land, so that the fertile island of St. Beveland might be joined to the mainland, and thus the kingdom would be greatly benefited. The Government did not, however, want to undertake the necessary works itself, but was ready to give up the whole of the old channel to any company that would construct the new canal above mentioned, as well as an embankment and road across the old channel, so as to connect the island of St. Beveland with the mainland. Accordingly upon these terms a concession was granted to M. Von Alstein and others, giving them the whole of the lands to be reclaimed, which would amount to a very considerable tract. The plan, upon due investigation, appeared to offer considerable advantages to any party who would undertake it. It was accordingly brought before the English public. A Belgian and English company was formed to carry it out. The money was duly subscribed, and Mr. Thomas Hutchings, at that time a large railway contractor, and considered to possess undoubted means, offered to take a large number of shares. He accordingly became the contractor for the work, and I was appointed the engineer-in-chief, while M. Von Alstein became the managing director, and a M. Dronker, a Dutch engineer and contractor of considerable experience, was appointed by Mr. Hutchings sub-contractor under him. I went over to Holland to examine the whole district and the works proposed, and it appeared to me that if they were properly carried into effect, it would turn out a very fair speculation, yielding considerable profit. My visit was made in December, 1851.

Things having, in 1852, been satisfactorily settled with the Government at the Hague, as soon as the weather would permit, arrangements were made for commencing the works, and in the month of June Mr. Hutchings, myself, and the English and Belgian directors being present, the first sod of the canal was turned with considerable ceremony by one of the Dutch Princes.

Whilst the canal was proceeding, the company was empowered to enclose as much of the land of the old channel as was considered advisable by a jury of Dutch experts, without whom nothing could be done. M. Von Alstein, the manager of the company, attended, and the jury marked out a space in the old channel of the Scheldt, which was always covered to a considerable extent at high water of spring tides, and even neap tides, including the green marsh as well as the sands. The Dutch jury marked a space of 3000 acres, which included about one-third of green marsh and two-thirds of sand—part of which was clayey; when this was done, I went over to examine it, and was much surprised to find that so large a space had been decided upon. I said to M. Von Alstein, the director, that I thought it was far too much to be taken in at once; that it would cost a great deal of money; that the sands were not worth the expense; and that I thought it would be far better to confine the intake to the green marshes for the present year, and that before the sands were taken in they ought to be worked up so as to become green marshes. For, in fact, in England, where I had reclaimed many thousand acres, I never thought of enclosing bare sands. The manager, however, told me that was totally contrary to the Dutch system, and as the jury had decided upon taking the sands as well as the green marshes to the extent above mentioned, it must be done, and he would not listen to my recommendation of taking in a less quantity. The Dutch engineers also fixed the dimensions and form of the embankments, which I did not approve of. In fact, I disapproved of the whole plan, and told M. Von Alstein my opinion, and that I could not take upon myself the responsibility. The manager said that the works must be carried into effect as decided by the Dutch jury and the Dutch engineers. I was therefore obliged to be silent, particularly as Mr. Hutchings, who was the contractor and also the leading shareholder, had consented to it. The works accordingly began under M. Dronker, Mr. Hutchings' sub-contractor; and Mr. Brown, one of Mr. Hutchings' partners, was sent over to superintend the works on the part of Hutchings, Brown, and Wright. I remained some time on the spot, and had a boat fitted up to live in at Barth, for there were no lodgings to be had. I soon saw by the manner in which M. Dronker was carrying on the works that they could not succeed, and I wrote repeatedly to Mr. Hutchings to come over, otherwise he would be ruined, as his partner—Brown—knew nothing about it, and allowed Dronker to proceed as he liked. It appeared to me as if it was nothing more nor less than throwing away money by handfuls, no adequate amount of work being done for it. I never saw such gross mismanagement in my life. There were from twelve to thirteen hundred men employed at 3s. to 4s. a day, with a number of assistants, and they were not doing work enough for half that number. I also expostulated with Dronker, the Dutch contractor, but he would not listen to me, and said that he would do as he liked. The works continued to proceed in this manner, and I wrote almost every post to Hutchings to come over himself without delay, or to stop the works, for it was impossible to complete them for any reasonable sum in the manner in which they were being carried on. I told him, moreover, that it was in vain to attempt to reclaim 3000 acres at once, and the better plan would be to confine his operations, in the first instance, to taking in the green marsh, which was about 1000 acres, and to wait until the sands were worked on the outside, which would be done much more rapidly when once the green marsh was enclosed. I told the manager and Dronker the same; however, they would not listen. At length, after great difficulty, I got them to divide the enclosure into two parts, each consisting of 1500 acres, and I hoped that I should be able to induce them to subdivide these two again into 750 each. By this means the green marsh in each division would have been reduced to about 500 acres, which would easily have been enclosed at a considerable profit, leaving the sands to be dealt with hereafter according to circumstances.

However, nothing would do but they must continue pushing on the outer bank over the sands, which I saw was impracticable at any cost within reason. At last, Hutchings came over himself, and when he visited the works held up his hands in astonishment, and saw that he was a ruined man. By this time money ran short, the workmen rebelled for wages and threatened destruction to Hutchings and all concerned; and the Dutch Government, being applied to, sent three hundred soldiers and two armed cutters to keep order. The men were paid, great numbers were discharged, and the works went on upon a much better system, and considerable progress was made, but still they would not confine themselves to the enclosure of the green marsh. After a great deal of difficulty they nearly succeeded in enclosing 1500 acres near to Barth; but before this could be done Hutchings' funds were exhausted, and he was obliged to stop payment. If my advice had been followed this never would have occurred. Some fresh parties then joined the concern, and, adopting my advice, confined themselves to enclosing the 1000 acres of marsh land.

At certain times of the year, particularly at the end of summer and in the autumn, it is difficult to conceive a more unwholesome district, as at those times the marsh fever invariably makes its appearance with the most deadly effect. I had several fine, strong, healthy young men as assistants with me, living in my ship, and at dinner it was by no means uncommon to see one taken ill and fall off his chair. The only remedy was to give him a strong stimulant of wine or brandy, wrap him up in blankets, and send him off as quietly as possible. Fortunately I had been so thoroughly cured of my fever when at Naples, as before mentioned, that I had become as it were acclimatized and never experienced the least attack. The island of St. Beveland, where these works were carried on, is one of the most rich, and fertile districts imaginable, teeming with luxurious vegetation of every kind, and abounding in beautiful little villages, the very models of cleanliness and comfort; but amongst them the fever lurks in the most insidious form. Mr. Brown, a fine, powerful man of about forty-five, whilst superintending these works for his partners and himself, had an excellent house in the village of Yersike, about four miles distant, supplied with every English comfort both of furniture and provisions; but after a time he caught the fever, which stuck to him for a considerable time, and ultimately he was obliged to leave the place, as otherwise he must have succumbed. The Dutch never go out in the morning without taking a cup of coffee and a dram of bitters, composed of gentian, quinine, and gin, and a pipe, which is scarcely ever out of their mouths, and they repeat the dose at night and not unfrequently during the day. In the English fens and lowlands we do pretty much the same, and good port wine and quinine are considered a specific; but I do not think that the fever is so bad with us as in Holland; the cause may be that these Dutch islands being surrounded by the sea, the tide leaves a large surface of mud exposed to the action of the sun, and thus a considerable amount of malaria is engendered; this I have always found to prevail most densely where there are trees, which prevent it from being dissipated; so that it is always safer to be in a boat, or in a house without any trees near it, so as to be exposed to the free circulation of the air, also to sleep at least 20 feet above the ground.

The whole of the seaboard of Holland requires to be remodelled. The numerous channels through which the tidal and fresh waters pass occupy a considerable surface which is comparatively useless, and only serve to deteriorate the main channels of the rivers, and thus prevent them from discharging their waters effectually, and so keeping them open, in the best state for drainage and navigation. If these superfluous channels were filled up, and the islands which they surround were united to the mainland, a great quantity of valuable land would be gained to the State; the extensive embankments which are now necessary to prevent these islands from being submerged, and which entail a great and constant tax upon the kingdom, would be considerably reduced; the main rivers and harbours would be materially improved, and the general surface of the water in the interior would be lowered; the drainage also might be improved in the same manner, so as to render a considerably less amount of artificial drainage power necessary, which would in turn cause a corresponding reduction of the taxes; while the land would be greatly improved, because the present general body of water is too near the surface, so that it is impossible to carry on agriculture to the greatest advantage; also the risk of breaking the banks would be greatly reduced. I think, moreover, that the vast multitude of shoals which skirt the coast might, by the adoption of proper means—and those not expensive, compared to the object to be obtained— be raised sufficiently so as to be converted into valuable land, and be added to the kingdom with considerable profit, while the navigation along the coast and in the interior might be greatly improved. Further, the whole of the Zuyder Zee, which is said to have been under cultivation in early times, might be reclaimed. As extraordinary spring tides seldom rise high along these coasts, and as vegetation generally takes place at the level of high water of neap tides, the lands on the coast, if properly managed, should not be above 4 or 5 feet below the level of high water of spring tides, although I believe it is a fact that many tracts in the interior are much lower, and this arises from their having been embanked too soon.

Now that these lands have been brought into cultivation at vast expense, it would be difficult to raise them without rendering them useless for a considerable time. Still, in all contemplated new enclosures, the lands should be raised by warp- ing, that is, accumulating the alluvial matter, which can always be done if properly managed up to the level of high water of neap tides. In fact, I repeat, a considerable extent of Holland has been embanked too soon, and what has been done it would be extremely difficult, perhaps not advisable, to disturb; although, if any of these very low lands are not sufficiently fertile, it would be well to warp them up with fresh soil. They might thus be sufficiently raised to do away with the greater part of the artificial power now required to keep them dry, and be rendered very fertile; thus a double advantage would be gained. This subject is well worthy of the seri- ous attention of the Dutch Government and its able staff of hydraulic engineers, who, nevertheless, go too much upon the old routine, and prefer patching up the old system in preference to striking out a new course. Probably this is not the fault of the engineers, as they would naturally attempt any new course whereby they might distinguish themselves. The Government, also, would most likely be in- duced to adopt any new method, provided that it could see its way to bettering the condition of the country. But, on the other hand, the Government feels great disin- clination to depart from an old system which, it may be said with some reason, has continued so many years, has answered its purpose very well, and has rendered Holland, under the circumstances, one of the most extraordinary countries in the world. This is very true; but then it must be recollected that hydraulic science has advanced like other sciences, and what would be considered good practice some centuries back would not be thought so now. The Dutch introduced their system into England in the reign of Charles I., when Vermuyden was considered the first hydraulic engineer of the time. He certainly was considerably in advance of the English engineers of that period; but at the present time his method would be wholly inadequate to deal with the circumstances to be encountered, and it was abandoned long ago. The combination system, and comparatively narrow channels, combined with warping the land, is what is required in Holland as in all other countries similarly situated.

In 1855 I went to Portugal, at the request of the Government, to make arrangements for a general system of railways and other works. As we passed the Fort Belem, at the entrance of the Tagus, the view gradually opened out, until you obtained a *coup d'œil* of the whole, which really is very fine and striking. Fancy a range of streets, houses, palaces, and churches rising rapidly from the water's edge until they crown the summits of the hills on which they are built, backed by the picturesque outline of the Cintra hills, and extending for upwards of three miles. We landed at the custom house, close by the Plaza, commonly called by the English Blackhorse Square, from the fine equestrian figure of the Marquis of Pombal in the centre. This is certainly a handsome square, surrounded by all the public offices, terminated in the centre, on the south, by a fine triumphal archway, which leads by a spacious street to the square of Dom Pedro. Beyond this are the public gardens, very tastefully laid out.

I was introduced to Dom Fernando, the husband of the late Dona Maria, Queen of Portugal, and at that time Regent during the minority of his son. He was of the House of Coburg, and brother to the late Prince Consort. He received me very graciously, and spoke English very well, although with a peculiar accent. His Royal Highness afterwards invited me to a ball and supper, at the Palace of Belem, which was extremely well managed, without magnificence or ostentation. I was there introduced to the Minister of Finance, Signor Mello de Fontes, a young man of great talent and an excellent orator. On the following day I attended a launch at the dockyard, which was honoured by the presence of the Prince Regent, and here I was introduced to the Duke de Saldanha, President of the Cabinet; the Duke de Terçeira and his amiable wife; Viscount de Bandiera, together with the distinguished men of all parties.

Before proceeding to Oporto I was obliged to remain several days until I could get my official instructions as to what they required. I employed the interim in visiting the different objects in Lisbon most worthy of the traveller's notice. I first went to the great aqueduct, which is really a very grand and magnificent work, the finest of the kind in Europe, and every part is readily accessible. It enters at the highest part of the northern portion of the town, and conveys the water to a large artificial reservoir, whence it is distributed by pipes to fountains situated in different parts of the town, and from these fountains it is carried to the various houses by means of carriers and water-carts. These carriers are chiefly Gallegoes or Spaniards from the Biscayan Provinces, who are allowed to charge so much per barrel.

It is singular to see them travelling about from one part of the town to the other with cries of "Agua," and it is more singular still that this practice should have been allowed to continue so long, to the great and unnecessary tax upon the inhabitants, where, if pipes were only laid to the houses, the water might be distributed at probably one-tenth of the cost. The Gallegoes are a hard-working, temperate race, who save every penny they can, and as soon as possible retire to their own country to enjoy the hard-earned fruits of their labours.

The water before it reaches the city is collected in reservoirs distributed about the valleys, at the base of the Cintra hills, but these reservoirs are upon too small a scale, and in dry seasons the water is very scarce. A Portuguese company has since been formed to enlarge the works, increase the supply, and deliver it direct to the houses, but, like most Portuguese companies of the kind, has failed from the want of understanding the subject and from mismanagement.

I visited the fine old Cathedral of Belem, built in honour of Prince Henry, the Portuguese navigator. The style is a species of richly florid Gothic, and the interior is spacious and imposing. There was an hospital for invalid sailors attached to it, which is now converted into a naval academy.

I also visited the Castle of Belem close by, where all ships coming to Lisbon are obliged to stop and show their papers and get pratique before they are allowed to proceed. The castle itself is a very small fortress, built in the Moorish style, and mounts a few light brass guns, and is manned by about fifty artillerymen. It is a very pretty object to look upon, but as a fortress it is good for nothing. There is another fort, called St. Julian, about five or six miles lower down, on the same side, which commands the channel passing close to it, but it also was in a dilapidated condition, with a few light guns. It is certainly much stronger than that of Belem, but wholly unfit to prevent the passage of any large vessel of war, or to resist a land attack. There is another small fort near the bar at the entrance of the southern channel. This fort is circular, and called the Bugia. In the centre is a lighthouse, with a revolving light. This also as a fort is good for nothing. I accompanied Dom Fernando and his ministers to examine the entrance of the Tagus, which was said to be deteriorating. In the northern channel, during the heavy south-west gales, which are very prevalent, there is a very heavy broken sea, which at times cannot be passed without great danger. The strength of these gales when at their utmost is extraordinary. The waves break with tremendous violence on the shore, and carry the sand in vast masses to a considerable distance inshore, so as to render the soil perfectly barren on the north side of the entrance. These sands extend for several miles, and the whole coast is covered with it. At low water, in some places, they are dry during the ebb for a mile and upwards.

The port of Lisbon properly speaking is at the custom house, which immediately adjoins Blackhorse Square. It consists of nothing more than an open roadstead, where all the mercantile vessels lie at their anchors, and are loaded and unloaded by means of lighters, at great delay and cost, and with frequent interruptions from heavy gales of wind; but it rarely happens that the vessels suffer any material damage, as the mouth of the river, although about seven miles wide, is never seriously agitated. Docks or landing jetties might be made here with considerable advantage to the trade. At the request of the Government, I prepared some plans upon a moderate scale. Docks also might have been made in the bay to the westward, although it would have involved the removal of the custom house, which would, however, be a great improvement. The naval dockyard and Admiralty are close by, and, in fact, form part of the west side of Blackhorse Square. It is a miserable place, and consists of a dry dock, two or three building slips, with a ropery, and some warehouses, all in the most antique fashion, and supplied with cranes, tools, &c., of the last century. In fact, it required to be wholly remodelled and removed to the bay above mentioned. The Government has since then parted with this bay to a private company, which has nearly filled it up, and propose to convert it into building ground. As the Great Eastern and Northern Railway terminates on the east side of the custom house, docks in connection with it might be made between it and the custom house, and as these are so obviously wanted, it is not improbable that they will be made some time hence; but, unfortunately, from some cause or another, the railways have been mismanaged, and the Government has no funds to undertake any great works itself.

The roadstead for large foreign vessels of war is situated about a mile and a half lower down, nearer to the southern shore, where there is ample depth of water, and where they are completely out of the way of the merchant vessels.

The channel, from the entrance at Belem to Blackhorse Square, is about three miles long, and from two to two and a half miles wide. It then branches out to about seven miles wide, and continues more or less of about the same width for nine or ten miles, when it contracts to a mile, so that it may be readily imagined that the harbour is one of the finest and most spacious in Europe, always excepting that of Vigo.

At Lisbon I visited the collection of royal state carriages, which, without doubt, is the finest and most extensive anywhere. I think I counted nearly fifty, gilt and decorated with the royal arms in the most elaborate manner. Upon the panels are very beautiful paintings executed by first-rate artists. Some of these carriages are above two centuries old; and it is curious to observe, that whilst the decorations are most elegant and finished in the highest style of art, the ironwork, springs, &c., in many of these are of the rudest description.

The Sunday after my arrival there was a grand bull fight, which Dom Fernando and some of the royal princes honoured with their presence. These bull fights are totally different from those of Spain. Here the bulls have their horns capped with large bosses of wood, so that they can do no harm, whilst, on the other hand, they are never killed. Instead of being tortured with barbed arrows and spears by a host of banderilleros and picadores, and when thoroughly exhausted and worn out killed by the matador, they are encountered by about a dozen stout and most active men, accustomed to the fight. These grapple with the bull, and master him by main force, and pin him down to the ground, and some of the most powerful and skilful will, of a sudden, seize the bull by the horns, and throwing their whole weight upon the animal, completely roll him over, when he becomes perfectly helpless. This is a great feat, and is most enthusiastically applauded. But notwithstanding their strength and activity, the bull-fighters frequently get knocked about a good deal, and receive heavy falls and bruises, but they never get tossed, and are seldom severely hurt. The sight is by no means disgusting, cruel, or barbarous, like those of Spain. Indeed, there is a good deal of fun in it, and at times you cannot help laughing heartily, and generally go away much pleased. After every fight, the men came forward, and were handsomely rewarded by Dom Fernando and the audience.

There still exist the remains of many fine convents and churches; of some the building had never been finished. When monastic institutions were abolished these buildings became the property of the State, and have since been sold to the public and converted to different purposes; some to manufactories, others to barracks, others to domestic residences. The inmates have been dismissed, with moderate pensions, and thrown out upon the world. The property brought little to the State; and as they were in a great measure founded by the charity of private individuals, to which the State had no right, it would have been better to have allowed them to die a natural death. Religion has gained nothing, the State little or nothing, while gross superstition has degenerated into scepticism and infidelity. All violent changes have an opposite result to that desired. The roots of old institutions once torn up, it is difficult to substitute a new plant to succeed them. It would be better if such important changes could be made gradually, but, unfortunately, the old generation, strong in possession, and forgetful of the times in which they live, will listen to no alteration, and cling firmly to the past, as if it must endure for ever. The new generation, born under different circumstances, have no respect for old institutions, and regardless of what is good in them, will listen to no compromise, and are never satisfied until they have rooted them out altogether.

There can be no doubt that the institution of convents and monasteries was of great service to the world in the middle ages—in preserving the learning of past times, in teaching the ignorant, in distributing alms amongst the poor, and in healing the sick. They curbed the ambition, and controlled the violent passions of the barbarous feudal lords, who considered that the world was made for them alone, and gratified their lusts at the expense of the unfortunate people over whom they ruled. No power but that of religion could have controlled them. They felt its wondrous influence, and were told in stern, unmistakable language that there was a hereafter, and that the great God who governed the world would surely visit their sins with condign punishment, unless they sincerely repented of their wickedness, and prayed for His forgiveness. We must therefore be grateful for the services which religion, under its peculiar forms and ceremonies, rendered in those days; and although abuses by degrees crept in, yet these might have been remedied by much milder measures. The monasteries and convents, however, served their purpose; but now they are gone, and it is to be hoped that as the world becomes wiser and more enlightened the great Creator will be worshipped in a purer and simpler manner.

Having heard so much of the filthiness of Lisbon, I was most agreeably disappointed to find it so much the reverse. With the exception of the old part of the town, to the westward of Blackhorse Square, and which was filthy enough, few cities, certainly not London, were cleaner than Lisbon at the time of my first visit in 1855. The streets were broad and straight, well lighted and watered, and the buildings regular and handsome. The police certainly was not the most effective, but still the town was safe.

I left Lisbon about four o'clock in the afternoon, in a fast steamer, full of passengers, and after a fine passage of sixteen hours, with only a moderate swell ahead—which, however, made most of the passengers sick—we arrived off the bar of the Douro at six the following morning, when a pilot came on board, and we crossed the bar and proceeded up to the town quay, about two miles from the entrance of the river; I was recommended to the Peninsular Hotel, situated in the higher part of the town. This was entirely a Portuguese establishment and extremely rough, where you boarded and lodged at so much per day. At the table d'hôte I made acquaintance with a fine old Peninsular veteran, Colonel Owen, formerly attached to the Duke of Wellington's army, where he was constantly employed on outpost duty, and he necessarily saw a great deal of hard service. He was a fine-looking man, six feet high, and seventy years of age, with a determined countenance, and full of fire and energy even at that advanced age. He possessed a good deal of talent, had studied much, spoke the Portuguese like a native, and was an excellent writer both in that language and his own. I found him a very agreeable, intelligent companion, and we soon became well acquainted with each other, which was the more agreeable as he knew well and had frequently served with Colonel Somers Cocks, a distant connection of mine, who was a distinguished officer and favourite of the Duke of Wellington. Colonel Cocks, much to the regret of the Duke, was killed at the siege of Burgos Castle. Owen knew his whole history, and at my particular request wrote a complete and very interesting memoir of him, which I privately printed. Owen was almost my constant companion, and having long resided at Oporto, knew all the principal merchants in the place, both English and native, and was much liked and respected. He introduced me to several, but being upon public business, namely, the improvement of the port, which all desired, I had no need of introductions, for the principal inhabitants called upon me and invited me to their houses; amongst others may be mentioned Messrs. Lambert, Sandeman, Herries, and others, the principal wine merchants of the place, who paid me the greatest attention and hospitality.

I lost no time after my arrival in proceeding to examine the port, and to consider what was best to be done. I soon discovered its defects, which were the exposed and dangerous nature of the bar at the entrance of the Douro, the depth of water over it being scarcely 10 feet at low water of spring tides, the tides rising only from 6 to 7 feet; the conformed and circuitous channel both within and without the bar, in which there were several rocks; and the prevalence during three-fourths of the year of strong westerly winds from south-west to north-west. Moreover, during heavy floods the outgoing current was so strong that it carried all before it, frequently tearing vessels from their anchors, driving them to sea, or wrecking them in the harbour. From these different causes the bar could only be attempted during the flood or ebb for entering or departing during the most moderate weather, and then only by vessels drawing 12 to 14 feet at spring tides. From all these combined causes the harbour of Oporto might be said to be hermetically sealed for three-fourths of the year, and frequent instances have occurred of vessels having made the passage to the Brazils and back again, whilst others have been beating about the offing waiting for a favourable opportunity to cross the bar; and equal time was lost by those vessels that were waiting, laden in the port, to go to sea. And the mails for Oporto, the second city in the kingdom, frequently had to be delivered at Vigo, 60 miles farther north, and to be transported thence by land, which occasioned the loss of one or two days. In consequence of these serious disadvantages the trade naturally suffered materially.

In order to remedy these defects, I drew up a plan by which I proposed to carry out two piers—one on the ridge of rocks on the south side, and another on the north—in such a direction that the entrance between them, which was to be 500 feet wide, should be least exposed to the prevailing storms; and the space between the piers on the inside of the entrance should be wider than the entrance itself, so that any swell which might enter would diminish as it proceeded up the harbour. I also proposed to remove the rocks both inside and outside, and to strengthen the channel. By this means I expected that the bar would be lowered about 5 feet; so that at high water of spring tides there would be about 20 feet, and a vessel drawing 18 feet could enter at springs, and 16 feet at neaps. The expense of these works was estimated at about 400,000*l.* This no doubt would have been a very great improvement, and worth the money, although at times even with this expenditure the bar would not be approachable either way.

Under these circumstances, I explored the adjacent coast in order to ascertain whether another and better place for a harbour could not be found, and at Mattozenhas, about three miles farther, I found that there was a detached reef of rocks, 2000 feet long, lying about a quarter of a mile distant from the shore, and between it and the shore there was a depth of from 25 to 50 feet at low water. This reef had been observed by others before me, and it had struck them that an excellent harbour might be made. As this also lay within my instructions, I made a design for a harbour; but upon maturely considering the subject, I thought it would be scarcely safe to make a close harbour at this point, as, from the great quantity of sand which lay along the coast, there was a chance of its filling up. I therefore proposed to make it open: that is to say, to construct a breakwater along the line of the rocks as far as they reached, and as much farther each way as might be considered advisable. In fact, this breakwater might be extended so as to make a magnificent asylum harbour for the largest vessels of any class; and by making an open landing pier within the breakwater, vessels could take in and deliver their cargoes, and send them to Oporto by rail, when vessels could not pass over the bar of the Douro.

As there was plenty of fine granite on the spot, and labour could be obtained comparatively cheap, I considered that for the sum of 500,000*l.* an excellent harbour might be made here with a double entrance, and as there would be a free passage for the current both ways, no sand would lodge within the breakwater. In fact, the current, being confined, would force out the sand and make the harbour deeper, so that at low water (spring tides) there would never be less than from 25 to 30 feet in the shallowest part of the northern entrance, and above 50 feet at the southern. This place appeared, and, in fact, was so much better and more advantageous in every respect than Oporto, that, considering the increased cost would not exceed 100,000*l.*, I strongly recommended it in preference to expending 400,000*l.* on the old entrance. Still, if the Government thought it advisable to improve the entrance of the Douro, to some extent this might be done advantageously by the expenditure of about 150,000*l.* Both the King and the Government approved of my plan for the breakwater and landing pier at the Secars Rocks, near Mattozenhas, and the Finance Minister—Fontes—gave me an order in writing to provide the necessary apparatus, and to commence the work immediately. But soon after, the Government being changed, the whole was stopped, and nothing material has been done since, although, I believe, some attempts have been made by the Portuguese Board of Works to blow up the rocks at the entrance of the Douro, with very little effect; and as funds are very scarce, it is probable that nothing of consequence will be effected for a long time, which is the more to be regretted, as meanwhile the trade of Oporto suffers most materially.

Having completed my investigation of Oporto harbour, I prepared to return direct to Lisbon, and report to the Government the results of my examination, when I received a telegraphic message saying that the Government wished me to examine the harbour of Viana, some thirty miles farther north, also that of Aveiro, twenty miles farther south, and that of Figuera, the mouth of the Mondego, about twenty miles south of Aveiro; and that they would send a small vessel of war to take me to those places. I therefore determined to await the arrival of the war-steamer, which came two days afterwards. It was not above 200 tons, mounting three or four small guns, and commanded by a lieutenant of the Portuguese navy, having with him about thirty men. I embarked and proceeded at once to Viana, which we reached about three o'clock in the afternoon. There being a heavy swell, we could not enter the port, and a pilot boat with six rowers was therefore sent out to take us ashore, where we were received by the authorities. As our time was limited, I immediately proceeded to examine the harbour, which was formed by the river Scina, the entrance to which was at times very dangerous, from the same causes as those already described as existing at the bar of the river Douro, namely, a shallow, much-exposed bar, little tide, tortuous channel, and exposure to a heavy swell during the prevailing westerly winds. I observed, however, that there was a long reef of rocks running parallel to the channel nearly as far as the bar, and that by erecting a breakwater upon the reef, for which there was plenty of stone adjacent, the current, instead of spreading over the rocks and losing its force, might be confined, and thus act more effectively upon the bar and deepen it several feet; and, further, by this means the entrance channel would be protected, and very little swell could get in. Thus the principal defects of the entrance would be remedied; and the small depth of water at the quays could be increased by dredging. I consulted the pilots and the practical men of the place, as I always do (for from experience they generally have a knowledge of the subject), and it was satisfactory to know that their opinions entirely coincided with mine. I accordingly made my notes, and returned on board the vessel in the evening about eight o'clock.

The days being long, and low water in the harbour, it gave me a very good opportunity of examining it.

I had occasion to visit Viana subsequently. It is a very pretty little commercial town, surrounded by a rich, thriving country; it does a good deal of trade, and is therefore very prosperous, there being some wealthy merchants and a good deal of property in the place. The valley through which the Scina flows is very beautiful and fertile, surrounded by mountains varying from 2000 to 7000 feet high. The large village of Scina, about twelve or fourteen miles above Viana, is particularly picturesque. It is situated in a delicious country, abounding in corn, wine, and oil, and apparently wanting in nothing.

As soon as I got on board the vessel weighed anchor. The night was fine, so that we reached the offing of Figuera, at the mouth of the Mondego, about noon the next day. I was met on landing by the captain of the port, who had received intelligence of our coming, and on account of the heat we were obliged to take refuge in his house, where he exhibited all his plans, and the measures which had been taken to improve the bar at the entrance, which, upon the whole, as far as they went, were satisfactory. After a couple of hours' rest we took a boat and sounded the bar, upon which the water was very shallow. There was a very extensive reservoir for tidal and fresh water within, which, under proper management I considered might be rendered very valuable in deepening and scouring out the entrance, and in lowering the bar; but at low water there was a narrow channel, which was constantly changing, so that there was no good entrance, and consequently neither the fresh nor tidal waters could be either admitted or properly discharged; thus the quantity of water was reduced, and the effect upon the bar considerably diminished. I therefore saw at once that it was necessary to confine the low water channel to some extent by slight works, composed of osiers and faggots, extending from two to three feet above the low water level, so that it should be confined always to the same course; thus the low water line would be lowered, more tidal water would be admitted and discharged, the surrounding lowlands, which were frequently flooded, would be better drained, and the bar improved. I observed also that the channel at the entrance was too wide, and that consequently it was much exposed to the swell during westerly gales. I therefore proposed to reduce the width of the entrance to a certain extent; by this means a greater quantity of water would be admitted and discharged, the bar would be reduced, the drainage improved, and a less quantity of swell would be admitted; when I explained my views to the captain of the port, who was a very intelligent person, he quite agreed with me.

Having finished my examination here, we embarked at sunset and proceeded to Lisbon, which we reached on the following morning, when I took my leave of the commander of the vessel and his officers, who had treated me most kindly, for which I felt very grateful; I wanted to make a present of some champagne, but this was most politely declined. I therefore renewed my thanks, and went on then to the Braganza hotel. I next day called upon the Finance Minister, Fontes, and Vicomte de Luz, the chief officer of the Board of Works, reported generally what I had done, and said that I would proceed to England at once, and send my plans, estimates, and reports as early as possible. I accordingly left two days afterwards, and reached England in the middle of June, after an absence of between six and seven weeks.

When I was in Portugal I had a good deal of conversation with M. Fontes about their railways. He said that they had already given a concession for them to an English company, Messrs. Shaw, Waring, and Co., of whom the Government was now desirous if possible of getting quit. M. Fontes also talked to me about their financial affairs, and said that they had had a good deal of trouble with our Stock Exchange. I did not know the precise cause of this at the time, but it was afterwards explained to me that the English Stock Exchange would not allow their funds to be quoted in the English money market in consequence of the Portuguese Government having reduced the interest of the last Portuguese loan without the consent of the subscribers, so that unless this was remedied, it would not be possible to raise another loan in England. I saw the full force of this, and told the minister frankly my opinion, which he appeared to take very well, for he then said that he wished me to lay out a line between Coimbra and Oporto, and gave me instructions accordingly.

In the month of August, Dom Pedro, the heir apparent to the throne, and his brother, Dom Louis, paid a visit to our Queen at Osborne House, Isle of Wight, and Count Lavradio, the Portuguese minister in London, recommended me to go there to pay my respects to his future majesty. I therefore went, and was honoured with an audience on board Her Majesty's yacht, which was lying off Osborne, and in which Dom Pedro and his brother were living at the time, as the Queen did not wish them to be on shore, on account of the prevalence of the smallpox in the vicinity. Dom Pedro and his brother received me very courteously, and we had a great deal of conversation about Portuguese affairs, particularly concerning the various engineering works that he wished to see carried into effect, such as railways, harbours, docks, &c., and he said that until these were done Portugal could not be on a par with other nations. He also very kindly invited me over to Portugal again. I took my leave and returned to Southampton, where I dined with Count Lavradio at Radley's hotel, and then returned to London. It should be added that Dom Pedro attained his majority on the 15th of September, when he was crowned.

As soon as I got my harbour plans and reports ready, I determined to go over and present them in person after the coronation. As I had been commissioned by the Government to make the surveys and estimates for the proposed railway between Coimbra and Oporto, and consult two other English contractors about making railways in Portugal, I went to Messrs. Peto and Betts, who said that they would be happy to undertake them, provided that they could make proper arrangements with the Government, and that they did not interfere with any other contractors or companies who were then employed; and that in order to ascertain the Government views, they would send over their agent, Mr. Giles, with me. As I had previously agreed with Mr. Cheffins to make the survey of the line between Coimbra and Oporto, he preceded me with his staff, while Mr. Giles went to Lisbon with me in the Royal Mail Company's vessel the 'Trent,' leaving Southampton on the 9th September, 1855.

After a fine passage we reached Lisbon on the morning of the 14th of September. I immediately presented myself to M. Fontes, the Finance Minister, and the Duke de Saldanha, President of the Cabinet, and delivered my plans, which were well received, and tickets were sent to us to visit the various ceremonies of the coronation.

It happened just before our arrival that the Government and the English contractors of the Great Eastern Railway from Lisbon to Santarem, who had been quarrelling for some time past, came to an open rupture, and the Government took possession of the whole of the works and all the materials, machinery, and plant, with an armed force, turned the Company adrift, and would have nothing further to do with them. The contractors complained to the British representative, Mr. Ward, Chargé d'Affaires, who was very indignant at this summary mode of treating his countrymen; and M. Fontes was equally indignant at the way in which the Company had behaved; but more of this hereafter. In the meantime the coronation took place, and a very pretty sight it was. We had the usual demonstrations of fêtes, reviews, illuminations, salvos of artillery, &c. To heighten the spectacle a British fleet of five line-of-battle ships was sent over to do honour to the occasion. These fêtes lasted for three days, and no business was done; but in the mean time I had the honour of being presented to the king at the great palace of the Ajude; and when M. Fontes returned to his office again, I called upon him and presented Mr. Giles, Messrs. Peto and Betts's agent. Afterwards, when Mr. Giles had left, M. Fontes commenced talking to me about their dispute with Shaw, Waring, and Co. He spoke very temperately on the subject, and said that justice should be done to them. I said that I did not wish for a moment to defend them; in fact, I did not know the merits of the case on either one side or the other. At the same time, I said that it would be far better to arrange with them amicably; and if they could not agree, they should settle accounts and dissolve the agreement; as until this matter was satisfactorily concluded, it would be impossible to get any fresh English contractors of respectability to finish the works, or form a new English company, or indeed to raise any money in England, which the Government at that time wanted to do. Moreover, it was desirable above all things to arrange matters with the English Stock Exchange, and until the affair of Shaw, Waring, and Co. had been settled this could not be done. These arguments appeared to have much weight with the minister, who replied, that he would think them over; that he was desirous of coming to an amicable settlement; that he had very little doubt but it would be ultimately arranged properly; and that when Mr. Griles and I returned from our examination of the country, he would be glad to see us again.

CHAPTER VIII.

Surveys in Portugal and Tunis.

After visiting, with a great deal of pleasure, Batalha, by far the finest ecclesiastical edifice in Portugal, we reached Coimbra, than which, with its environs, it is difficult to conceive a more beautiful prospect. The town, with its numerous churches and academical buildings rising from the opposite shore of the Mondego, and situated in a rich fertile plain, backed by the lofty and picturesque hills of Busaco, clothed with wood to their summits, with a fine old bridge in the foreground, while the Mondego is seen winding through the valley until it is lost in the sea, which forms the distant horizon, altogether made a picture which, for loveliness, was unsurpassed by anything I have seen in Portugal. We found here, a rare thing at that time in Portugal, a clean, comfortable little hotel, near the Rocio, where Mr. Cheffins and his assistant had arrived before us. We accordingly made preparations to start next morning at six o'clock, on horseback, to explore a line along the coast to Oporto.

We started punctually at six, and crossed the valley of the Mondego, which is here about three miles wide, and is quite flat. It has a rich and fertile alluvial soil, but is frequently inundated by the floods of the Mondego, so that the roads, such as they were, viz. extremely bad, were formed by rough causeways raised 3 or 4 feet above the level of the adjacent land. In fact, at the time of my visit there were no roads worthy of the name in Portugal, except the road above mentioned, from Cairegado to Coimbra; and upon this, what was still more extraordinary, a regular four-horse mail-coach, like those of England, had been just established for the first time to run from Lisbon to Coimbra, at the rate of about seven miles an hour. Another similar road had been commenced from Oporto to Braga, and had been carried only a few miles. In fact, unless you had seen and visited Portugal you could hardly have conceived such a state of things possible. All the other roads were nothing but mere horse-tracks, in the most wretched state imaginable, full of holes and great stones, so that you could seldom go faster than a walking pace without danger of breaking your neck at every step, except in those parts where the track lay through forests and open plains. Over these the rude bullock carts jolted up and down at about the rate of half a mile an hour, and the creaking of the axles might be heard two or three miles off. As for inns, for the most part there were none, and the wearied traveller had to carry everything with him, and take refuge in some miserable cottage full of filth, which he was obliged to clear away before he could establish himself with anything approaching to comfort. I had heard of this before leaving England, and I took care to provide myself with a light travelling-bed, and a canteen with the necessary provisions.

Leaving Coimbra, we reached Aveiro the next evening, after a most fatiguing journey of fifteen hours in the saddle; during a great portion of our ride we were exposed to a terrific storm of rain, and lost our way, and narrowly escaped passing the night in a pine forest. Not far from Aveiro, Messrs. Pinto, Basto, & Co. had established a large manufactory for porcelain and common crockery ware, which was in a very thriving state, and the imitations of the English porcelain were very creditable, as well as the artistic decorations; as they were well protected by high import duties on foreign porcelain, they enjoyed a monopoly for a time, and made considerable profits. These duties have latterly been greatly modified, and it has not been found worth while to continue this establishment. It has, however, I understand, answered very well for the proprietors, who have, by this and other enterprises, realized considerable fortunes. In fact, it was quite evident that such an establishment, which had to import all the materials and the fuel from England, could not compete with the superior talent and industry of England without extraordinary protective duties, which are nothing more or less than a heavy tax upon the country. Portugal is not a manufacturing country. She has a most genial climate and generally a rich soil, and can produce a variety of raw commodities which England and the north of Europe cannot. Portugal therefore, by exporting these articles, such as wine, oil, fruits, cork, salt, cattle, &c., can always realize a good profit, which will enable her to import all other articles required at a far less cost than they can be produced at home, besides contributing materially to the State revenue. Messrs. Pinto and Basto were not at the establishment when we visited it, but their manager had received notice from them that we were coming, with orders to show us everything, and to give us a handsome entertainment, which he did to the very letter, and we returned to Aveiro much delighted with our reception. Indeed, the manager regretted greatly that we had not stayed there the previous night, and if we had known it, we certainly should have done so. For although, considering all things, we were not badly off at Aveiro, still we should have been much better off at Messrs. Pinto and Basto's establishment, and should have been saved the severe fatigue of the last ten miles of our journey.

The next day we started for Oporto, and passed the thriving little town of Ovar, situated at the north end of the Aveiro lagoon, but having a separate entrance from the sea, formed by a small stream passing through the town. Our railway passed along the base of the hills, which was very favourable. From Ovar we continued near the shore through a forest of pines for several miles; these pines would furnish excellent sleepers for the railway at a very moderate cost. From the forest we continued along the shore through most favourable ground, until we reached the entrance of the Douro: the distance from Ovar being about twenty miles. From the entrance of the Douro there was a favourable line for the railway for two miles along the banks to Oporto, where it would have been necessary to cross the river by a rather expensive bridge to reach the chief part of the city. This, however, must have been done in any case, and as all the great wine depôts or lodges, as they are called, are on the south side, and as the wine trade of Oporto is the most important, this would have been a great advantage; moreover, there is on the south side, near the entrance, a much better position for docks.

Between Ovar and the Douro stands a rock surrounded by extensive sands. On this rock is a chapel, reputed to be of peculiar sanctity, which is annually visited by immense numbers of pilgrims. It is very curious to see the extent to which these devotions are carried, and the occasionally strange results. In going one day, on foot, from Mattozenhas to Oporto during one of these festivals, I was surprised to meet a man walking, or rather shuffling, along in an extraordinary manner, in what appeared to me to be a sack. I immediately stopped to look at him, and gazed on him with astonishment. Several persons, apparently friends, accompanied him, and I inquired what it meant; I was informed that the person whom I saw enveloped in a sack had, during a severe illness, prayed to the Saint of the church of Mattozenhas to deliver him from his illness, promising that if he recovered he would annually on the Saint's fête-day walk to the church in his grave-clothes, to return his thanks for his delivery from death. The poor fellow seemed to be much fatigued, and no wonder, as travelling for two miles in such hot weather over a dusty road in such an uncomfortable costume must have been very severe work.

Next day we were off before daylight, and reached our old quarters at Aveiro the same night. I soon saw that it would be difficult to get a good line for the railway over this district; it was too hilly and expensive, and would have required long tunnels through granite rocks and heavy embankments. I therefore gave up this line, that is as far as Ovar. After leaving Aveiro we took a new route nearer to the hills, and with the exception of a part of the river Vouga, it was preferable to the coast line formerly mentioned. The summit near the Busaco hills was a little higher, but there the line was shorter, and on the hill there was plenty of fine limestone fit for bridges and any other works. This line evidently therefore was the best; I ordered it to be minutely surveyed and levelled, and the result proved that my anticipations were correct.

We reached our old quarters at Coimbra, and having examined the city again were much more pleased with it than before. The University is a very handsome building upon an extensive scale, with an excellent library, museum, extensive lecture rooms, and a competent establishment of professors and lecturers. The costume of the professors and students, although totally different from that of our own, was very appropriate; and the method of teaching and conferring degrees was extremely good, although a little too much savouring of sacerdotalism, but still greatly improved in this respect to what it was only a few years before. The city moreover was comparatively clean, and there appeared to be a degree of outer tranquillity and prosperity about it which pleased me much. I have seen many foreign universities, but none delighted me more than Coimbra. After leaving Coimbra we passed through Thomar and reached the valley of the Tagus; this we descended until we arrived at Santarem, which is but an inconsiderable place; it is situated on the highest point of the ridge, and bounded by precipitous rocks which abut on the Tagus, there being a small valley on the north, also on the east and west. Being thus isolated, it occupies an admirable position as a fortress, and might easily be made impregnable; commanding, as it does, the country all around, if it were properly defended, the approach of an enemy would be extremely difficult; and if fortified according to modern ideas, with a sufficient garrison and provisions, it might hold its own against all comers for almost any length of time; in fact, it may be considered as one of the keys of the kingdom. When I saw it, its works were in a very unfinished and dilapidated state. From Santarem we proceeded down the river to a small place termed the White House, where we embarked on board the steamer for Lisbon. The hills on either side of the Tagus alternately advance and recede; but the whole scenery is very rich, the soil being highly cultivated to the very tops of the hills, and in places abounding with cattle. In the flat country, however, bordering on the river, malaria and intermittent fever are very common, and indeed the whole course of the Tagus requires regulating.

We reached the Braganza at Lisbon in the evening, after a rough but satisfactory journey. Mr. Giles was now anxious to go to England, and a few days after our arrival he sent in a general proposal to the Government, on the part of Peto and Betts, for making railways, and then started in the packet for England, leaving his assistant, Mr. O'Neil, with me, to make any further necessary explanations. The season was getting late, and we had no time to lose, for it was now nearly the middle of October. I saw M. Fontes several times, and he always received me very civilly, and said that he was determined to meet Messrs. Shaw, Waring, and Co., fairly.

Mr. O'Neil and myself then started to survey the country for lines suitable for railways in the east and south. After being detained for a time by heavy rains at the White House, we reached Salvatierra, a wretched place on the left bank of the Tagus, about seven o'clock in the evening, fully two hours after dark; it was a nasty pestilential hole, close by the marshes; however, bad as it was, we were very glad to get there. There was no inn in the place, but after a little inquiry we found a shopkeeper who was willing, for a consideration, to accommodate us. We accordingly agreed with him, and took possession; we found the rooms full of corn, beans, &c., which we soon cleared out, swept the floors, walls, and ceiling as well as we could, and unpacked our things. I sent out my man for some provisions, we got a good wood fire lighted, and after about an hour and a half had a tolerable supper, turned into bed, and slept soundly until the morning. In fact, I never grumbled; I took the people as I found them; they were generally very civil, and did everything in their power to make us as comfortable as possible. I always carried plenty of cigarettes and cigars, and distributed them liberally, not only to the men attending the horses, but also to the people in the house and to any casual visitors that might come in; as I had by this time acquired some knowledge of the Portuguese language, I made myself as agreeable as possible, and in consequence every one did the best they could to help me. The Portuguese are a very civil, quiet people, and if you only treat them properly, as I always did, they will do everything they can to accommodate you. I of course also took good care of the horses and men, and they were so much pleased that they never grumbled, and would do anything I required. Throughout the whole of my journeys I never had a single squabble; we all went on merrily together, and whenever I could get a good laugh I always did, and this was not seldom. We left Salvatierra at daylight, and proceeded over a gently undulating country, and in three days reached the old fortress of

Estremoz. The people of the district through which we were now travelling were extremely primitive, and in one large village, where we halted, as usual, at midday, the whole population, it being Sunday, turned out to have a look at us.

The country here is elevated and very well cultivated. The olive-trees are planted in regular rows, and carefully attended to, a thing I had not before observed in Portugal, as they are generally in forests, and left entirely to nature. The vines also were more looked after, and some excellent wine is made here. Upon the whole this was certainly a thriving district. We took a ride to the ridge of mountains about ten miles to the south, which, extending east and west, separates this district from the southern part of the Alentejo. They are covered with verdure to the top, afford excellent pasture for sheep and cattle, and form a very picturesque prospect. Estremoz is in a commanding position, but its fortifications were in a dilapidated state. After having spent a day here, we proceeded over an elevated, fertile, cultivated country, to Elvas, on the frontier. The Government had been employed for some time past in making a good road here, and a considerable portion had been finished, so that we made tolerable progress, and reached Elvas soon after midday, and found a very clean, comfortable little hotel.

Elvas is a very strong fortress, situated upon the summit of the ridge of hills forming the western boundary of the valley of the Guadiana, which winds through the extensive valley below, about seven miles distant, and the boundary or frontier line between Spain and Portugal is about half way between Elvas and the Guadiana.

The fortifications are very extensive, and upon the whole well laid out and tolerably well kept up. There are several large outworks, particularly that on a hill about a mile to the north, which may be said to be the citadel. It commands the town, being situated upon the highest part of the ridge. The view from this is very extensive, overlooking the dreary and bare valley of the Guadiana, with Badajoz, the Spanish frontier fortress, about nine miles distant; and farther on the bare bleak hills of Estramadura bound the horizon. On the west you have a fine view of this part of Portugal, whilst north and south you command the view of this elevated ridge, which bounds the Guadiana as far as the eye can reach.

Having explored this part of the country, and having satisfied ourselves that it would be difficult to get a good line here, we turned our attention to the southward, and again leaving Estremoz, reached the poor but considerable village of Oçana, not many miles from the frontier. I therefore determined to stop there for the night, and to my surprise found that there were two or three small inns. I selected the best, and bad was the best. I was informed that Oçana was the resort of the worst characters of all sorts, both Portuguese and Spaniards, who made this their head quarters, from whence they could easily cross the frontier of either country, according as either Spain or Portugal became too hot for them. I was told that we ought to be upon our guard, for it was not improbable that we might be attacked in the night, or waylaid on our route the next day, for the country through which we should have to pass was very wild and lonely. Therefore, as is always the case, I thought the best plan would be to put a good face upon it, and to show that we had no fear, put confidence in the people around us, and make them friends. I therefore invited as many as I could get, men and women, to a sort of ball, inquired for some musicians, and found half-a-dozen with guitars, castanettes, and tambourines, and as the inn where we were would not hold all the guests, I engaged the other two inns also. The whole population crowded to the balls at these inns, and I visited them alternately, danced with the women, and made myself as agreeable as I could. Wine and refreshments were not wanting, everything passed off in the best manner, and nothing but mirth and good humour prevailed. Soon after midnight I had had enough of it, and the guests below, excited by wine, became very noisy. I told my servants, guide, and horse-keepers to have everything ready by four o'clock on the following morning, the horses saddled and the baggage packed, outside the town, and having previously paid the reckoning, we quietly departed, without beat of drum, leaving our delighted guests in the midst of their revelry, and in no disposition to follow us. By daylight we had travelled five or six miles without encountering any person on the way—for road there was none. We laughed heartily at our adventure, and I was afterwards congratulated at having succeeded so well, for we had had a narrow escape of being robbed, and probably something worse. We passed through a wild but not unpicturesque country, and reached Abrantes, on the Tagus, without further adventure, on the afternoon of the second day. This is one of the most considerable fortresses in the kingdom, situated in a commanding position of great strength; but the works were very much neglected, and there was scarcely any garrison. The town itself is a poor place.

As regards the country through which we had passed being adapted for a railway, there was no doubt that a good line could be made through it, although at considerable expense; the difficulty was how to cross the main ridge which separated the valley of the Guadiana from that of the Tagus. It certainly was most desirable to connect Abrantes with Elvas, but the descent from it to the Guadiana valley was too rapid; still, this was only a matter of expense. I thought that some more practicable gap through the ridge might be found farther southward, but this I was obliged to leave for future investigation.

I now determined to proceed northwards, in order to ascertain whether a line for a railway could be obtained preferable to that which we had previously examined when returning from Coimbra.

Having slept at our old quarters at Thomar, we proceeded to the banks of the Zezere, which we found to be extremely lofty and falling rapidly to the river. We proceeded for some miles along them, but the higher we went up the river, the higher we found the hills; in fact, there was no valley except that occupied by the river, so that we had to give up all hopes of a line in that direction. However, in the course of a few days' explorations we were fortunate enough to find an easily practicable line for the northern railway, in the direction of Oporto.

In one of these expeditions we met a Portuguese gentleman dressed in the native costume, and attended by his servant. He courteously saluted us in good English, and then asked where we were going, and what we were doing, which, of course, we told; he then asked, "Where do you sleep to-night?" and when we told him, he said, "Nonsense; you cannot and you shall not sleep there: it is not fit for a dog" (in which he was not far wrong). "I live close by. Yon must come and sleep at my house, and leave as early as you like to-morrow morning." We gladly accepted his invitation. We returned to the cottage, dressed ourselves in rather better costume, while our new friend rode forward to receive us. His house was something like a good farm, surrounded by a courtyard and farm-buildings, situated in the midst of a kind of rough park. The night was very dark, and we had some difficulty in making our way through the woods, and if it had not been for the barking of the dogs, we should scarcely have found it. However, we got there, and were most cordially received, shown into a comfortable room with a blazing wood fire, and in about half an hour after sat down to an excellent supper, with plenty of good wine.

Our host was a most excellent and well-informed man of about sixty, of the middle size, rather stout and well-made, with a fine open countenance. His name was Don S. de Silva. He had served in the Portuguese army under the Duke of Wellington, and had been engaged in several battles with the French, of which he was very proud. He took the greatest interest in the railways, offered every assistance in his power, and insisted upon our making his house our home whenever we came that way. He said that after the war he had retired to his estate, and devoted the whole of his time to farming, and that he passed a very active, pleasant time. I think he said that his wife and family were at Lisbon, where they frequently went, as they had many friends and connections there. We were shown into a comfortable and clean double-bedded room, and the only thing we required to make us thoroughly comfortable was more blankets, for the night was excessively cold. However, I slept soundly, and awoke next morning thoroughly refreshed. We had an excellent breakfast, and took leave of our worthy host with many thanks for his most hospitable entertainment.

We reached Thomar about midday, having travelled over a rather better road than usual. Here we halted at the Old Inn for a couple of hours, and then made the best of our way over the old road to Basquenha, before mentioned, where we slept. The next day we hurried on by Santarem, caught the tugboat and the steamer at the White House, and reached the Braganza, Lisbon, the same evening, having been absent about eighteen days; fortunately, with the exception of the first day, we had tolerably fine weather, without which we should have had a good deal of difficulty in making such a rough journey.

I forgot to mention that while Mr. Giles and myself went to Coimbra and Oporto, we sent Mr. O'Neil to explore the line of the Tagus beyond Abrantes as far as the frontier, in order to ascertain whether there was any practicable line for a railway in that direction. Mr. O'Neil, however, gave a very unfavourable account of it, for a few miles beyond Abrantes the Tagus passes through a deep chasm, hemmed in on both sides with steep, precipitous granite rocks, swarming with eagles, and wild and desolate to a degree. There were scarcely any inhabitants save a few shepherds, clad in goat-skins, who lived in small cottages stuck in the clefts of the rocks, or built upon some small patches of ground close by the margin of the river, with a few vegetables and fruits growing round them. They were thoroughly surprised at seeing Mr. O'Neil; they said they were never visited by strangers, and could not conceive why he had come there. Nevertheless, they were civil, and gave him freely the best they had, which was very little indeed, and would scarcely take anything in return; when pressed, they said they would keep the money for his sake, or would buy something in remembrance of him.

The eagles appeared quite at home there, and were seldom molested, except when they had carried off some favourite pet-lamb or kid, or had paid a visit to the poultry. Mr. O'Neil's report was so unfavourable that we did not think it advisable to pursue our investigations farther in that direction.

We had now pretty well explored the country between the Tagus and the Douro for the north, and the Tagus and the Guadiana for the east line to connect Portugal with Spain, and it only required that these should be investigated to ascertain which was best. Upon my arrival at Lisbon, I called upon M. Fontes, and reported what I had done. We had also a good deal of conversation about their finances; arranging affairs with Shaw, Waring, and Co.; getting good contractors for making further lines; raising a loan in England, &c. I said with all deference that the whole of these things might be satisfactorily settled; but that the first and most important of all was to arrange matters with the Stock Exchange in England, for until this was done a loan could not be obtained (and the Government was much in want of money) without paying very high interest, which would soon consume the revenues of the country. For example, their stock of 50 was only nominally quoted at 18 to 20, for which they paid 3*l.* interest, or 15 per cent.; whereas, if they settled with the Stock Exchange, their funds would at once rise to 40 or upwards, and thus they might borrow at 7 per cent. I further ventured to hint that by free trade—that is, lowering their import duties—the revenue would be much benefited, and that their exports would be increased in the same ratio. M. Fontes listened with much attention, but said nothing.

Having finished my business, I prepared to return to England, and was told by M. Fontes that I should have an audience of the King. I accordingly presented myself, and had an audience of his Majesty Dom Pedro, who was an exceedingly well-informed, sensible young man, of about twenty-one, and whose sole object and desire was the advancement and improvement of his country, to which he devoted himself day and night. He was exceedingly amiable and modest, and much beloved and respected, not only by his own subjects, but by every person who had the honour of being presented to him. His Majesty received me in the most condescending manner, entered into the subject of railways, the great advantage they would be to Portugal, the making of common roads to join them, the benefit of free trade, and the necessity of improving agriculture and extending the cultivation of vines, olives, corn, fruits, and all native produce, as well as the improvement of the harbours, and everything else which would advance the prosperity of the country. He had studied all these subjects thoroughly, and was perfectly at home in them; and with regard to the finances, he quite agreed with me. I took my leave of his Majesty, and he shook me by the hand, and bade me farewell, trusting that he should soon see me again.

I then called upon the Duke de Saldanha, the President of the Cabinet and Commander-in-Chief of the Army, a very estimable man, who spoke English very well; he had served under the Duke of Wellington during the great Peninsular war.

He spoke English perfectly, and was very popular amongst his countrymen, and always happy to see the English. When I told him that I was about to leave the next day for England, he said that he was very glad to hear it. He further remarked that a Cabinet Council had been held the day before, and that I should hear more of it before the day was over.

I returned to the Braganza, packed up my things, and was ready to leave by the packet the next day. At nine o'clock the same evening I received a message from M. Fontes to come to him at his office. I went there immediately, and he said he had been ordered to proceed to England with me, to endeavour to settle the matter in dispute between his Government and the English Stock Exchange, and the quarrel with the railway contractors.

We started together next morning, and on arriving in London I had the pleasure of being partly instrumental in bringing about an arrangement between the principal holders of Portuguese stock in England and the Portuguese Government, to their mutual satisfaction and advantage; also an arrangement between the Portuguese and Messrs. Shaw, Waring, and Co., which, as we shall see, was not subsequently ratified. M. Fontes returned to Portugal with considerable *éclat*, having re-established the credit of Portugal with their English stockholders, and settled the railway difficulty, and it was thought that he and his ministry had a long and prosperous career before them. Unfortunately, however, this turned out quite the contrary; whether it was envy at his success, or some other political cause, I do not know, but in the spring of the following year, 1856, Fontes and his ministry were obliged to resign, and the Marquis de Loulé, a connection of the royal family, succeeded him. The Marquis de Loulé was a remarkably fine, handsome man, a thorough gentleman, with considerable talents, of imperturbable temper and great good humour; I know no man with whom he might better be compared than the late Lord Palmerston. Loulé, like Fontes, was equally desirous of completing the railway system of Portugal.

I continued the surveys for a line between Coimbra and Oporto, and when they were completed I received a tender from Messrs. Peto and Betts, saying they would execute the line for 10,000*l.* per mile, without the land.

As his Royal Highness the Prince Consort took considerable interest in the improvement of Portugal, and as I had heard indirectly that information respecting the progress of the railways would be gratifying to him, I communicated with his Royal Highness's secretary, and received a command to go to Windsor. I went, and was received in the most affable manner. The Prince was thoroughly acquainted with everything going on in Portugal, and took great interest in the construction of railways in that country. He thanked me for the information I had communicated, and wished me every success.

Towards the end of August all the plans, estimates, and specifications of the line between Coimbra and Oporto were completed, and in the month of September I went over and presented them in person. I was very kindly received by his Majesty and by his ministers. When I arrived, I found that M. de Soveral, the very clever Chief Secretary of the Portuguese Embassy in London, was home on leave.

He also took the warmest interest in the establishment of railways. The dispute with Messrs. Shaw, Waring, and Co. had not yet been settled, and consequently no new arrangement could be made. I had frequent meetings with the Marquis de Loulé and M. de Soveral on the subject, and fully explained to them that it would be impossible to get any English contractors of influence to form a new and powerful company of English capitalists to make railways in Portugal, unless this matter could be honourably adjusted; and I added that I thought the terms agreed upon with M. Fontes were, upon the whole, the best that could be adopted.

The Marquis listened with much attention to all my arguments, and finally gave me instructions to see Messrs. Peto and Betts and their friends, upon my return to England, in order to ascertain upon what terms they would be willing to form a new company to complete the Northern Railway to Oporto; and I was to inform him of the result, but not to commit the Government in any way. He requested me, before I left Portugal, to examine the port of Setubal, or, as the English call it, St. Ubes, a little to the south of Lisbon.

I accordingly proceeded to reconnoitre the port, and found that it was formed by a river which ran a considerable distance into the interior. St. Ubes was situated on the right bank of it, about four miles from the bar, but the river here was so wide and shallow that it was only an estuary. The opposite side being covered at high water, the returning ebb and flood waters were expanded over such a wide space that the current had not sufficient force to maintain a good channel, nor did it act upon the bar powerfully enough to maintain a good passage over it. Moreover, the estuary was exposed to the full effect of the south-westerly gales, which, sweeping across the Atlantic with great violence, send a very heavy swell across the bar, rendering it extremely dangerous to approach, and driving a great quantity of sand into the mouth of the river; this would materially reduce the depth, and render the navigation both dangerous and difficult; and there was too little fresh water coming down to be of much service in assisting the operation of the tide. I soon saw what was necessary to be done to remedy to a great extent, if not entirely to remove, these evils; this was confine the channel on the opposite or sea side by a low line of wicker or fascine work, in the first instance to the level of a little above half tide, and then to raise it as required; the sand would soon accumulate behind it, and thus the tidal and fresh waters, being always confined to the same channel, would greatly improve it, and considerably lower the bar. I also thought the channel for some distance above the town should be deepened, embanked, and improved by dredging, and that below the town to the bar the channel should be dredged also to assist the operation of the current. I embodied these observations in a plan and report on my return to Lisbon, and delivered them to the Marquis, with which he was much satisfied.

I now prepared to return to England, in company with my friend de Soveral; but as the latter was desirous of breaking his voyage at Oporto, for the purpose of ascending the Douro and visiting his estates, I agreed to accompany him. We reached Pescoa de Ragosa, a very important place in this district, being the centre of the port wine trade. Here is held the market where all the port wine growers assemble at the end of the vintage, and here come all the wine merchants from

Oporto, and make their purchases for the season. The veritable port wine district is a very remarkable country. It is situated on the banks of the Douro, which are here composed of lofty, undulating hills, clothed with rich soil, rising rapidly from the river, about forty miles above Oporto, where the granite formation, on which no vines will grow, ceases. The district consists of the transition rocks above the granite; it extends about forty miles, as far as the cataracts of St. Joao de Pesqueira and is about five or six miles wide, including both sides of the valley of the Douro; one side is exposed to the morning and the other to the evening sun, and the reflected rays make the valley like an oven during the summer months, so that no place could be more advantageously situated for the development of the vine. No trees are allowed to grow upon the hillsides. The vines are like so many gooseberry or currant bushes, and they are cut down almost to the roots so that the nutriment applied to the plant may be developed in producing the fruit to the greatest possible extent. The vines are planted in rows, and the intervening space is carefully weeded, so that the whole power of the sun is concentrated upon the vines; and it is wonderful to see the quantity of fruit borne compared to the extent of branches and leaves. When the vintage has been made, there are a certain number of tasters who are appointed by the Government. These men, who neither smoke, snuff, or drink, then examine the wine, which they do by a saucer, much indented in the bottom, of pure silver, which enables them to judge of the colour; then they smell it, and taste it by the touch of the tongue without drinking any. By these three tests they pronounce the quality of the vintage, and this, combined with the quantity, determines the price. It is said that the wine is prepared according to a certain standard for the English market; that is, the wine is made to resemble as near as possible a vintage that has pleased the English, who are the great consumers of port wine; this is done either by mixing some old approved wine with the new, or adding brandy. There has been a great deal of argument and doubt about adulteration. I will not take upon myself to decide, but I will say this, that a good vintage of port wine requires no doctoring.

At Ragosa, after the vintage, all the great wine merchants of Oporto assemble, and there make their purchases for the ensuing season. All transactions are made in cash, for the wine growers are very important and wealthy personages. The merchants are equally so, and take their cash with them. I heard an anecdote when there, which says much for the honesty of the Portuguese. M. Sandeman, one of the wealthiest and most important wine merchants of Oporto, went up to the wine district after the vintage one season, with his sons and friends—for it is generally made a party of pleasure—to purchase wine for the ensuing season, and they took 15,000*l.* in gold to make their purchases, the gold being carried in sacks on mules, the whole party riding, as usual, for there were no roads for carriages. The first night, having arrived at their usual halting-place, just as they were sitting down to their supper, old Sandeman said to his son, "George, I hope that you have taken good care of the mules and the money?" "Yes, father," said George; "it is all right." "Well, George," said old Sandeman, "you had better go and see." George immediately went out to the stables, and after some minutes came back in great consternation, and said, "Father, I can't find the mules with the money." At this old Sandeman got in a great rage, and said, "George, you shall have no supper until you find them." The whole party got up at the same time, and took lanterns and torches to look after the lost money and mules. Away they went back several miles on the road they had come, and there they found the mules, which had slipped their halters, quietly feeding by the roadside, with the bags of gold untouched. They returned with them to the inn with much satisfaction, got the mules comfortably installed in their stables, having previously relieved them of the gold bags, which they took to their bedrooms. Now perhaps in no other country could you have found such an example of honesty as this. It was well known before they started from Oporto that the Sandemans were going up to the wine country to purchase wines for the ensuing season; and as they were the most wealthy merchants of the place, it was equally well known that they would buy the largest quantity of wine, and would take the greatest amount of hard cash to pay for it; therefore, if the natives wished to make a prize, here was an excellent opportunity to do so. But quite the contrary, the natives were honest, and had not the least idea of robbing, which they might have done, if so disposed, most easily.

From this place we proceeded leisurely to Guimaraens, one of the ancient capitals of Portugal. The natives appeared different from any I had seen. The better classes were dressed in jackets, ornamented with large silver buttons, knee-breeches, leggings, and silver-buckled shoes; having over all a loose, blue cloth cloak, and a large, conical, broad-brimmed black felt hat. It seemed to be a place wholly *sui generis*, and totally separated from the rest of the world. The well-to-do people seemed to have nothing to do but to walk leisurely about conversing with their friends, and smoking their cigarettes. There was no trade, no bustle, nothing, as it were, moving; it was one of those places where a man disgusted with the world would wish to bury himself. The repose was something surprising to a person accustomed to active life. We found a pretty good inn, and a convent with a few old nuns, who were still allowed to live there, and who employed themselves in preserving in a particular manner a very fine sort of plum that grows in the neighbourhood. They fetch a high price in all European markets; I bought a few to take to England. There is a splendid old Moorish castle outside the town, which, although in ruins, was not in a very bad state. Upon the whole I was much pleased with the place, so totally different from anything I had seen either in Portugal or Spain. I should have liked much to have spent several days there, but the inexorable time would not permit; we therefore set out the next day for Braga, and passing through the beautiful environs, which are a perfect garden, we continued our way through a well-cultivated and picturesque country, and reached Braga soon after midday.

Braga, also one of the ancient capitals of Portugal, is a handsome town, more modern than Guimaraens, also much more lively, and surrounded by a rich, fertile, picturesque country. It is well built, with handsome streets, and squares upon a moderate scale, with an appearance of quiet imposing dignity. It is the residence of many old families, who seem to enjoy their *otium cum dignitate*, and to be wholly independent of the world. They appeared to be above the grovelling pursuits of trade, for there was certainly little appearance of business about, yet withal we could not help admiring its clean, comfortable, well-to-do appearance.

From here we proceeded over a tolerable road towards Viana, the sea being about a mile distant; and the intervening space, composed of rich alluvial soil, was well cultivated. We passed several granite crosses, erected near the village churches, one of which was extremely elegant, beautifully worked in a species of Corinthian style. I never saw granite so elaborately worked as in this part of Portugal, namely, the district between the Douro and the Minho; in fact, they have no other stone but granite, and no other people work it so well; but then the natives are the most docile and industrious in Portugal. They are never idle; constantly at work, either abroad or at home, and saving money. Large numbers emigrate to the Brazils, where by their energy and industry they may acquire ample fortunes, and return home to their native country, where they buy an estate, build a handsome house, and spend the remainder of their lives in tranquillity and enjoyment.

On reaching Vigo, after passing through Viana, I found that I had to wait two days for the steamer. I could not help remarking the superior beauty of the Spanish as compared with the Portuguese women, a thing I could not well understand, as the personal appearance of the men is about the same in both countries, although the Spanish peasants are apparently more robust than those of Portugal.

Immediately on my return I set to work to carry out the instructions of the Marquis de Loulé, and endeavoured to obtain powerful contractors and wealthy and influential capitalists who would form a company to undertake the railways in Portugal. I frequently had communications with Messrs. Peto and Betts, who with their friends were perfectly ready to come forward to form a company, provided that Messrs. Shaw, Waring, and Co. were satisfactorily settled with. In these negotiations M. de Soveral was indefatigable, and of immense service, as he was perfectly acquainted with the views of his Government, and knew what would be acceptable to them, and what they would not agree to. At last, towards the latter end of December, certain conditions stating the terms upon which they would form a company and find the capital were drawn up and signed by Messrs. Peto and Betts, and seven other well-known capitalists, and were submitted to the Portuguese Embassy, to be transmitted to the Government for their approval. Towards the end of January, 1857, they answered that they generally approved of the conditions proposed, with one or two exceptions, which Peto, Betts, and Co. agreed to alter. The Government then sent an official letter to Count Lavradio, Ambassador in London, requesting that Sir Moreton Peto and myself would come out to Lisbon *immediately*, to finally conclude the arrangements, in order that a Bill might be prepared without delay to be laid before the Portuguese parliament. In March, 1857, I accordingly proceeded to Lisbon.

It is not my intention to enter into the unfortunate disputes between Peto and Co. and the Portuguese Government—disputes which terminated in the abandonment of what would have been for Portugal a great national work. Doubtless, as in all these cases, there were faults on both sides; and I believe that one of the main causes of the failure, on behalf of Messrs. Peto and Betts, to carry out the works, was that their resources were swallowed up by a great variety of speculations, some of which, as we have seen, did not in the long run turn out very profitable, and they were really unable to undertake them. I will proceed to relate briefly the only other occasion on which I was connected with that firm.

But first I may mention that I completed my work on 'British and Foreign Harbours' (which had occupied what little leisure I could command during some years) in 1854, having previously, in 1847, published a monograph on Plymouth Breakwater. I may also add, that conjunction with the late Mr. John Plews, I constructed a considerable extension of the Cardiff Docks for the trustees of the Marquis of Bute; as it is fully described in the above work, I need not here further refer to it.

In the early part of 1859 I was asked to proceed to Tunis, in company with one of Messrs. Peto and Betts' agents, to examine into the feasibility of constructing a railway from the Goletta to the city. I accordingly started in March, and having embarked on board a French steam-packet, reached Tunis after a four or five days' passage, including a stoppage, for some unexplained reason, of two days at Philippeville.

The view of the Bay of Tunis cannot be compared in picturesque effect with that of Bona, which we had just left; still there is something wild and striking about it. On the east the bay is bounded by a lofty ridge of bare irregular hills, with a narrow strip of level marshy land bordering the bay. On the west it is bounded by the celebrated peninsula of Carthage. In front, to the south, there is the Goletta, or channel to the Lagoon, surrounded by the custom house and a small town, and beyond is the Lagoon, extending about six miles, at the farther end of which is the city of Tunis, seated upon a gentle eminence, bristling with minarets, and a lofty chain of hills in the background, the whole having a wild, uncultivated appearance, so that at first sight you were puzzled to conceive whence supplies could be obtained for feeding the metropolis of the kingdom. We got clear of the Turkish customs after a good deal of delay, not from any fastidiousness of the officials, for they were easy and good-natured enough to let anything pass, but from the confused and blundering manner in which all business is transacted. Everything was then bundled into a large boat, which was also laden with merchandise of all kinds, as much as it would hold. We then entered, along with a Colonel West, who had come out upon a shooting excursion, and set sail for the capital across the Lagoon, which was about six miles long and four or five miles wide. The channel was nearly in the centre, and had five or six feet of water. We passed numerous flocks of wild geese, ducks, cranes, and flamingoes, disporting themselves in the water. We reached the landing quay of Tunis, outside the walls, about two or three in the afternoon, and immediately proceeded to the only hotel in the place, kept by a Frenchman, and, upon the whole, it was very clean and comfortable; but before we could get to it we had to wade through a sea of filthy mud in a narrow lane that was scarcely 12 feet wide, bounded by the city wall on the one side, and a row of miserable buildings on the other, showing little more than bare walls, the windows looking into small courts on the inside, which were approached by solid well-barred gateways. It was nearly dark before we got installed in our new domicile, and then we dined at not a bad table d'hôte in the French fashion, passed the evening agreeably, and went to bed early, rather tired after the kind of knocking about we had had during the day. Fortunately the weather was fine after the great quantity of rain which had lately fallen. The next day was fine also, and after an early breakfast we got a carriage with a couple of horses, and drove along the west side of the Lagoon to the Goletta in order to select the line for the railway; nothing could be more favourable, the country being—to use a homely phrase—as flat as a pancake, and therefore required no particular exercise of the engineer's art.

Having so far completed our investigation we adjourned to the examination of the ruins of Carthage, of which scarcely anything remains, except the cisterns for supplying the city with water, which are of massive masonry, the walls being lined on the inside with a thick coat of stucco, which was glazed, and presented an excellent, smooth surface. These cisterns were covered with arches, so that the water was preserved from the action of the sun, and was thus always kept cool and in the best state for use. The water was brought, by means of an aqueduct, from a fine and plentiful spring close by the mountain of Kegouan, about 40 miles distant, and was carried with the requisite inclination by means of tunnels pierced through the hills, and extensive lines of arched aqueducts across the intervening valleys, some of these aqueducts being above 60 feet high, and the total length of the tunnels several miles; in fact, the whole aqueduct was a series of tunnels and bridges about 40 miles long, and is certainly a most extraordinary work, not to be surpassed by anything of the kind in existence at the present day. The conduit for the water was about 2 feet 6 inches wide and 3 feet deep. In this single example we have nearly all the improvements of modern times, namely, excellent water, an ample supply, and covered storing reservoirs. The water required no filtering, but it has not been ascertained whether it was distributed to each house; probably not; most likely it was delivered to the fountains, where the natives sent for it. As baths are known to have existed at Carthage, it is probable that the water was supplied direct to them as well as to the houses of the more wealthy citizens and to the palaces. Iron pipes were not then known, and consequently they were obliged to carry the water on aqueducts, otherwise they had no mean of resisting the hydraulic pressure.

Besides these cisterns, one cannot make out distinctly any other remarkable buildings; but there are plenty of remains of foundations of walls, some of them of masonry and some of them of brickwork, showing great solidity; there are also great quantities of pottery and fragments of marble sculpture lying about. As to the celebrated arsenals and docks, it is still more difficult to point out satisfactorily their extent, form, and position; but from the accounts we find in ancient writers, and from the well-known recorded fact that the Carthaginians were a great commercial as well as warlike nation, it is evident that they must have possessed the means of building, sheltering, and repairing both classes of vessels, those for commerce and those for war. These docks and arsenals must have been on the sea-shore; the peninsula is composed of comparatively high land, and they could not have built them anywhere else, for the low lands which border the peninsula on the south side were not, so far as we can learn, included within the walls of the city, and it was not likely that they would have left such important establishments as these, upon which in a great measure their power depended, unprotected. I repeat, therefore, and I believe it is confirmed by most authorities, ancient as well as modern, that these docks and arsenals were on the sea-shore; and as they would not have built them on the northern side of the peninsula, which is so much exposed to the strong northerly gales, they must have placed them on the eastern shore, which is tolerably well protected by the opposite sides of the bay. Indeed, I walked round the northern shores of the peninsula and carefully examined them, and could find no traces of any works having been executed there; but upon the eastern shore I could discover traces of considerable works. The ships of those days were comparatively small and drew but little water, and by running out moles or breakwaters of loose stone into the sea, a sufficient space might have been enclosed to answer the required purposes. We know that the ancestors of the Carthaginians did this to a great extent at Tyre, and we can have no reason to doubt that they adopted the same system at Carthage. This is a question still open to discussion; but I think, after what has been stated, that the arguments are in favour of the eastern shore; and until more decisive remains have been found elsewhere, I must adhere to my conclusion, for we must not forget that this is the weather shore, where all such works should be carried out, so as to afford the greatest facility for egress and ingress.

As the Bey was not then in Tunis, I determined to make use of the interval by going to see the remains of the ancient city of Utica, about twenty miles distant, west-north-west. My two companions were not very well, and therefore thought that they had better remain at Tunis, in case anything connected with our business should occur; so I determined to go alone, as I was told that I should meet with no difficulty, for the country was perfectly safe. I accordingly hired a carriage with three excellent horses, and engaged a clever Frank servant, an Italian, half Jew, half Mohammedan, who had lived many years in Tunis, and besides English, French, and Italian, spoke the Arabic very well also. He was a clean fellow, and was well recommended by our vice-consul. Being told that I should find nothing on the way, I took a good provision basket and plenty of cloaks. We started soon after noon over a wretched road, or rather open track made by the peasants' carts, and as the weather had been very wet the wheels of the carriage were frequently half-way up to the axles in mud. The country through which we passed was wild and lonely in the extreme, not a creature to be seen. After having driven about eight or ten miles we came to a kind of village, or cluster of about half-a-dozen mud huts whitewashed, where there was a sort of café of the roughest kind, and close by it there was a sort of château belonging to some aga or district chief, surrounded by trees and a rude wall, the whole having a most solitary and gloomy appearance. There we halted about half an hour to refresh the horses, which were tired enough, and at length we reached the caravanserai, a solitary building two stories high, surrounded by a high mud wall.

It would be difficult to conceive a wilder or more desolate spot. In front was an extensive marsh, half covered with water, through which the river pursued its devious course, the banks being covered with rushes; at the back lay the dreary country through which we had passed; to the right the marshy plain extended to the sea, some 10 miles distant, and to the left it stretched as far as the eye could reach, bounded by blue hills of considerable elevation. There was not a soul in the house but the kanghè or master, and another man and a boy. He welcomed me, however, very civilly, and showed me to the upper floor by an outside staircase; here I found two rooms with bare walls, brick floor, a trestle for a mattress, a wooden table, and a couple of rude chairs; to my great delight there was a chimney-place, in which I soon lighted a wood fire. The kanghè brought up a couple of tolerably clean mattresses, two oil lamps, and some bread, and water, which was all he had. This, however, was of no consequence, as I had come provided, and after a good supper, in spite of the loneliness of the place I slept soundly. As there, was no road any farther, it was necessary to provide horses to go to Utica the next day, which the kanghè said he would do. At daybreak we mounted, and were just about starting when we were joined by an aga, a fine handsome fellow, exceedingly well mounted. He was accompanied by two or three servants, also well mounted and armed. The aga saluted me very courteously, and said he was proud to see Englishmen, and that if I was fond of sporting he would be most happy to see me at his castle, and would show me some capital sport, wild boars, deer, partridges, quail, &c.; for this kind invitation I thanked him, but was obliged to decline on account of want of time. We rode together along the marshy plain, our horses at times being up to their knees in water, and crossed the river by a rough stone bridge, immediately beyond which we reached the high land which bounds the river on the west. Here we parted in the most courteous manner, and I continued along the left bank for two or three miles farther, when I reached the ruins of Utica, situated at the foot of the hills, at a height of about 30 feet above the level of the bottom of the valley, and about a mile distant from the left bank of the river, and 5 miles from the sea-shore, with nothing but a dreary flat marsh intervening. Utica was originally seated on the sea-shore, and was a port of considerable importance; the marshes that now exist have been gradually formed during the lapse of centuries; and in this there is nothing remarkable, as there are many similar examples, even to a much greater extent, at the mouths of rivers in different parts of the world.

Of the city nothing now remains but portions of the broken down walls, a square tower near the middle, and some remains of foundations scattered about. Nothing could be more desolate and solitary than the whole scene, which afforded a most striking contrast to its former magnificence; and reflecting upon its ancient compared with its present state, one could not help remarking, "Alas, poor human nature!"

On my return from Utica I found that His Highness the Bey had consented to receive us at his palace, distant about 12 miles from Tunis, on the following day at eleven o'clock. We accordingly started off about eight o'clock, and got there in good time, and were presented to the Bey—who was seated in great state and surrounded by his officers—by our consul, Mr. Wood, and were most graciously received. The Bey was about fifty-five, of the middle size, rather stout, with an open, frank, kind countenance; he conversed in Arabic with the consul in the most genial manner, said he was glad to see us, and was much pleased at the idea of having railways in his kingdom, about which he had heard so much; he said he would give us every assistance in his power, and hoped that this would lead to other European improvements, that would tend so much to the prosperity of his country. After about twenty minutes of very agreeable conversation we retired, very much pleased with his Highness's courteous reception. In front of the palace there was an encampment of about 2000 troops of all kinds, horse, foot, and artillery, whom the Bey was in the habit of reviewing daily.

The remainder of the story is soon told. We knew perfectly well that the French were very jealous of any foreign capital or enterprise being introduced into Tunis, that the consul was somewhat suspicious of us, and that he and his spies were always on the watch: we were warned of this, and advised to be very cautious. So well was our secret kept, that the French consul had come to the conclusion that we were nothing but a party of ordinary travellers, and was just getting rid of his suspicions. On the other hand, the Bey and his Divan had agreed to grant a most favourable concession, and there can be no doubt that the line would have succeeded admirably in every respect. But in an evil moment Messrs. Peto's agents discovered that they could not make a proper report to their employers without taking levels. In vain I represented the extreme imprudence of such a proceeding; that it was a mere surface line, without the slightest engineering difficulty, and which did not present the least obstacle; moreover, that the land was to be conceded free of all cost, and that, in short, there was nothing which men of their experience could not estimate accurately without taking the smallest measurement. They persisted in their intention, and, of course, the moment they got out their instruments the French consul understood the whole affair, and in an audience with the Bey threatened him with the vengeance of France if the concession were granted. The poor Bey had no alternative but to submit, and there the whole matter ended.

Before leaving Tunis I saw all the different sights; amongst others, serpent charming, which is quite a profession. The charmers display considerable dexterity; indeed, it is a profession of long standing. The northern portion of Africa, on account of its sandy and sultry climate, is peculiarly well adapted for the breeding of serpents of the most venomous kinds, and in consequence they attain an enormous size and secrete a large amount of poison. The bites of many of the species are fatal, and hence it has become for ages past the object of certain of the natives to study the habits and characteristics of these reptiles and the antidotes to cure any unfortunate individual who may by chance have been bitten by them; indeed, the Roman armies during their campaigns in this part of the world were always accompanied by a certain number of these serpent charmers, and the profession exists at the present day. You find them in Egypt and throughout the whole of the northern part of Africa, and they not only practise the cure of serpent bites, but also collect numbers of them for exhibition to the multitude, to show their power over them, and by this means realize a considerable amount of money. At Tunis I saw several of these serpent charmers, who were always surrounded by a large crowd. They appeared to have complete control over the reptiles; they always kept their eyes constantly upon them, and regulated their movements by a wand in their hands, whilst an attendant boy kept time by beating a small drum and blowing a pipe with a low monotonous sound.

I was much surprised to find the natives so very civilized; we walked about the town, in the suburbs, and in the country, without experiencing the smallest molestation or incivility. I was told that robberies were very rare, and that frequently large sums of money were sent by a single messenger on a dromedary from Tunis to Tripoli without the least danger of being robbed. There were scarcely any palm trees to be seen, but we had plenty of the finest dates, which were brought from the interior, I think Tafilet.

Upon the whole I was much pleased with Tunis. It is a country possessing great natural resources of all kinds, by which, if only properly developed, this kingdom might be again rendered very powerful, as it was in the time of the Carthaginians and Romans, when, it is said, it contained above ten millions of inhabitants, whereas at present there are little more than two millions.

CHAPTER IX.

Surveys at Odessa and Vienna—Harbour at Ponta Delgada—Ramsgate—Dagenham.

During the latter part of the year I had some correspondence with the municipality of Odessa about paving their streets and making a complete system of sewers for the city; and in the spring of the following year I was requested by the municipality to come to Odessa, and to confer with them as to the best means of carrying these works into effect, and the cost of doing so.

I accordingly started for that city; but when I arrived at Vienna I could get no idea as to when the steamer would leave for Odessa. So to pass away the time I determined to visit the Sömmering Mountain, over which the Imperial Elisabeth Railway passes on the great line from Vienna to Trieste. As that part of the line which crosses the Sömmering Mountain was said to be a great feat of engineering, we determined to visit it, and certainly it was a very creditable performance. The inclines were very steep, and the curves very sharp and very numerous. Galleries were cut through the rock, high embankments made, and bridges thrown across the ravines, and the railway, having to wind round the sides of the mountain, was very tortuous. The works, however, upon the whole were not badly designed or executed, though I will not say whether a better line might not have been found. The engines used were of the most powerful kind.

Having returned to Vienna we descended the Danube and reached the Sulina mouth. This mouth of the Danube was selected as the most capable of improvement, and considerable sums of money have been expended upon it, under the direction of Sir Charles Hartley, an English engineer of considerable reputation. The works designed by him consisted of two embankments or moles carried out from the shore, one on each side, and nearly parallel to each other, in an E.S.E. direction, for a considerable distance seaward, having a good opening between them for the entrance. These moles, by keeping the current of the river to one course, enable it to act more effectually in deepening the channel and lowering the bar. The operations of the current are assisted by dredging. The works, I understand, have been very successful so far, although by no means completed. An increased depth over the bar and in the channel has already been obtained, which is a very great improvement, for on account of the continual shifting of the channel, and its shallow and tortuous course, it was at all times uncertain and frequently very dangerous. When we passed, the new channel had not been completed; the captain of our steamer therefore thought it advisable to come to anchor for the night and wait until the following morning. There is a wretched place called the town of Sulina, on the right bank of the river, near the commencement of the moles; it consists of wooden houses, stores, shops, and cafés, scarcely two feet above the level of the water, and surrounded by marshes, that send forth the most pestiferous malaria. I was told that fevers constantly prevail there, and I am not surprised at it, for a more abominable, uninviting place I have seldom seen. Next morning we started for Odessa soon after daybreak, and after a tolerably smooth passage reached the western harbour at about four o'clock in the afternoon.

The view of Odessa, at about two or three miles' distance, as you approach it from the sea, is rather pretty and imposing. It is situated upon a calcareous sand-stone cliff, about 40 feet above the sea, with a very fine row of majestic stone buildings running the whole length from north to south, having a spacious road, terrace, and garden between them and the edge of the cliff. The town behind is for the most part well and regularly laid out, with wide, spacious streets, at right angles to each other, and some handsome shops and buildings, the residences of tradesmen and merchants. A great number of wealthy nobles and landowners reside here in winter, and houses worthy of the name of palaces, such as those of the elegant and high-born Countess of Urrenzoff, the Governor-General Prince Manukebè, Count Tolstoy, Mr. Maas the great banker, and numerous others. The surrounding country, although the soil is rich, is open, with very few trees, and has therefore a bleak, uncomfortable appearance. The custom-house officers were very civil, to my surprise, and gave us very little trouble, so we soon got permission to land, and immediately went to the Hôtel de Londres, a very fine extensive building, situated in the grand row of buildings already mentioned facing the sea. Here we obtained handsome, spacious, well-furnished rooms, and lived very comfortably at a moderate charge.

The next day I called upon the Governor, Count Strogonoff; the Mayor, Count Tolstoy; the English Consul, Mr. Grenville Murray, and several members of the municipality. Having paid these formal visits of ceremony, at which I was courteously received, I immediately began to inquire into the best mode of paving and draining the city, where the best materials were to be obtained and their prices; in fact, everything connected with them.

Before proceeding further it may be proper to describe the state of the place as regards paving and sewers, which may be summed up in a few words. There was neither one nor the other; and it is difficult to conceive how such a fine and wealthy city could have been built, or could have existed so long without them. First, with regard to the streets. The soil is composed chiefly of sand mixed with clay, which during fine dry weather makes a tolerably good road, but the moment it becomes saturated with water, which is the case for a considerable portion of the year, it is converted into one vast puddle, and the large and constant traffic cuts it up into deep holes and gullies, so that in a comparatively short time the road is a sea of mud, and almost impassable. During my stay there were only a few days' rain, but even this gave me a tolerable idea of it, so that I could easily believe the account given to me by the authorities and other inhabitants was by no means exaggerated. During the worst state of the streets, the obstruction of bullocks and horses trying to drag the loaded waggons and carriages was something dreadful, and many a waggon was left irrevocably fixed in the mud with numerous carcasses of horses and bullocks lying beside it. It was a singular thing that nearly all the corn warehouses were in the upper parts of the town, about a mile from the harbour where the corn was to be shipped, instead of being close by. During the wet season it costs as much to get the corn from the granaries to the harbour as to take it from Odessa to London. The butchers' shops were at the upper end of the town, from half to three quarters of a mile from the eastern part, where all the principal people live, and when provisions were required for the family they were obliged to hire a carriage with three or four horses to get them; and unless a family kept a regular store of provisions they ran a great risk of being starved. The roads in the country round are not a bit better than those in Odessa; during wet weather they are almost impassable. About twenty-eight miles from Odessa, at a place on the River Dneister, is a kind of depôt for the vast quantities of corn brought down from the interior. Immense heaps or hills of corn were lying there when I visited it. In wet weather these cannot be removed, and I was informed that a great deal was burnt or allowed to rot because it could not be taken away, either on account of the badness of the roads or the dangerous state of the bar at the entrance of the river.

The sewers were very simple. Gullies had been made along each side of the streets, into which all the filth was thrown, so that in dry seasons it accumulated there, creating the most offensive effluvia, and in wet weather it would not run off, on account of the gullies being blocked up with mud. It was impossible to find a city in a worse state, and it was astonishing that such a great, wealthy, and luxurious city could have so long existed in such a condition. As there was no stone in the neighbourhood fit for paving the streets, granite or a similar hard stone being the only kind fit for the purpose, the next question was where it was to be found, how to get it, and the cost of doing so. Upon inquiry, I heard that excellent granite might be obtained in any quantity from a quarry situated on the river Bug, and on proceeding there I found that very good stone might be got with great ease, and at a comparatively trifling cost. Having made my report, I returned to England.

In 1862 I was appointed chairman of the Civil Engineering Department of the International Exhibition, assisted by the Marquis of Salisbury, M. Bommart, M. Koch, of Berlin; M. Lelere, Belgium; M. Loehr, Austria; Cesare Valerio, Italy; the Baron Baude, M. Mille, Mr. C. Manby, Mr. Kelk, and Mr. Page.

Our report, I believe, was entirely satisfactory.

After this I was asked to examine the water supply of Vienna, and accordingly, having reached that city, I turned my attention to the following objects:

First, to the mode there adopted for supplying the water; this I found to be by means of steam pumping engines of inadequate power, which forced the water through iron pipes to fountains in the different streets, whence it was obtained and delivered into the houses by carts and carriers. The supply was not enough for the wants of the town, neither was the water sufficiently pure nor properly filtered. It contained a good deal of vegetable matter at the best of times, as the water was admitted through the porous soil adjoining the river into very small reservoirs, and it had no time to deposit the alluvial matter with which it was charged before it was delivered for use. Moreover, this method of supplying the water by means of steam pumping engines was a constant expense, and the more water that was required the greater would the expense be. I found, also, that an English party had proposed to extend and amplify the existing system, by erecting more powerful engines, and by making receiving and filtering reservoirs upon a much larger scale about three or four miles higher up the Danube, on the same side, where the water was clearer and more free from the sewage of the city. Now, when I considered that the population was even then between five and six hundred thousand, and that it was daily increasing, this pumping system appeared to me to be the worst plan possible to effect the desired object, unless no other means could be found. I therefore determined to explore the environs of the city, as I felt convinced, from the geographical features of the neighbourhood, that there must be numerous streams amply supplied with water, with their beds sufficiently elevated, and with reservoir room to any extent, to afford, by gravitation, an abundant supply of the best water to Vienna, not only for the present number of inhabitants, but for three millions and upwards. In other words, I proposed to conduct the water from some of these sources in a covered aqueduct, simply by its own natural inclination, to a reservoir situated above the tops of the loftiest houses, in the highest part of the city; thus all pumping would be done away with, and a vast yearly expenditure would be saved; the first cost of these works would not exceed the first cost of the extended plan on the old system above mentioned, as proposed to be made higher up the Danube. Further investigation completely established the correctness of my opinion.

On my return I took the railway to Trieste, passing by Baden, Neustadt, up the line or valley of the Leitha, as far as the base of the Sömmering ridge. Here I first visited the Fischa Dagnitz river, one of the tributaries of the Leitha; it is a splendid stream, about twenty feet wide, three to four feet deep, the water as clear as crystal, and flowing over a gravelly bottom. During the height of summer the temperature seldom exceeds 40 degrees of Fahrenheit. It is impossible to imagine a finer stream; it has been analyzed and highly approved by some of the first chemists, who pronounce it to be exceedingly pure. This water could be conveyed by gravitation through a covered conduit to the top of a hill overlooking Vienna, and from thence it could be delivered to the highest parts of the city in ample quantity, without pumping or filtration. The water of the Danube, even after filtration, cannot be compared with it. The Fischa Dagnitz turned a number of mills in its course, and this was the only objection to taking the water. But this was a loss that could easily be compensated for by making reservoirs, to be supplied by the surplus waters of this and other streams. The Fischa Dagnitz, therefore, appeared to me to be decidedly the best source for supplying Vienna with pure water, for any reasonable number of inhabitants. I next examined the Leitha above the Dagnitz, and here I found that there was a superabundant supply, although the water was by no means so good as that of the Dagnitz. I examined several other tributaries to the Leitha, passing over an extensive mass of *débris*, called the Steinfeld, where I found that by damming some large reservoirs could be made, from which an ample supply could be obtained of the same quality as that of the Leitha, although by no means so good as that of the Fischa Dagnitz. It was quite clear, therefore, that plenty of water could be obtained for Vienna now and for all future time from this quarter. But not wishing to overlook any source from whence a supply of water could be procured, I took the railway from Vienna to St. Polten, a town on the road to Linz.

Here I found a very fine and copious stream, though when I examined it, it was nearly in its lowest state. The ground between it and Vienna was very high, so that to a certain extent pumping must have been resorted to, if this source of supply was adopted. Another objection was the much greater distance as compared with the streams up the valley of the Leitha. I felt therefore satisfied that it would not do, and that the question lay between the Fischa Dagnitz and the tributaries of the Leitha on the Steinfeld above mentioned. All these observations I embodied in a report to the municipality of Vienna, and recommended the Fischa Dagnitz; for although there might be a few feet less fall, still, taking into account the far greater purity of the water, it was the best. The municipality received my report, and returned me their thanks for it. At all events, I had decided the question against taking the water from the Danube.

Having made these investigations and sent in my report upon them, it now remained for the municipality to decide. They said they would take time to consider, and accordingly appointed some of the most able and scientific officers of their own body, as well as those attached to the Government, to investigate the subject further. After above two years' examination these persons made their report, recommending that the principle of gravitation as proposed by me should be adopted, and that the supply should be taken from the tributaries of the Leitha in the Steinfeld: this was in fact adopting my plan, although I preferred the Fischa Dagnitz as the source of supply, as none could or did dispute the superior quality of its water to all others.

The municipality have been deliberating ever since on the best plan of carrying this great work into effect, whether by a private company, or whether they shall execute the works by contract and supply the water at their own cost. This question, as far as I know, has not been settled, although it has been now nearly six years in agitation; meanwhile the city suffers materially from the want of a good supply of pure wholesome water.

In the spring of 1861 a Mr. Parkenscholz, and M. José de Conté, a member of one of the wealthiest and most respectable families of St. Michael's, the principal island of the Azores, called and informed me that the Portuguese Government had decided to make a harbour at Ponta Delgada, the chief town of St. Michael's, the cost of which was estimated at the sum of 134,000*l.* I replied that I should be willing to undertake the superintendence and construction of this harbour, provided that I was not compelled to serve them beyond the term of four years, and that I was not to be responsible either for the plan or for the amount estimated to complete it. To these terms the Portuguese Government consented, and I started for the Azores in September, 1861.

We reached Ponta Delgada on the 20th September, and were very much delighted with our first view of it. The town rose rapidly from the sea, and presented a most interesting appearance. The spires of numerous churches, starting up from the level of the surrounding houses, pierced the blue sky, while here and there were gardens filled with the gayest of flowers, and groves of orange, lemon, and olive trees, the whole embosomed in a picturesque bay, backed by evergreen conical hills, reminding one a good deal of Naples.

On landing I was received with great ceremony, and was waited on by the Junta at my hotel. The next day I attended a meeting of the Junta at the Governor's house, when we discussed all the various preliminary operations that were necessary preparatory to commencing the harbour.

It appeared desirable that the first stone should be laid before I left the island, and preparations were ordered to be made accordingly. I gave Mr. Plews full instructions as to what was necessary for this, and having two days to spare I determined to accompany my friend, Mr. Thomas Ivens, on an excursion to Furness, a celebrated watering-place, situated near the eastern end of the island, about 27 miles distant. We started on a couple of good donkeys, with another carrying some provisions, and proceeded along the south shore over a very fair carriage road, for about five miles, through some neat villages embowered in orange and lemon orchards, passing also by comfortable villas and country houses, and then struck in a north-easterly direction across the island over an undulating well-cultivated country, chiefly growing Indian corn and other cereals and green crops. When we reached the summit of this part of the island we got a good view of the mountains to the eastward and of the sea on the north and south sides. We halted for about an hour, and then proceeded to a large town situated on the northern shore, surrounded by rich gardens and fields, having every appearance of prosperity. From thence we proceeded along the north shore over some very steep hills and cliffs overlooking the sea, the road still good, until we got to a little roadside inn about dark. There was, however, light enough from the stars to find our way, so off we set again, and going over a very wild hilly country got to the top of the pass which descended to Furness about ten o'clock at night. Here the road ceased, and we found it necessary to get off the donkeys and walk, for fear of being tumbled over their heads, although they were very sure-footed. At a subsequent visit to this place I crossed over the pass by daylight, and the view from the summit is very fine, looking into what was formerly the crater of a volcano, which is nearly two miles wide, surrounded by lofty rugged hills between 2000 and 3000 feet high, the bottom of the crater valley being now covered with rich verdure and gardens and tropical trees and plants of various kinds. Near the crater was the little village of Furness, with its church and white houses dotted about, and close by were the sulphurous hot baths sending forth volumes of steam, the whole forming one of the most picturesque and agreeable scenes imaginable. We got to the inn at the bottom of the valley at about 12 P.M., thoroughly tired, having ridden and walked for nearly eleven hours. The inn, considering all things, was by no means bad. I got a good bed and supper, and soon fell asleep, while my companion, Mr. Ivens, went to his family, who were stopping there. After a sound night's sleep I got up in the morning and took a warm sulphur bath as hot as I could stand it, which was about 96°; but you may have the water at almost any temperature, as it issues boiling from the spring. Close by, within a few yards, is a cold chalybeate spring, and not far distant is a vast extent of rich alum deposit, from which great quantities may be extracted, and a manufactory had then commenced operations. I walked about until midday, quite enchanted with the beauty of the place, dined with my friend Ivens and his amiable family, and about one o'clock in the afternoon mounted my donkey and started off with the guide for Ponta Delgada, being determined if possible to get there that night. We had a tolerably stiff hilly road fit only for mules, donkeys, or horses, before we could get out of the valley. Having ascended the summit of the first pass, we had a delightful view behind over the delicious valley of Furness which we had just left. We then came to a rather extensive lake, surrounded by evergreen hills of much less elevation than those surrounding the valley; and on the opposite side of this lake, which was two to three miles long and about a mile wide, we observed the country house of our Consul, Samuel Vines, Esq., seated on the side of a hill about 200 or 300 feet above the lake, embosomed in woods; and at the foot of the hill, close by the water's edge, was a strong sulphur spring, of the same temperature as those of Furness, and like them covered with clouds of steam. There was not a house near it all round the lake. It was singular that he should have chosen such a solitary spot. I continued along the south and west side of the lake, which is bounded by hills covered with underwood and evergreens, but not a house nor even a shed to be seen; only a few solitary cattle here and there with a shepherd boy. Nothing could be more still or lonely; but at the same time there was a degree of quiet and repose which gave to the place a certain undefinable charm not to be resisted.

Upon leaving the west end of the lake we ascended the hill over a rugged path, passing through a wild, bare district, and from the summit enjoyed a fine view of nearly the whole of the island, which was very beautiful whichever way you looked. We now descended a very steep path, the view changing at every turn. At last, about five o'clock, we reached the clean, pretty town of Villa Franca, where I halted nearly an hour for refreshment.

Villa Franca, which is situated in a small bay on the sea-shore, was formerly the capital of the island, but an earthquake having occurred near, it was abandoned for Delgada. It is still a thriving little place, with a rich surrounding country. There is a small island, a few hundred yards from the shore, where a good harbour might be made.

After leaving this town we had to take a rough road along the sea-coast for a couple of miles, partly through deep sand, and partly among scattered rocks, for there was no regular road. We then left the shore and travelled over the cliffs by an equally bad path, sometimes over deep chasms and sometimes up narrow glens, until we reached the high road again, at which I was very glad, as it was now dark and very difficult to find our way. We pushed forward with confidence, and the donkeys went on very well; after passing through numerous villages, sometimes lying on the sea-shore, at others a little distance inland, at last, much to my satisfaction—for I was very tired—we reached Mr. Rodrigue's comfortable hotel about midnight. Old Rodrigue was surprised that I had made the journey of between 50 and 60 miles in so short a time, for although it was the month of September, the sun was very powerful, and the road for many miles was very bad.

Next day we had another meeting of the Junta, and they made all the arrangements for laying the first stone of the new harbour, which was to take place with every possible ceremony. I found that Mr. Plews had got a couple of large stones well dressed for the purpose; the captain of the port had provided the sheers and tackle for hoisting them; and the Junta had procured in the town a very pretty silver trowel, a mallet, and mortar holder. A commodious gangway had been prepared from the shore to the west end of the old mole, fronting the area where the new pier or mole was to commence, according to the plan approved of by the Government.

The whole town was in motion at an early hour, and great numbers of people came in from the neighbouring towns and villages, all dressed in holiday costume. The town was decorated with the flags of various nations, amongst which the Union Jack was particularly conspicuous. In fact, it was considered a great national fête; the more so, as a work such as this was intended to be had never been previously undertaken in this island, or in the kingdom of Portugal itself. The procession was marshalled at the Town Hall, and consisted of the band of the militia of the island; then the governor, his secretary, and the Junta or committee that was to conduct the work, followed by the principal officers, merchants, and deputations from the chief towns in the island, with their respective banners, closed by a number of the most respectable inhabitants of the place; the lower orders, clad in their best, lining the way by which we passed. Upon arriving at the place we were met by the chief priest of the island, who, in a short prayer, invoked a blessing from the Almighty that the enterprise might prosper. Then, upon a signal being given, coins of the realm, together with a printed paper, containing an account of the proposed work, the names of the governor and Junta, the engineers, officers, &c., were placed in a glass bottle, and deposited in the cavity of the lower stone, which had previously been prepared and set. The governor having placed the glass case in the cavity, I handed to him the silver trowel, with which he spread out the mortar. The stone was lowered into its place, and the governor, having previously adjusted it, gave the usual three taps with the mallet, and the ceremony was finished with a discharge of guns from the fort, and numerous showers of rockets from the town, amidst the cheers and vivas of the bystanders, the band playing the national anthem. In the evening a very handsome entertainment was given at M. José de Conté's villa, on the outside of the town, to which the Junta and principal officers and merchants of the place were invited. The Portuguese band played admirably during and after dinner, and we all retired much pleased with the success of the day's proceedings. The governor presented me with the silver trowel, which I respectfully declined, and requested him to keep it as a memorial of the happy day; which he, after some hesitation, accepted, and I contented myself with the polished mortar holder, and the next leading man of the Junta accepted the mallet.

The next day I made final arrangements for my departure by the packet, which was expected on its return from Fayal the day after. In the mean time I had been considering the danger of the harbour as laid down or rather approved by the Government, and found that if the west mole was commenced at the east end of

Fort San Bray, as proposed, the fort would be exposed to a much heavier swell; but that by making it commence at the western end of the fort this would be avoided; moreover, the sum of 600*l.* would be saved, and the harbour would be made so much larger. This recommendation was afterwards adopted by the Government. I now took my leave of the governor and all the authorities, and my other friends, with my grateful thanks for their kindness and attention.

The island, taken in a direct line from north to south, is about 40 miles long, and from 7 to 9 miles wide. It is entirely volcanic. The east and west ends exhibit the most powerful effects of the volcanic force. In the former we see mountains raised to the height of about 3300 feet above the level of the Atlantic, in the centre of which lies the valley of Furness, the bottom being occupied by a lake that still sends forth sulphurous vapours; and in the latter or west end we find mountains of about 3000 feet, the centre of which is occupied by a large lake, without any exhibition of existing volcanic action. Near the centre of the island, which is the narrowest part, there are numerous minor conical-shaped hills of less elevation, but all more or less showing their volcanic origin. The island enjoys a most genial climate, and frost or snow is of rare occurrence; but during the autumn and winter it is visited by heavy gales from all quarters of the compass, which extend over a distance of about 100 miles. During this period a great deal of rain falls, and the climate may be said to be moist, much resembling that of Madeira. Formerly a good deal of wine was produced here, but since 1855, when the oidium disease made its appearance, the vintage has been very unsatisfactory; and although a certain quantity of wine, resembling that of Madeira, is still made for home consumption, none is exported. The principal productions of the island now are oranges and lemons, of which vast quantities are annually exported to the north of Europe. The district where these are produced is restricted to the centre of the island, commencing at Ponta Delgada, and extending eastward about 7 or 8 miles. On the north it is bounded by the central ridge of hills, and on the south by the sea, the width being about 3 or 4 miles, so that it enjoys the full rays of the southern sun. These orange and lemon orchards are cultivated with the greatest care, and wherever they are exposed to the east, west, or southern gales are protected by high stone walls. On the north the hills alone afford sufficient shelter.

Towards the latter end of October the season commences, and continues until about the end of February, during which time the harbour of Ponta Delgada is continually crowded with shipping, whilst on shore the inhabitants are busy packing the fruit in boxes; these boxes are made from the wood brought chiefly from the forests that clothe the mountains at the eastern end of the island. The vessels in which the fruit is exported are principally small schooners, built rather short, capital sea boats, and manned by the best of captains and sailors, who thoroughly understand their profession. They go to sea in any weather, which is at times most severe, rarely if ever meet with any accidents on the voyage, and make the passage to England in from eight to ten days. It is true they are sometimes driven ashore when they break from their moorings in the exposed roadstead of Ponta Delgada during heavy gales from the south-west to the south-east, to which it is exposed; but this will be obviated by the new harbour, and was one of the objects of its being made. Sometimes, during the prevalence of these gales, they are obliged to slip their anchors, and run for shelter to the northern side of the island, where they occasionally ship their cargoes, which is always done by means of lighters from the shore. Immediately they are laden they start with the first fair wind, however strongly it blows; they never wait for weather, but as soon as laden they put to sea, and generally make good, indeed, the best of passages.

I met a young botanist who had come out in one of these vessels to Ponta Delgada, for the purpose of making botanical researches in this and in the neighbouring islands, in the month of March; they had an excellent passage until they got within about a hundred miles of St. Michael's, when he said to the captain, "We shall be there to-morrow." The captain, an excellent sailor, looking at the signs of the weather, replied, "I don't think so. We are going to have a hard gale from the southward." He immediately ordered his mate to well batten down the hatchways, shorten sail, and make all as snug as possible. The captain was right; the gale from the south came a few hours afterwards, with a very heavy sea.

The young botanist frankly confessed to me that he began to be terribly afraid at seeing the tremendous sea running after them, and asked the captain if there was any danger; when the captain coolly replied, "Never fear: it is only a little loss of time. Go to your berth, and lie there quietly. We shall get there safe enough, with a little patience." The captain then ordered the vessel to be hove-to, and there she lay as comfortably as possible, never shipping a single sea, although the waves were running mountains high. After about five or six days she entered the bay of Ponta Delgada without having sustained the least damage. She got her cargo of fruit aboard and returned directly, and made one of the quickest passages that season to England.

Besides oranges and lemons, St. Michael's grows large quantities of Indian corn, wheat, barley, potatoes, and other articles, of which she exports largely to Portugal, and is considered the granary of the kingdom; her exports are much larger than her imports, and she receives back in return wine, oil, and manufactured goods. In fact, St. Michael's is looked upon by the Portuguese as their most productive and wealthiest island. The population is extremely civil, hard-working, and industrious, and the upper classes are wealthy, enterprising, and energetic, and send some of the most talented deputies to the Cortes, who by their ability and perseverance attain the highest offices in the State; for example, the late Minister of Finance, Senhor Avila, who, although a rough subject, was yet possessed of great talents and integrity. The island also furnishes some very fine hardy sailors and soldiers; and I was told that the other islands of Pico, Fayal, and St. Mary's, although not equal in extent or wealth, do the same.

On the 9th of October (1863) I again left London for Lisbon by one of the Royal Mail steamers, and reached it on the morning of the 13th following, and on the 15th, in the evening, I started in the packet for Ponta Delgada, which we reached on the morning of the 20th, after an agreeable and tolerably smooth passage. I found that some considerable progress had been made with the preliminary works; although these were not so far advanced as they might have been, in consequence of the Junta not having found the necessary funds. I had previously explained to them in written reports, that the more complete the preliminary works were made, and the larger the scale on which the operations were conducted, the sooner the harbour would be completed, and the greater would be the economy. I have already observed that the Government had approved of a certain plan for the harbour without consulting me, and that it had simply confided to me the charge of carrying the design into effect. However, I felt it my duty to consider the plan more maturely, so as to ascertain how far it was likely to answer the object intended; as, for example, whether the estimate made was sufficient for the purpose, and whether it could be done within the time stated. I could not at first do this, because I had not had time sufficient to investigate the local circumstances; however, upon my second visit, the experience of my former one, combined with the observations which Mr. Plews had made in the meantime, enabled me to master the subject; and upon carefully considering the plan adopted by the Government, and comparing it with the local experience which I had now obtained, I felt convinced that the design would not effect the object proposed. Vessels would neither be able to enter nor depart during the most dangerous and prevalent winds, without the risk of being shipwrecked; neither could those vessels lying in the harbour be considered safe. The plan was deficient in all the qualities necessary for a good harbour; as regards the expense, it would cost at least more than double the estimate; and as to the time, it would be extremely difficult to state when the works would be completed. Having clearly satisfied myself upon these points, I felt it my duty to inform the Junta, so that they might report the same to the Government. The Junta received my remarks very cautiously, and said that several objections to the plan adopted had occurred to them; however, they did not pretend to give any opinion upon the subject, and they requested me to make a full report upon the plan adopted, together with all my objections, and the cost of carrying it into effect. They also requested me to prepare a new plan, according to what I conceived best adapted to the local circumstances; also an estimate of the cost of carrying it into effect; and said that as soon as they received them they would send them to the Government, and would communicate to me their decision as early as possible. This I accordingly did upon my return to England.

This report was submitted by the Junta to the Government, and it was decided that the plan I proposed was the best, and the Government ordered it to be carried into effect. This decision was very gratifying to me, for I was strongly convinced that I was right, and if the Government had decided otherwise I felt that there was no alternative but to resign my situation. The works were therefore ordered to be proceeded with according to the new plan that I recommended. The Junta before I departed arranged with me that the contract for my services should be limited to five years from 1861, although I was previously informed that it should only be for four years, the sum for my remuneration being the same for five as for four years; this was certainly a loss to me, but I did not wish to make any difficulty about it, as I was anxious that my plan should be adopted.

My design consisted simply of one mole or breakwater carried from the west side of Fort St. Braz, and in such a direction that no eastern pier would be required, as the opposite shore of the bay would answer that purpose. The mole consists of two arms, one at the shore end, 2000 feet long, and the outer end from 800 to 1000 feet long, with ample depth within from 40 to 50 feet, and covering a water space of nearly double the extent of the old plan. This new mole when finished will have a strong promenade stone parapet 20 feet above the level of high water, and a roadway below 40 feet wide, lined by a quay wall on the inside, alongside of which the largest vessels may approach and take in and deliver their cargoes at all times. Railways will be laid along the quays, and cranes worked by steam will travel along them for loading and unloading the vessels.

This mole is now advanced outwards about 1600 feet, and if the Junta had only followed my advice, it ought to have been finished by this time. I always calculated that after the first year, when the whole of the works were in full operation, from 1000 to 1200 tons of stone should be daily deposited, whereas, upon an average they have not done half that quantity, in consequence of their not employing sufficient plant in the shape of waggons, trucks, cranes, locomotives, tools, &c.; however, that is their fault, not ours. I have constantly made reports pointing out these deficiencies.

The quarries having now been opened, railways laid, and a sufficient number of locomotives, waggons, and trucks having been provided for the present, I proposed to the Junta that they should commence depositing stone on the line of the great mole. The laying of the first, as mentioned before, was simply a matter of ceremony, as they were really not then in a position to commence the actual work. The Junta approved of my proposal, and accordingly, two days before my departure, this ceremony took place. About six waggons laden with blocks from two to five tons weight were drawn by one of the locomotive engines from the quarries to the end of the staging or platform in the line of the mole, and were there deposited with great *éclat*. The locomotive then returned with the empty waggons and brought six more, which were deposited in the same manner, and this operation continued throughout the day. As this mode of conducting the harbour works had never been before seen on the island, it created much interest and astonishment amongst the natives, and the ladies were particularly amused by taking a ride upon the locomotive engines. I simply observe once for all that this mole was to be constructed by depositing, from open staging in the line of the mole, blocks of rough stone varying from a quarter of a ton to 10 tons in weight, when they could be obtained. These blocks being deposited in the sea, the waves would soon drive and consolidate them together, until after a time the mass becomes immovable; in fact, the sea is the workman or mason to arrange the stone deposited in such a manner that it shall become fixed in its place; therefore, during the operation the more and heavier the storms the better, the great point being that the stone shall be carried out and deposited in such masses that the sea shall not break through it, but merely act upon it, by drawing down the exterior or sea slope to such an angle that it will stand after the heaviest storms. Now my father, who commenced this system at the breakwater in Plymouth Sound, and in other places, found that the sea slope of a mole or breakwater constructed in this manner would stand an inclination of about 5 or 5½ to 1 for every foot perpendicular, and 1 to 1 on the land side, as I have mentioned in a former chapter.

Throughout all my experience I have found the same, therefore the breakwater at Ponta Delgada was founded upon this principle. Five lines of railway of the 7-feet gauge have been carried out upon the staging, so that the top has a width of fully 50 feet; and as the works proceed outwards another will perhaps be added, if circumstances require it, which will make the width 70 feet. By keeping up a constant supply of stone, there will always be sufficient for the waves to act upon until the sea slope has attained its ultimate point of repose. As fast as the large blocks of stone are deposited and washed into their place, great quantities of quarry rubbish are supplied to fill up the minor interstices and render the whole mass more solid, until the slopes are in a fit state to be regularly formed and paved for receiving the parapet.

For some years there had been constant complaint from merchants and ship-owners that they were taxed for Ramsgate harbour when their vessels never did or could use it. These continued complaints, so often repeated, at last had effect, and Parliament decided, in the year 1861, that the passing toll of Ramsgate harbour should cease, and that only those vessels that used it should pay, according to a certain tariff. The trustees under whose direction the harbour had been made and maintained, complained to the Government that without the passing toll they did not see from whence funds could be derived to maintain it in a proper state of efficiency, and therefore they requested to be relieved of their responsibility, and tendered their resignations, which were accepted by the Government, and an Act of Parliament was passed in the year 1862 relieving the old trustees from their trust, repealing their Acts of Parliament, and vesting the harbour, all its funds, and responsibilities, in the Board of Trade. I succeeded my father in December 1821 as engineer-in-chief to the harbour, at the same salary, namely, 210*l.* per annum, which included travelling and office expenses of every kind, the trustees paying the salary of the resident engineer themselves. The harbour was in a very dilapidated state, in fact, it almost required rebuilding, when it came under my direction, and monthly visits of two and three days were necessary, besides attending the Board in London once every fortnight examining accounts, correspondence, reports, and plans, all of which required a good deal of labour and responsibility, and which, if paid for according to the usual professional scale, would have amounted to at least treble the sum of 200*l.*, or more; but considering the appointment to be permanent, and that upon retirement I should be entitled to an adequate pension according to my years of service, I thought it better not to decline. In the year 1822, a committee was appointed by the House of Commons to investigate Ramsgate harbour and everything connected with it. Mr. Wallace, afterwards Lord Wallace, was chairman of this committee; he made a searching inquiry, and found nothing wrong. I was examined at great length touching all works, the mode of managing them, and what would be the total cost of doing so. The cost it was extremely difficult if not impossible to state, for the greater part of the works were under water, and failures and accidents were constantly occurring without any previous warning, so that what was sound one day was in ruin the next.

As an example, I may mention the west pier-head: this upon examination carefully from above appeared quite sound; not a crack could be seen above low water. It is true that the projecting basement floor of the lighthouse, which was not properly connected with the lighthouse tower itself, and was little more than a mere shed, showed a slight subsidence, but this outer part of the lighthouse was in no way connected with the outer walls of the pier head; in fact, it rested upon the chalk filling between the outer walls and the lighthouse. Now this chalk filling might have subsided, by some leaks through the outer wall of the pier head, without denoting any serious defects in the pier-head walls; however the walls fell, with little or no warning, and the consequence was that it was necessary to rebuild the whole pier-head, and the lighthouse also, at a cost of nearly 20,000*l*. To a certain extent the same thing occurred at the east pier-head; but this I observed in time, and completely secured it at a cost of 2000*l*. The whole of the inner walls of the east and west piers were completely undermined, although from above they showed no signs of failure, and I was obliged to underpin them to a depth of from 10 to 12 feet below low water of spring tides. The old wooden sluices were worn out; and these it was necessary to replace. The main entrance from the harbour to the basin was in such a dilapidated state that it was necessary to take it down and rebuild it; this cost 18,000*l*. The quay walls of the inner basin were fast going to decay; and I was obliged to take them down and rebuild them. All the filling in between the outer and inner walls of the outer harbour had sunk in numerous places, and it was unsafe for any person or carriage to go over it; it was therefore imperative to take the whole of this out and refill it with proper concrete. The pavement also was completely worn out, so that we had to renew it. There was no means of supplying the ships with fresh water, or of extinguishing fires. It was therefore necessary to lay pipes with stop cocks and hoses round the piers and basin, connected with the main water-pipes in the town. The whole of the sewage of the town was discharged into the basin, and at low water it created a most offensive effluvia, which rendered this part of the town unhealthy; I therefore recommended that an intercepting sewer should be made round the basin, so that all the sewage should be diverted from the town and harbour into the sea, to the westward, and by this means it was carried away by the tide, and a great nuisance taken away from the town. There was no regular tide gauge to ascertain the exact rise and fall of the tide, so that the harbour light at night could not be shown at the proper time, neither could the day signal be hoisted to show when there was sufficient depth at the entrance. This defect I remedied by establishing a self-acting tide gauge in a well within the lighthouse at the west pier-head. The pole of this tide gauge was connected with a cylinder and a clock hand; on the cylinder was a roll of paper, and to the hand of the clock was attached a pencil, which, as the rod or tide gauge rose and fell, marked it on the paper; thus the rise and fall of every tide was indicated upon the paper, and the clock showed the time, so that the rise and fall of every tide was regularly registered and kept in a book from year to year. I also established barometers at different parts of the harbour, under lock and key, the latter of which was kept by the harbour-master; these barometers were set every morning, and the rise and fall was registered in a book, so that all the captains of vessels in the harbour could ascertain as nearly as practicable the state of the weather; rain gauges were also established, a regular account of the rainfall being registered. Before my time the trustees had got an admirable time clock, by Moore, for which they paid 200*l.*, and Mr. Turner, one of the chairmen of the trustees, got another clock from Dent's, which cost 105*l.* These two clocks were kept in repair by a competent person, and corresponded to a second with the Royal Observatory time at Greenwich, hence every captain of a vessel starting from Ramsgate could carry the correct time with him. Thus Ramsgate was provided with all the instruments for ascertaining the correct time, the state of the tides, together with the barometer, thermometer, the wind and rain gauges, and everything necessary to indicate the probable state of the weather.

All harbours ought to be provided with these instruments, and a regular journal should be kept, forming a careful record of the observations made from them. After the harbour was transferred to the Board of Trade I still continued as principal engineer, at the same salary, viz. 315*l.* per annum and travelling expenses.

In the middle of the last century a breach was made on the left bank of the Thames, near the village of Dagenham, and many thousand acres of the adjacent lowlands were inundated. The most skilful engineers of the day tried long and in vain to close the breach, but at last it was effected by Captain Perie, at a cost of 20,000*l.*; but although the breach was closed, and nearly the whole submerged lands relieved from the water, still a space amounting to about 100 acres, where the breach took place, has ever since remained covered, and is called Dagenham Lake at the present day.

The position of this fine sheet of water being on the London side of the Thames, its depth varying from 4 to 20 feet below low water of spring tides, the great depth and width of the river in front of it and its proximity to London render it admirably adapted for wet or floating docks. For a long time it passed unnoticed, until, the trade of London increasing, other docks were established on both sides of the Thames at and close to London; the enormous cost of these and the high rates which they were necessarily obliged to charge in order to get anything like a remunerating dividend for the capital expended, induced enterprising people to look out for some situation lower down the river where docks could be established upon more moderate terms, and where consequently the rates would be much lower. Amongst other places Dagenham Lake attracted their notice, and very naturally so, for it possessed all the requisites for making a complete establishment of the kind at a most moderate cost, far below that which had been expended upon any of the great dock establishments in London. Who were the first persons who originated the idea of converting Dagenham Lake into a great dock establishment I do not know, but amongst others, I am told, was Mr. George Burge, the well-known contractor, about 1845. Subsequently Mr. Crampton took up the idea, and proposed to convert Dagenham Lake into a great dock establishment nearly twenty years ago, but the project never came to maturity. At last Mr. George Remington, a well-known projector, entered into it in 1854, and asked me to join him. On investigating the subject I was satisfied of its intrinsic merits, and agreed to co-operate in the undertaking. A Bill was therefore obtained in the year 1855 for this purpose. It was simply proposed in the first instance to connect the Dagenham Lake with the Thames by means of a lock, together with some small warehouses, landing wharves, and a railway to connect it with the London and Tilbury line; the whole estimate of what it was proposed to do there being confined to 120,000*l.*, that is to say, 90,000*l.* subscriptions, and borrowing power of 30,000*l.* This, it must be observed, was merely to commence the undertaking upon a moderate scale; and it was intended to extend the quays and warehouse room in proportion as the increased trade required it, for the floating basin accommodation was equal to that of the largest docks in London, and the depth of water in the river and in the dock was greater. The dock, if it could be commenced upon this moderate scale, could not, it is true, have been considered as a powerful rival to the other dock establishments, but it would have relieved them from the greatest part of the lumber trade, which they could not accommodate without great inconvenience and even loss, such as the timber, guano, hemp, flax, coal trades, &c.; moreover, it would give accommodation for laying up in ordinary the great number of vessels which are always more or less unemployed in the port of London. It is computed that of unemployed vessels there are generally about 150,000 tons; now these vessels, at a penny per ton per week, would alone return 7000*l.* per annum, and as they require no superintendence except from their owners, they alone would have paid 5 per cent. upon the total capital of the Company, and all the other trade would have added so much more to their income. It was quite clear that it would be to the interest of all the unemployed vessels to lay up there, because they could do so at half the expense compared with the other docks; for even if they were to lie at their moorings in the river, although they would be charged nothing, still their expenses would be a great deal more. There was, besides, another trade open to these docks, that could not be accommodated in any of the others, namely, the foreign cattle trade, which is every day increasing, and which must continue to increase with the population of the metropolis.

The first Act, as I have already said, passed in 1855, and although several attempts were made to form a company to carry it into effect, they all failed. In 1862 another Act of Parliament was obtained, as the original one had nearly expired. In the new Act the powers were enlarged, and the works were extended to 300,000*l.*, with power to borrow 100,000*l.* more; and it was again attempted to form a company to carry it into effect, but failed. In 1865 a third attempt was made to form a company, and by the aid of Messrs. Rigby, the well-known contractors of the great Admiralty harbour at Holyhead, a company was at length formed. Those gentlemen contracted for the works at a certain price, and agreed to take a large number of the shares as well in payment. The works commenced under my direction, in the month of May 1865, and proceeded very well until the end of March 1866, when the Messrs. Rigby got into difficulties, and were unable to complete their contract, and the consequence was that the whole of the works were stopped. The state of the money market ever since has been so depressed that it has hitherto been impossible to find the money to carry them on, and thus this really valuable concern remains still in abeyance.

In 1866 another Act of Parliament was procured, enabling the Company to obtain more land and to increase the works, so that ultimately, when times become favourable, it is very probable that this great undertaking will be carried out, and will form one of the largest and most important dock establishments on the banks of the Thames.

During the year 1866 it was attempted to obtain an Act of Parliament for making a railway between Romford and the docks. It passed the House of Commons, but when it got into the House of Lords its supporters drew back and the Bill was abandoned.

CHAPTER X.

Retrospect—London Bridge—Sheerness Dockyard—Plymouth Breakwater and Victualling Yard—Steam Vessels for the Navy—Harbours—Railways—Broad and Narrow Gauge—Atmospheric
Railway—Water Supply and Sewage.

I have thus endeavoured to give, in the foregoing narrative, an account of my
professional and private life as near as my memory would serve. I have not had
a single date, or note-book, or journal to refer to; so that many inaccuracies may
have occurred, particularly with regard to the dates, although the facts and circumstances are, I believe, pretty fairly narrated.

In my professional career I consider that I have executed the following works:

I. London Bridge. This was designed by my father, as far as the general outline
and proportions, but he did not live long enough to design any details, such as
the depth of the arch-stones and those of the inverted arches between the main
arches, or the adjustment of them, so that the whole might be placed in a perfect
state of equilibrium, not only as regards the individual arches, but also with each
other; neither was the width of the foundations of the piers and abutments given,
nor the extent of piling necessary, the cornice and parapets, stairs, pilasters of the
piers and abutments, the construction of the cofferdams and centres; the specification as to what materials should be used, and how they were to be put together;
the approaches to the bridge on both sides, or how they were to be designed and
put together; all these had to be worked out and executed by myself. It is true that
my brother George gave me his advice when I required it, but still I was the sole
engineer, and the whole responsibility rested with myself. The execution of these
works was rendered much more difficult than intended by my father, for at his
death the site was that of the old bridge. But the Committee of the Corporation of
London insisted that the new bridge should be built immediately above the old
one, the latter to be left standing during the construction of the new bridge. I was
therefore obliged to build it in the deep hole above the old bridge, which was from
25 to 30 feet below the level of low-water mark of spring tides.

II. The completion of the great works of Sheerness Dockyard. These, as I have
said, had been wholly designed by my father upon an entirely original and novel
plan of hollow walls, which he first carried into effect at Great Grimsby Docks, in
the year 1786. These walls, though composed of a mass of materials of the same
weight as ordinary dock walls, were distributed over a wider area, and pressed
less heavily upon that surface in proportion to their extent, and therefore the soft,
sandy foundation upon which they were built was able to bear them without

yielding; the increased friction also produced by the increased surface of their base enabled them to withstand with greater effect the lateral pressure of the earth behind them; thus a double object was gained, namely, security against both vertical and lateral pressure.

When my father died, on the 4th of October 1821, the northern half of the new dockyard, including the sea wall, the great basin, the three large dry docks at the west end, and the mast ponds and locks, had been nearly completed; so that it only remained to fix iron gates for the dry docks and those of the mast and boat ponds, which had been already designed and ordered, and were put into their places under my direction. This portion of the dockyard, although comprising the most extensive and costly part, was not the most difficult. The most arduous task still remained, namely, the construction of the northern portion. Here was the greatest depth of water, varying from 25 to 30 feet at low water of spring tides, the worst foundation, and the situation was much exposed to northerly and easterly winds. These obstacles were felt so strongly by my father, that he originally contemplated carrying out the works by means of the diving bell; but as so much experience had already been obtained by the employment of cofferdams in similar constructions, where they had been very successful, it became a question for my serious consideration whether it would not be better to use cofferdams for the northern portion of the dockyard, instead of employing the diving bell, which would necessarily require much more time. After consulting with the enterprising contractors, Messrs. Jolliffe, Banks, and Nicholson, who had completed the works already made, and Mr. John Thomas, the experienced resident engineer, we came to the unanimous conclusion that it was perfectly practicable to construct the remainder of the works by means of cofferdams; and although it would be rather more expensive, nevertheless they could be done much better and far more speedily than by the diving bell; and, indeed, they told me that my father had expressed the same opinion before he died; and that there was little doubt but that if he had lived he would have recommended cofferdams instead of the diving bell. I consulted my brother George upon the subject, and he was of the same opinion. We resolved to recommend that the remainder of the works should be completed by cofferdams, and the Admiralty approved of our recommendation. Messrs. Jolliffe, Banks, and Nicholson therefore undertook the contract for these works at the sum of 845,000*l.*, and gave ample security; and they were most successfully finished for the sum of 854,000*l.* in round numbers, or at about 9000*l.* beyond the contract price, our estimate being nearly 900,000*l.*; so that they were actually completed for about 45,000*l.* below our estimate, and fully three years sooner than they would have been if the diving bell had been used. Of course the real merit of these works is due to my father; but I claim some credit for having successfully carried them into effect, for if any failure had taken place—and there was very great difficulty and risk—I should have been blamed for it, and probably been ruined at the outset of my career, as the whole responsibility rested with me; my brother never went near them.

III. I finished the Chatham dry docks, commenced by my father, at the cost of 100,000*l.* In these there was nothing remarkable; after those of Sheerness they were much less difficult, although of a somewhat similar kind.

IV. The next great work was the finishing of the great breakwater in Plymouth Sound. The chief merit I claim for this is in adding the benching or berm on the outside, at the base of the sea slope, which breaks the sea before it reaches the slope and prevents it from acting injuriously upon it. I also claim a certain portion of the credit for arranging and executing the paving of the upper surface, and the dovetailed masonry of the two ends of the breakwater.

V. The design and execution of the Royal William Victualling Establishment, at Stonehouse, near Devonport, I claim entirely as my own, with the exception of the machinery, for which my brother George is entitled to an equal share of credit with myself. This establishment, including the cost of the land, amounted, I believe, to between 600,000*l.* and 700,000*l.*

VI. The great basin, two building slips for first-rates, mast slip, and the river wall in front, at the Royal Dockyard at Woolwich, costing 340,000*l.*

VII. In company with Mr. Joseph Whidby, Mr. Walker, and Captain Fullerton, of the Trinity House, I made a report for removing the bar, by means of dredging, at the entrance of Portsmouth harbour, upon which there was only 13 feet at low water of spring tides, which we estimated at 55,000*l.*; and it is singular that this important work was never carried into effect until many years afterwards, when it proved to be completely successful as far as it went. The bar was lowered 5 or 6 feet, and it might be lowered 8 or 10 feet more, so as to enable the largest class of vessels to enter and depart at low water of spring tides, which would be of the greatest possible advantage to the public service; and although the Admiralty have not carried the dredging far enough, still there is now 18 feet at low water of spring tides, which enables the largest class of vessels to pass the bar at half tide, instead of only at high water as before. This fully proves the value and correctness of our joint report; it only now requires that our recommendation should be carried further, and there can be little doubt that it will be successful. This great national harbour will be rendered accessible at low water, and it ought to be, particularly after the enormous sums that have been expended upon it, for unless the depth over the bar is increased all improvements will be comparatively valueless. Mr. Murray and myself wrote a joint report to the Admiralty, recommending that, in order to assist the dredging operations over the bar, a sluice should be erected across the entrance to Langston harbour, with the gates or doors of the sluice pointing inwards, so that at high water they might be shut, and all the water, or so much of it as might be required, should be sent through Portsmouth harbour at ebb tide, to assist in scouring down the bar. Of course, in order to render these works effective, it would be necessary to enlarge the connecting channel between Portsmouth and Langston harbour, so that all the Langston tidal water should flow out through Portsmouth during the time of ebb.

The Admiralty up to the present time have not adopted this report. They must, however, in order to preserve the requisite depth over Portsmouth bar, do either the one or the other, or both; that is to say, they must increase the dredging operations, or send more tidal water over it, and the latter can only be obtained from Langston; as this harbour is of little commercial value, supposing that any partial silting up should take place, the depth could be restored by dredging; but if both the dredging of Portsmouth bar and the additional quantity of tidal water from Langston harbour should be resorted to, the bar might be kept down to the depth required, and Langston would not be injured. If these two operations are skilfully conducted, so as mutually to assist each other, the result will be successful, and this success is the more necessary, in consequence of the quantity of land which is now being reclaimed from Portsmouth harbour for the new works.

VIII. The great flour mills and biscuit machinery at the Clarence Victualling Yard, Portsmouth, were designed and executed by my brother George and myself. The idea of the bread apparatus was proposed by M. Grout, and worked out by ourselves. The great flour and biscuit mills at Deptford were also designed and executed by my brother and myself.

IX. The Thames Tunnel shield; the rolling machinery of the Bombay, the Calcutta, and the Mexican mints; the machinery at Constantinople for manufacturing small arms; numerous locomotive engines and tenders for different railways, amongst them the 'Satellite,' for the Brighton Railway, which was one of the first that travelled at the rate of 60 miles an hour. The engines and machinery for several of Her Majesty's vessels of war, amongst which may be mentioned the 'Bull Dog,' the yacht 'Elfin,' and others; four iron vessels, engines, and machinery for the Russian Government for the Caspian Sea, the first that were ever placed there; two yachts for the Emperor Nicholas; the 'Vladimir' frigate; two large screw vessels of war for the Baltic; three also for the Black Sea; several for the Danube Company; cranes, sugar mills, diving bells, and machinery; gantry cranes for the mahogany roofs of the West India Docks; spinning and all kinds of machinery, from the year 1821 until the year 1852.

X. The first sea-going screw vessel that was constructed, namely, the 'Archimedes;' and also the 'Dwarf,' 1839, the first screw vessel of war that was introduced into the navy.

XI. I recommended that the use of the Cornish high-pressure condensing system should be introduced into the steam-vessels of the Royal Navy. At that time they were entirely upon the system of Boulton and Watt, when steam was only employed to the extent of 5 lb. pressure upon every square inch. Now it was well known that the intensity of the power of steam increased in a much greater ratio than the additional quantity of fuel required to raise the temperature, so that high-pressure condensed steam was much more economical than low pressure. There was a good deal of prejudice against it, in consequence of the decided objections of Boulton and Watt, and therefore it was not adopted at the time, but by degrees this prejudice has been overcome, and now steam of 25 to 30 lb. is employed in the Royal Navy, with great advantage and economy.

XII. I may also say that I was the means of introducing oscillating engines into the navy. These I believe were invented by a Mr. Witté, of Hull, but in consequence of the extreme accuracy required in making them, and some degree of prejudice against the vibratory action of the cylinder, this very valuable invention was laid aside. The able and ingenious Mr. Maudslay took it up, but was dissatisfied with it, and abandoned it. Mr. John Penn, who had a small establishment for making machinery at Greenwich, then adopted it, and commenced manufacturing these engines upon a small scale for the steamboats on the Thames. He improved on the idea, acquiring the greatest experience in constructing the engines, and he was convinced that they could be made upon any scale with equally successful results. It happened about this time that the Admiralty required new engines of greater power for their official yacht, the 'Black Eagle,' whose speed averaged little more than 8 knots an hour, and they applied to Boulton and Watt, who had made the old engines for the 'Black Eagle.' They said they could easily make more powerful engines, but that these would necessarily be heavier, and sink the vessel lower in the water, when the resistance would be so much increased that very little additional speed would be gained, and therefore it would be better to have an entirely new vessel. The Admiralty did not wish to incur the expense, and the matter was likely to fall to the ground. Penn heard of this, and, quite uninvited, sent in a tender to make new engines for the 'Black Eagle,' double the power of the old ones, of the same weight, and occupying the same space, for a sum, not, I think, exceeding the cost of engines of the same power on the old method. He further offered, if the Admiralty officers were not satisfied, to take them out, and replace the old engines at his own expense. I happened to be present upon other business with the Comptroller of the Navy, Sir Thomas Byam Martin, when Penn's tender was sent in, and after reading it he threw it to me, and said, "Rennie, what do you think of that; should I accept it or not?" I read Penn's tender carefully, and knowing something about the oscillating engine, and having a good opinion of it, I said I thought he should accept it. "Then," said he, "I will do so, and if it turns out badly you shall have the blame." "Very well," I replied, "if it turns out badly I will take the blame." Penn's offer was accordingly accepted. The engines were made and fixed on board; all the conditions of the tender were fully complied with, and the Admiralty were perfectly satisfied with their bargain. From that time forward Penn became one of the chief manufacturers of the Admiralty engines, and has continued to be so up to the present time.

The harbours which I made are described in my work on 'British and Foreign Harbours'; they were a portion of Kingstown, in Dublin Bay; Donaghadee, Port Patrick, Port Rush, Warkworth, Sunderland, East Hartlepool, Whitehaven; nearly rebuilding Ramsgate harbour; Ponta Delgada, in the Azores. I designed harbours for Oporto; the Mattozenhas; Viana, Aveiro, Figuera, and St. Ubes, for the Portuguese Government; also for Douglas, Castleton, Peel, Ramsey, and Laxey, in the Isle of Man, for the local authorities; and Redoubt Kalé, in the Black Sea, for the Russian Government.

XIII. London Bridge; Hyde Park, Kensington Gardens, and Staines Bridges, besides finishing those at Crammond and New Galloway, designed by my father.

I laid out and carried through Parliament the Brighton Railway and the Blackwall Railway, in 1838; also the Manchester and Liverpool Railway, in conjunction with my brother George, in 1827. In 1838 I designed the Central Kent Railway, which, by passing through the centre of the county, connected all the leading towns on the main line, besides reducing the distance between the metropolis, Dover, and Folkestone to the minimum.

I also projected a line for the Great Northern in the years 1844-45, which was admitted to be the best and shortest line; but it unfortunately failed in consequence of the late Mr. Francis Giles not having completed the parliamentary surveys. I laid down a railway between Leeds and Carlisle, that would have materially shortened the distance between the important manufacturing town of Leeds, Carlisle, Glasgow, and the north of Scotland; a line between Leeds and Bradford, and another between York and Scarborough. Another, called the North Wales Railway, between Bangor and Port Dyllaen, where I designed a capacious harbour and docks, that would have been of the greatest advantage to Liverpool, avoiding the dangerous navigation between that place and Port Dyllaen, and affording an excellent point of departure for Ireland. I also made a design for a new port for Holyhead, upon the principles laid down by my father, that would have answered the purpose far better, and have saved in a great measure the expense that has been incurred by the present ill-contrived harbour, and which has not answered the object intended.

In company with my friend Mr. George Remington, I designed the direct London and Manchester Railway in the years 1844-45; this line would have reduced the distance between London and Manchester to 176 miles, besides affording railway communication to a number of the intermediate towns, such as Bradford, Burton, Leicester, Congleton, and other places that had not hitherto received the benefit of direct railway accommodation. This line was pronounced by the Board of Trade to be the most important and best laid down line that had been brought before Parliament, and was strongly recommended by them; and it would have been carried, but unfortunately there was another competing line by Mr. Rastrick, that was ultimately abandoned by its promoters, who, before doing so, united with us; but in doing this the reference books containing the names of the owners and occupiers along both lines became mixed, and the result was, that seven miles of the reference of the competing line was substituted for seven miles of our line, and *vice versâ*. This was fatal, and the Bill was consequently lost; and this valuable line, almost the best of any in England, could never be resuscitated. The North-Western Railway, thinking that they were safe from all competition, declined taking up the line, though their interest imperatively called upon them to do so, and further, would not unite with nor buy up the Midland from Leeds to Rugby. The Midland Company then determined to make an independent line to London, and took the identical course laid down by Remington and myself. They have become a very formidable rival to the North-Western, and this is precisely a similar case to that between the South-Eastern and the London, Chatham, and Dover Companies. If the South-Eastern Company had only adopted my Central Kent line, which was laid down in 1838, before they had commenced their present line—and they promised to do so—the London, Chatham, and Dover Railway would never have been made, and the county of Kent would have been better served, many millions would have been saved, and many thousand unfortunate shareholders would have avoided ruin.

I laid down lines for the kingdoms of Sweden and Portugal, which have been more or less adopted, and projected a line from Odessa to Moscow. Also the London, Brighton, and South Coast as far as Salisbury, and from thence to Warminster, which has since been adopted. A line from London to Birmingham, Leeds, and Carlisle; Leeds and Bradford; Dumfries and Port Patrick; Newry and Enniskillen, in Ireland; Bangor to Port Dyllaen, North Wales; Cannock Chase line, in Staffordshire, through an undeveloped coal district, another of my lines which has since been carried into effect. The East Lincoln, from Lynn to Great Grimsby; the direct London and Norwich, from Bishop's Stortford to Thetford, which would have shortened the distance between London and Norwich and Yarmouth. All these lines were laid upon the direct principle, that is, taking the shortest distance that the nature of the intervening country would permit between the two termini; this principle is now proved to be the correct one, and if it had only been acted upon before, we may readily conceive the vast amount of capital which would have been saved, while the counties through which railways have been made would have received a much greater benefit; whereas, by the system which has hitherto been adopted, a great number of unnecessary lines have been constructed, and a constant competition and rivalry have taken place between the different companies, and now, with reduced dividends and increased charges, they find out their error, when it is too late to be remedied.

Another most important error has been committed by a too narrow gauge having been adopted. My brother and myself, when we carried the Bill for the Manchester and Liverpool through Parliament, in the year 1826—and this may be considered almost the very commencement of the railway system—after investigating the width between all the various carriage wheels, whether for goods or passengers, we decided that the width of gauge from centre to centre of the rails should be 5 feet 6 inches or 6 feet. When Mr. George Stephenson became the engineer for executing the line, he decided that the gauge should be only 4 feet 8½ inches from centre to centre of the railway, for no other reason than that the gauge between the old colliery rails was 4 feet 8½ inches; hence arose all the subsequent difficulties. It was quite clear that 4 feet 8½ inches was too narrow. Brunel, seizing on this mistake, proposed at once to make the gauge 7 feet from centre to centre of the rails for the Great Western Railway. This was as evidently too much as Stephenson's was too little. The power of a locomotive engine is in proportion to its weight, and the greater the weight the greater the power, acting as it does by its adhesion to the rails; and to increase the power of an engine upon the narrow gauge could only be done with safety by increasing its length; for if it be done by increasing the height, the centre of gravity would be raised also, and the motion of the engine would be rendered unsteady; and by increasing the length the engine would be less adapted for going round sharp curves. Now in the ordinary traffic of goods, such as coals, &c., extraordinary velocity was not required, and therefore the width of the gauge was not of so much consequence, but when it came to carrying passengers the case was wholly altered. Latterly the coaches and mails had travelled at the rate of 10 and 12 miles an hour, whereas goods were seldom carried at the rate of more than 3 miles an hour. If passengers were to travel by railway it would not be less than 12 miles an hour, and therefore it was at length necessary to provide for this velocity, and more; otherwise, as there was a certain prejudice on the outset against railway travelling, the latter could not expect to have the preference. But when it was ascertained, as it was at the trial of engines upon the Rainhill plane of the Manchester and Liverpool Railway, that the imperfect locomotives of that day could go at the rate of 30 miles an hour, the whole case was changed; the carriage of goods, which at first was most important, gave way to that of carrying passengers, and it was evident that the whole system of locomotion, whether of goods or passengers, must be absorbed by railways. It was therefore more especially necessary that the question of the gauge should be most carefully considered. I may be answered, certainly, that the improved locomotive engines upon the narrow gauge realize a speed of 50 to 60 miles an hour, and this is fast enough for anything; but then this cannot be done without incurring greater risk than upon a broader gauge. The Great Western realize a speed of 45 miles an hour without the least risk, i.e. including stoppages, whereas the narrow gauge does not do more than 35 to 37 miles an hour, and that probably with a greater wear and tear of the rails. A medium therefore between the two gauges, that is 5 feet 6 inches or 6 feet, instead of 4 feet 8½ inches or 7 feet, appeared to my brother and self the proper gauge; and if such had been adopted we should never have heard of the 7-feet gauge, and the 5 feet 6 inches or 6-feet gauge would have been universally adopted, to the great advantage of all.

Before leaving railways, it may be proper to say something about the atmospheric system. When an experiment was made on a large scale and succeeded very well, it was subsequently reduced to practice upon the Dublin and Bray Railway, between Kingstown and Dalkey, a length of about 3 miles. Here it succeeded perfectly; the steepest incline was completely mastered, and the smoothness and luxury of travelling were unequalled. Brunel afterwards took it up, and employed it upon the South Devon Railway. There it succeeded also perfectly as far as speed and luxury of travelling were concerned. The difficulty however of making the valve in the exhausting tube was so great that it was ultimately abandoned, after having incurred great expense, and the locomotive system was again resorted to. The Croydon Railway also adopted it, but gave it up for the same reason as the South Devon. My brother and myself were much taken with this system, and made several of the steam engines for it, that answered their purpose perfectly, and we thought that by a little more perseverance in it, the difficulties complained of might have been overcome, but the proprietors would not listen either to Brunel or ourselves. The Stephensons made a dead set against it, and, taking the facts at the time, perhaps they were right; but it is very rarely that a new invention succeeds at the first or second trial: it requires time to ascertain the defects, and to study more minutely the remedy, and, after a little while, the cure for the evil is found out. I should not be surprised if ultimately the atmospheric system comes to life again: indeed, the very strongest opponents of it have already adopted it in London, with certain modifications, for conveying the mail bags in London from the General Post Office to some of the railway stations, with considerable success, and Mr. Rammell made an experimental line of this kind at the Crystal Palace. The defects in the original lines were principally those of workmanship, and can be remedied by degrees, as is always the case whenever a principle is sound, for it only requires perseverance to achieve ultimate success.

XIV. Drainage of lowlands upon a large scale I have carried into effect in several instances already described. The completion of the Eau Brink Cut, the designing and making the Norfolk Estuary Cut below Lynn, and the Marshland works, by means of which from 350,000 to 400,000 acres of land are drained; the Nene Estuary Cut, by which about 150,000 acres of land are drained; the improvement of the Witham between Boston and the sea, by which the drainage of about 250,000 acres has been materially improved; the Ancholme drainage, by which 50,000 acres of lowlands have been well drained; altogether amounting to between 800,000 and 900,000 acres.

XV. I may also say that I have embanked from the estuaries of the Ouse, the Nene, and the Witham, about 6000 acres of fen land, which is now more or less under cultivation. I have also laid down a plan, at present being carried into effect, by which 32,000 additional acres will be embanked from the estuaries of the Ouse and Nene; and another plan for embanking 45,000 acres from the estuaries of the Welland and Witham; indeed, my original plan of 1837 was for embanking from 150,000 to 200,000 acres of land from the estuaries of the Ouse, Nene, Welland, and Witham, and the Great Wash; and I have no doubt that in time this will be effected, and another large and most valuable county—all rich agricultural land—will be added to the kingdom. I also obtained an Act for embanking 32,000 acres from the north side of the estuary of the Thames, near Shoeburyness. I believe that, in addition to this, three times the amount may be taken from this and other parts of the Thames estuary. Let to these be added the lands which may be saved from the estuaries of the Humber, the Forth, the Tay, the Clyde, the Solway, Morecambe Bay, and the Mersey, altogether from 500,000 to 600,000 acres of land may be reclaimed, or three large new counties may be added to the kingdom, capable of producing annually an additional supply of 3,500,000 quarters of corn, which, at 3*l.* per quarter, would, after deducting 20*s.* per quarter for the cost of production, add a revenue of about 6,000,000*l.* a year to the country. A great deal may be done in this way also in Ireland. We should, however, deduct a million a year for the first fifteen years to cover the cost of embankment. The clear annual gain would be 5,000,000*l.* a year to the country; or, putting it in another light, the land so acquired would maintain an additional number of inhabitants. Besides this, large tracts of lowlands adjacent to these estuaries might be greatly improved in their drainage, in connection with the reclamation works, which would add considerably to their produce.

The execution of all these works, besides draining the quantity of land I have stated, and more than doubling its value, has also very greatly improved the navigation.

I also extended the Newry Ship Canal nearly two miles, which has a depth of 16 to 18 feet, and is 130 feet wide, with an entrance lock 50 feet wide. I deepened the old canal to Newry, so that large vessels, drawing nearly 16 to 17 feet, can come up to the town.

XVI. Soon after my father's death, in 1821, when I may be said to have entered my professional career upon my own account, I began to consider the water question; that is to say, the best mode of economizing water, so that those districts where it might be most required could be supplied, as far as the physical geography of those places would render it practicable. Generally speaking, there falls a certain quantity of rain in every district during the year, and this, with more or less regularity, at particular seasons and times. In some places the rain is periodical, and falls in the course of three or four consecutive months; in other countries it falls at different times, principally, however, in the winter and autumn months. Now after the periodical rain is over, the whole country is deprived of water throughout the remainder, or about three-fourths, of the year. The remedy for this is to construct reservoirs in the most convenient places, upon such a scale as the wants of the country may require; in these reservoirs the surplus waters should be stored during the periodical rains, to serve as a supply in the dry season, not only for domestic purposes, but for irrigation, navigation, &c.; the reservoirs should, in some cases, be covered, and in others open, even to the extent of making them large lakes. They should be provided with proper sluices and culverts, open or covered, as may be required, and best adapted for distributing the water in the most beneficial manner.

Having obtained a sufficient supply, the next point to be attended to is, to take care that the water shall not be polluted: in order to effect this, in all thickly-peopled districts the sewage should not be discharged into the river or watercourses, but into separate, isolated, and well-ventilated tanks, and then be deodorized by mixing it with earth, or subjecting it to any well-known process for this purpose, and the refuse should be distributed for manure; thus the sewage, instead of being a nuisance, will become valuable for agricultural purposes.

By these means, regulated according to the particular circumstances of each case, the whole question, viz. economy of water, which is so very important in every respect, is solved. I have long endeavoured to make it clearly understood, but in England we are slow to move in a new direction. The enemy must be at our doors before we are prepared to meet him, and then we begin in earnest. Such has been the case with the water question: we carried drainage almost to the utmost extent, so that the rainwater was discharged into the adjacent watercourses and rivers with the greatest rapidity and was carried off to sea, and we thought not a moment that the day would come when we should want it. The universal cry was, "Only get rid of the water, and all will go on well." At the same time all the sewage matter was discharged into the watercourses, the cry being, "Only get rid of the sewage, and our cities and towns will be healthy, and we shall hear no more of it;" little thinking that the streams would be polluted, and that water when most wanted would not be forthcoming, and that even the moderate quantity that could be obtained would be unfit for domestic purposes. The Thames and all the great rivers and streams were converted into common sewers, threatening to spread pestilence around them. The water that was to be obtained for domestic purposes was polluted to such an extent, that the malaria caused by the foul state of the watercourses was increased by drinking the contaminated water that we fondly expected we had got rid of. At last the public opened their eyes, and asked how all this had arisen; then commissioners of all kinds were appointed by the Government to investigate these important questions; and what is the result? Precisely that which I mentioned years ago, namely, 1. That means must be established for economizing water and for affording an ample supply at all times. 2. That all sewage matter must be diverted or be prevented from being discharged into the watercourses. 3. That as far as practicable the sewage matter must be utilized for manuring the land. All these three propositions, which constitute the whole elements of these important questions, are now being carried into effect by Acts of Parliament; better late than never, for if these terrible evils had been allowed to exist much longer the consequences would have been most fatal.

About four years ago I wrote two letters to 'The Times,' which were printed in that journal, embodying my views upon this subject in a detailed manner, according to the principles above described. I am extremely glad that at the eleventh hour the subject is beginning to be thoroughly understood, and it is to be hoped that now the proper remedy will be employed; it is contained in the principles that I have recommended for the last forty years. I may not perhaps claim the merit of the whole; but this I must say in justice to myself, that I have contributed in some degree to direct attention to the subject, and I most sincerely trust that, having been made conscious of its importance, the public will not be content until the question has been thoroughly sifted, and the evils complained of successfully remedied. Up to the present time neither compensating reservoirs for the due supply of water during the dry seasons have been made, nor, with a few solitary exceptions, has the sewage been excluded from the rivers, nor have the watercourses been properly improved so as to prevent inundations of the adjacent lowlands. In fact, the authorities have only just begun to get an idea of what is required to obtain an ample supply of good water; but the more they investigate the subject, the more they will find that only upon a right understanding of the principles above recommended can this supply be procured. Sewage matter has now been recognized as a fertilizing agent, and the only points undecided with regard to it are the best modes of deodorization, so as not to injure its manuring value, and the most suitable method of applying it to the land, whether in a liquid or in a solid state.

With regard to water for domestic use, considerable progress has been made: the water is conducted into covered reservoirs, where it is excluded from the action of the atmosphere; it is also filtered, so that all the alluvial and tangible vegetable matters are excluded; and the best method of separating from it those injurious ingredients with which it is chemically combined has made great progress. These, no doubt, are considerable advantages gained, but unless the means of obtaining an ample supply be used, the other advantages will be comparatively of little service. It is true they will be valuable as far as they go, but if there be a deficient supply of water, there will remain a great deal to be remedied, therefore it will be necessary to secure an ample supply by means of open reservoirs.

CHAPTER XI.

The Formation of Natural Breakwaters—The Society of Civil Engineers—The Education of a Civil Engineer—Some Hints on Practice—Estimating.

In the introduction to my work on 'British and Foreign Harbours,' I have suggested a method by which shoals formed by alluvial deposits in the open sea might be converted into effective breakwaters, so as to become harbours of refuge; or the means of removing them altogether. It is well known that many existing shoals form, to some extent, safe roadsteads at certain times of tide, e.g. the Goodwin Sands, the banks outside Yarmouth Roads, the banks off the coast of Holland, and many other places. These are generally formed off alluvial shores, where the meeting of opposing currents causes an eddy or line of stagnation, and the alluvial matter held in suspension is deposited, forming a bank, according to the extent, width, and direction of the eddy. In some instances, as in the case of deltas of rivers, and along coasts where the waters are densely charged with alluvial matter, these shoals, by continual deposit, are raised to the level of high water of neap tides, when a succession of marine vegetation appears on the surface, finally becoming a rich grass marsh; except under special circumstances, the land is seldom raised higher, and where there is no flow of tide the same result takes place at the medium level of the waters.

In other cases, as in the open sea, where the waters are exposed to violent agitation by the wind, these deposits not only rarely reach the level of high water, but, except under particular circumstances, seldom exceed the level of half-tide, and often the banks remain many fathoms below low water, though even in their lowest state they are far above the bottom of the sea. As all these banks are composed of alluvial matter, we can only ascribe the different levels, first, to the variable quantity of alluvium with which the waters are charged; secondly, to the degree of agitation to which the waters are exposed; and thirdly, to the velocity and extent of the opposing currents which produce the banks. Having thus stated generally the causes that produce these banks, I now come to my proposition, namely, the best mode of utilizing them for making harbours of refuge, or the method for clearing them away where they may be injurious.

With regard to the first, it is only necessary to increase the power of the depositing eddy by means of artificial works, to raise the banks to any height required; by this means they may be rendered permanent breakwaters at the least expense. Secondly, where these shoals are injurious they may be removed by diverting the course of one or both currents, so that the line of stagnation shall be destroyed; the action of the sea will then gradually remove the shoal. Thus we have the means in our power of converting these sandbanks into most valuable harbours of refuge, or of removing them where they are found to be injurious. This I do not pretend to call an invention, but simply an idea, and I am not aware that it has been suggested before. Modern engineers have not sufficiently directed their attention to the construction of harbours. It is a very simple affair to build piers or breakwaters of any extent, provided the requisite means are forthcoming, but it is a totally different thing to ascertain whether, after these works have been constructed, they will answer the purpose originally intended.

When President of the Institution of Civil Engineers, during the years 1845-6-7, I drew up detailed reports of the history of the profession from its commencement in Great Britain up to that time, showing what had been done in every department, by whom, and at what date. These reports are published in their 'Transactions.' Subsequent presidents have to some extent adopted a similar course; but with all due respect to them, they have not taken that large and scientific view of the profession of a civil engineer which it is imperatively necessary to adopt in order to keep the profession up to that high tone which its importance requires, not only for its own credit, but for the benefit of the world at large. Perhaps there is no profession (with all proper respect to others) that has conferred so much benefit upon mankind as that of the civil engineer. Its objects are clearly defined in the two mottoes belonging to the Smeatonian Society of Civil Engineers, which was the first of the kind established in this country, having originated with Smeaton, Mylne, and my father, namely, "*Omnia numero pondere et mensurâ;*" Ὧν φύσει κρατοῦντες τέχνῃ νιχώμεθα. Up to that date the profession of a civil engineer may be said to have been unknown in Great Britain; previous to that time we were simply known as "*vulgar mechanics*" —men who toiled with their hands, as masons, bricklayers, carpenters, blacksmiths, &c. But those who so called us would have entertained a very different idea of the "mechanics" if they had been forced to dispense with their services. Let me ask how could the various and complicated operations which alone render modern trade, and therefore modern civilization, possible, be carried on without the aid of the mechanic, *alias* the civil engineer?

The object of the Smeatonian Society was merely a social gathering in the form of a club, to assemble the members at dinner at certain times, when they could discuss in a friendly manner the various subjects connected with their profession, and to endeavour to obliterate all those rivalries and jealousies which unfortunately are too common amongst professional men of all classes. The society was to serve as a rallying point for the profession, and it was believed that when their members increased sufficiently (for there was little more than a dozen engineers in the kingdom at the time who were counted as such) the society might extend its usefulness by reading papers, discussing them, and publishing them regularly to the world, in the same manner as the already established scientific societies; this has since been done by the Institution of Civil Engineers. But I think the time has now arrived when that Institution should be enlarged, and take a wider sphere. It has hitherto been confined too much to the class practising purely engineering works; but the mechanical engineers now form a body which must be treated with every deference. It is very true that the latter are freely admitted into the institution, but there seems to be a tacit understanding amongst the former that they should not attain the honour of becoming presidents and vice-presidents. It is true that the late Mr. Field, a most distinguished mechanical engineer, was elected president, and served his time; but this, I believe, arose more from his having been one of the earliest members of the institution than from any respect due to the particular class of the profession to which he belonged. Now there cannot be a greater mistake than this. Every member of that institution, to whatever class he belongs, from the moment he is elected should be in every respect upon precisely the same footing as those who are now considered the governing class, and the ablest man should be chosen from each grade as president or vice-president alternately, so that each department should successively occupy the chair. Also, instead of choosing the president and council by rotation, according to seniority, the acknowledged best men in every department should be chosen as officers. And further, the institution thus regulated should have the power of giving certificates of competency after the candidates for admission have been duly examined by independent examiners; and until they have received these certificates they should not be allowed to practise. This is the rule in every other learned profession, and there can be no reason why it should not be adopted by the engineers. It is the only method by which it can take rank amongst the learned professions; and as no other requires more scientific knowledge, or is entrusted with a greater portion of responsibility or a larger amount of trust, or where failure becomes more disastrous, it is quite clear that no man should be allowed to practise it unless he has passed a proper examination, and has received a certificate of competency from proper authorities.

Against this proposal it may be argued, that many illiterate men, although of great original genius, would be excluded if their competency were tried by such a test. My reply is, let them not be tried only by the ordinary rules of scientific books, but also by the general principles which the candidate professes, and let those principles be tested, to prove how far they are in accordance with sound philosophy. A man like Stephenson or Brindley, although illiterate, may understand these principles perfectly, and yet may not be able to explain them. Nevertheless, let him be examined, but in a different manner from the ordinary routine, and it will soon be discovered whether his profession and his practice are founded upon true mechanical and philosophical principles.

If these examinations are properly conducted every possible objection will be abolished, and no scientific educated engineer, or any illiterate person of true scientific genius, will be prevented from pursuing the profession, whilst only the speculator and charlatan will be excluded. By this means the public will be assured that the works for which they subscribe the funds will be conducted in the best manner, and most probably to a successful termination. At present, the system upon which public works are carried on is wholly wrong. There is no system. Any man without business, competent or not, dubs himself engineer, starts a project, well or ill founded, as the case may be, *generally the latter*, and issues a prospectus to the public, to obtain the necessary funds to carry his proposal into effect. Next he gets a contractor to back him by taking a certain number of shares, provided that he has the contract at his own price. The shares he looks upon as good for nothing, and therefore adds so much more to his ordinary profits, so that instead of receiving 10 or 12 per cent. upon his cost price, which is the usual rule of the trade, he gets double, with the shares into the bargain, all of which is added to the capital of the undertaking; and in order to carry into effect this wasteful policy, the contractor generally stipulates for two or three of his own nominees to be placed upon the board, to "*look after*" his interests, so that, in point of fact, he pays himself pretty nearly what he likes. The engineer, who ought to be his master, loses all control over him, and in many cases becomes little better than his servant. This is certainly a most discreditable state of things, and has been the cause of the most wasteful expenditure, and the ruin of many valuable undertakings, and it will always continue to be the case so long as the present system prevails.

The real object of the civil engineer is to promote the civilization of the world, by the proper application of all the great mechanical means at his command, and to take a high, independent position as a scientific man, thoroughly versed in his profession both theoretically and practically, and wholly independent of contractors, and all sinister influences. Unless he can do this, he never will be held in that esteem and respect, or take that high position without which no professional man can properly discharge the duties that he owes to himself and to the public.

Against what I have said it may perhaps be urged that I assign too high a place to the profession to which my father and myself have had the honour to belong; but I think that when the subject has been calmly and fairly considered it will be generally admitted that I have not done so without reason. Without wishing for a moment to depreciate the merits of any other body of men, I think it will be conceded that the objects proposed by the engineer, and the acquirements, knowledge, and experience that he must possess before he can practise successfully, are at least equal to those of any other profession, particularly after the practical examples exhibited to the world of the great benefits that engineering has already conferred upon mankind. Therefore are we entitled to be ranked amongst the most learned professions, and to receive all the honours they have most justly earned; and I trust the time is not far distant when this justice will be accorded to them.

Before concluding this sketch of my career I will offer a few observations as to what I consider, from my experience, the best plan of education for the profession of a civil engineer. Hitherto there has been no regular system. A youth leaves school about the age of seventeen or eighteen, without any previous training, and his parents, thinking that he has got a mechanical turn, as it is termed, decide at once to make him a civil engineer, whether he likes it or is fit for it or not. They then send him, with a considerable premium, to an engineer of some standing and practice, who, unless special conditions are made (and very few engineers will make them), will not undertake to teach him the profession. The pupil is sent into the office, and placed under the direction of the principal assistant, who directs him to do whatever is required, if he can do it, whether drawing, writing, or calculating, or anything else; and if he wishes to learn anything, he must find it out himself: neither the principal nor assistant explains the principles or reasons of anything that is done. If he prove to be steady, intelligent, and useful, keeps the regular office hours, and evinces a determination to understand thoroughly the why and wherefore of every kind of work that is brought before him, and by this means acquires some practical knowledge, he will soon attract the notice of his employer, and will be gradually transferred from one department to another, until the expiration of his pupilage, which varies from three to four years; then, if he really has acquired a competent knowledge of the profession, and the employer thinks his old pupil can be of further service to him, he is engaged at a moderate salary, to be employed in such capacity as he is fit for. If during his pupilage he has made but little progress, nothing beyond mere routine, he is discharged with a certificate according to his merits, and sent into the world, to find his way forward as best he can.

Now it should be understood that the pupil only learns one part of his business, such as the construction of railways, canals, improvement of rivers, docks, drainage, harbours, and waterworks, and the buildings connected with them; but there is another and very important part of civil engineering, namely, *the mechanical department*, of which he remains totally ignorant. Nor will he gain any insight into the raising of coals, iron, or any other geological product. Now, in order to form a good civil engineer, in my opinion it is absolutely necessary that he should be well acquainted with all these different branches. To this it may be replied, that it is not necessary an engineer should be acquainted with all departments of the profession, but only with the one to which he intends more particularly to devote himself. Now this is a very great mistake, for they are all so intimately connected, that without having a general knowledge of the whole you cannot practise in any one department with complete success; for whenever you have to rely upon the resources of another department you can never make sure of being thoroughly well served, unless you are yourself a tolerable judge of work. I repeat, then, that an engineer who has studied only one department cannot be termed properly educated. And the question arises, what is the best mode of education for a pupil to obtain this multifarious, and, as I contend, absolutely necessary, information, to enable him to practise the profession of a civil engineer in the most enlightened, scientific, and practical manner? My answer is this: Let him first get a sound elementary education in the several departments of arithmetic, algebra, geometry, natural philosophy, geography, geology, astronomy, chemistry, land and hydrographical surveying, as well as grammar, English composition, history, French, German, and Latin, according to the improved system of modern education; every youth of ordinary talents has a tolerably fair knowledge of these at seventeen or eighteen. What then should be the training for an engineer? First let him go through the best course of modern education at his command, including the elements of geometry, mathematics, and the physical sciences, not excluding Latin and Greek, in spite of the prejudice against them now frequently expressed. Then let him be apprenticed for two or three years to some good steam engine and machinery manufacturer, where he should learn to make drawings and calculations, handle tools, make models, steam-engine machinery, and put machinery together. By this means, if he applies his mind to it properly, he may become a practical as well as theoretical mechanician, which is the soundest basis for good engineering; indeed, without this it is impossible for an engineer to be thoroughly successful, but being well grounded in this most important knowledge, all the others will become comparatively easy. Having gone through this apprenticeship, let him bind himself for three or four years to some well-known civil engineer, of large practice in railways, docks, harbours, waterworks, canals, drainage, rivers, &c. In this office the pupil will learn everything connected with these departments, and as they are founded more or less upon practical mechanics, he will soon find that from his previous mechanical education he has already acquired considerable knowledge of them, and it will only be necessary to apply those principles, modified according to the particular circumstances required: in fact, the principles are the same, although applied upon a larger scale.

In laying down a railway the young engineer will have to consider the particular local, geographical and geological features of the country through which the line is to pass, and the extent of mechanical power that will be necessary to work it effectually, consistent with the cost of making the cuttings and embankments. Here is a purely mechanical question, which the pupil's previous instruction will enable him to decide, and which he could not do without this instruction.

If it be a question of improving a river, the quantity of water flowing through it, the inclination of its bed, the extent and levels of the district which it has to drain, will reduce themselves to the laws of the pressure and movement of fluids, which are explained under the general theorems of hydraulics and hydrostatics, supplemented by certain rules derived from practical experience, such as friction, &c.

Again, if it be the making of a harbour, the student must first thoroughly examine the nature of the locality, that is, its geographical position and geological character. As regards the former, whether the harbour is to be at the mouth of a river, whether that river discharges its waters into a bay, or through a projecting exposed line of coast where the main tidal currents run continuously and rapidly past it. With regard to the latter, whether the adjacent coasts be flat and alluvial; or elevated, but still composed of soft alluvial or sandy and calcareous soil, easily abraded or worn away by the passing currents; or whether they be composed of the harder or primary rocks. He must also carefully consider the strength and the direction of the currents. All these various conditions must be carefully weighed before coming to a decision.

In constructing close harbours, the same observations must be made. Each of these cases requires a totally different kind of treatment, and the correct method can only be ascertained by a thorough investigation and knowledge of the local circumstances, such as winds, tides, currents, coasts, &c., so that the harbour when constructed may afford every facility for ingress and egress, safety when within, and not be liable to any deposit.

In order to give the requisite supply of water to canals it is imperative that sufficient reservoirs should be established chiefly at the high level if possible, also at each intervening ascent and descent; but it is most desirable that there should be only one high level, and generally speaking this may obtained; but when, from particular local circumstances, this cannot be done, then the high levels, even at considerable extra expense, should be reduced to as few as practicable. The same may be said with regard to railways, but in the case of canals it is always absolutely necessary that there should be reservoir space to supply the greatest amount of lockage that may be required during the season when there is the least quantity of rainfall. The rainfall in any given district may always be ascertained by proper rain gauges; and whenever it has been found that there is no probability of obtaining a sufficiency of water to pass the amount of trade that may be expected over any given length of canal, then the high level must be lowered sufficiently to obtain the required supply. When, from peculiar local circumstances, this cannot be done, then it will become necessary to erect steam engines of the requisite power to pump back the water from the lower to the higher levels. But as a rule it will be found, that by laying out a canal properly, and by storing sufficient water to answer all the required lock supply at proper places, pumping back will only be necessary in extreme cases. This, however, is a question of detail that will be governed by the local circumstances of each particular case. With regard to the construction of canals, that must be regulated by the quantity of trade to be passed, and the charges that it will bear; but, within certain limits, the larger the canal the better. In the case of ship canals for seaborne vessels, it is advisable to construct them wherever they can be made at a reasonable cost, and there is traffic enough to pay a fair interest upon the capital.

In the drainage of extensive districts of lowlands, whether bordering upon rivers or otherwise, it is the better plan, with some exceptions, to divide the lowland from the highland waters, and to discharge them by separate outfalls; because if they are both discharged by one outfall, the highland water, coming from a higher level, and naturally having the greatest velocity, will force its way first to the outfall, and until it is discharged the lowland water cannot get off, but will accumulate upon and inundate the adjacent lands. Again, if only one outfall be provided, a much more extensive system of main and interior drains will be required, as these latter must serve as reservoirs to contain both waters until they can be discharged by the common outfall; but by keeping them distinct from each other, the highland water may readily be discharged into the upper part of the rivers or watercourses, whilst the lowland water may be made to discharge itself at the lowest point the outfall will admit of, and will get off before the highland water can reach it. Moreover, the highland water, being discharged so much higher up the watercourses or rivers, will scour out their channels as well as the outfall, prevent them from filling up, and preserve them in the best state both for drainage and navigation. These catchwater drains for the highland waters will also be found very useful for supplying the lowland districts with fresh water for cattle, domestic purposes, and irrigation during the summer and dry seasons, when fresh water is so much needed for the lowlands. This system was first introduced by my father, in 1805, in the drainage of the extensive district of lowlands bordering upon the river Witham, between Boston and Lincoln, amounting to about 150,000 acres.

Generally speaking, before attempting to improve the interior drainage of any lowland district, it is necessary, in the first place, to examine the state of the outfall, and how far it is capable of improvement; before this is ascertained it is impossible to lay down any effectual plan. In order to make the outfall effective it should be improved to the greatest extent practicable, so that the low-water line or level may be reduced to the lowest point. Having done this, the interior drainage may be laid out accordingly. When this is combined with the catchwater system above described, the drainage may be rendered as complete as possible, as far as it can be upon the natural principle of gravitation. When the water cannot be discharged from the outfall at all times by gravitation, we must enlarge the main and tributary drains, so that they may serve as reservoirs to contain the drainage water during the time that the outfall sluice is closed in consequence of the water in the river or the sea, where the outfall sluice may be placed, being higher than the level of the water in the main and interior drains. No land can be considered as properly drained unless the surface of the water in the adjacent drains can be kept from 2 to 3 feet below the surface of the adjacent lands at all times. There must be no stagnation of water; at the same time there must always be the means, as far as practicable, of supplying the land with that proper degree of moisture necessary for nourishing the soil, either from the direct rainfall or from the water discharged into the catchwater drains from the adjacent highlands; and if these be not sufficient, then they may be supplemented by reservoirs of the proper dimensions attached to them. The best mode of arranging this is, of course, a matter of detail, keeping always in view the great principle of a thorough drainage and an ample supply of fresh water. The system that I have above explained is based upon the soundest principles of theory and practice, and therefore I feel no hesitation in recommending it.

With regard to the sewerage and drainage of towns, the same principle may be adopted, modified according to local circumstances. The drains here will require greater fall or inclination. The sewage should not be discharged into the watercourses, but into separate depôts at a proper distance from the dwellings. These depôts should be thoroughly ventilated, and the sewage deodorized by mixing it with earth, or some other suitable substance, that will not impair its value, and then it may be sold for manure; and thus instead of becoming a nuisance it may be turned to profitable account.

All rivers in densely populated countries should have their flood waters stored in capacious reservoirs, with proper sluices, in the main or adjacent subsidiary valleys, so that during the dry seasons there may be always an ample supply of good water for domestic and agricultural purposes, irrigation, and navigation. The reservoirs will also be advantageous in preventing the too frequent inundations and consequent devastation caused by floods.

In waterworks gravitation should be adopted wherever practicable, so that the source of supply shall be placed at such an elevation that it may command the highest part of the buildings to be supplied, thus all artificial power for pumping will be avoided. But in most cases, except where natural lakes can be found, it will be necessary to make settling or filtering reservoirs, from which the water when sufficiently pure may be delivered into the supply reservoirs, and both of these should be capacious enough to contain a sufficient supply for a month, more or less, according to the particular local circumstances. Last, but not least, the quality of the water for the proposed supply should be thoroughly tested chemically, in order to ascertain its purity; it should be as soft as possible, and be free from vegetable as well as all other matter prejudicial *to health*; and it must be obtained in sufficient quantity to guarantee a supply of thirty gallons a day to each inhabitant of the town, with the means of augmenting the supply at the same rate for any increase of inhabitants. The conduit which is to supply the service reservoir should be covered throughout, as well as the service reservoir, which of course should be occasionally cleansed; the other, or settling reservoir, near the fountain head, need not be covered if made large enough; that also should be cleansed as often as is necessary.

Where the water cannot be supplied by means of gravitation, then the artificial method of pumping by steam engines or water-wheels, or other means, must be adopted; but in this case also settling, filtering, and service reservoirs must be employed, as already described. It is unnecessary to remark that in all cases the reservoirs and conduits should be made thoroughly water-tight and impervious to any drainage water from the adjacent districts.

Docks may be divided into two classes, viz. floating and dry docks; the former may be designated as enclosed spaces filled with water, penned up to such depth as may be required for floating vessels of all classes. These docks or basins must be rendered water-tight, and in most cases it is necessary to surround them with nearly vertical walls, to economize space and to enable vessels to come alongside and discharge and receive cargoes.

With regard to the situation of these docks and designing the plans for them, this depends upon the local circumstances and the requirements of the particular class of vessels that they are to accommodate, and the trade that is to be carried on in them. Without a thorough knowledge of all these circumstances it is impossible to give anything like a correct opinion as to their dimensions, mode of designing them, or any other particulars. I may say generally, however, that as these docks are always situated contiguous to some river or harbour, either with or without the tidal ebb and flow, the position and direction of the entrances to the docks become of the greatest importance, in order that they may not be too much exposed, and that vessels may be enabled to enter and depart with the greatest facility; and in such part of the river or harbour where there is the greatest depth of water and the best channel outwards and inwards. There should also, as far as possible, be the means of supplying the basins with clear water, in order to diminish the amount of deposit within; there should also be a smaller or entrance basin adjoining the outer lock, the level of water in which can be more readily adjusted with that of the adjacent river or harbour, so that vessels may be taken into the docks with the greatest despatch out of the reach of the currents in the outer harbour, and without the necessity of lowering the surface of the water in the inner basin.

Floating docks in general should have dry docks attached to them, for the purpose of repairing vessels; and these dry docks should communicate by means of a tunnel or culvert with the tidal river or harbour.

With regard to the warehouse accommodation for receiving and delivering the different classes of merchandise brought to or taken out of the vessels frequenting the docks, these should as far as possible be made fire-proof, and should be properly adapted for the reception of the different articles placed in them, so that they may be stowed away in the most convenient manner and be readily accessible. Where space will permit it is desirable to keep the warehouses as low as possible; by this means the damage in the event of fire will be greatly reduced, and the expense of taking in and delivering goods considerably diminished, and the cost of construction lessened also.

Between the warehouses and the edge of the dock there should be sufficient space for a road all round the warehouses; and between the road and the edge of the dock there should be landing-sheds, so that the cargoes of vessels, when discharged, may be placed there, to be examined and sorted, and from thence taken away to their destination, or delivered into the warehouses, as occasion may require. All inflammable articles, such as oils, naphtha, turpentine, tar, pitch, jute, hemp, flax, &c., should be stowed away in low warehouses or covered sheds, completely isolated, and with the interior divided into distinct compartments, with access round each. These compartments should be no larger than necessary. Railways should be laid along all the quays, and should be carried through the ground floors of the whole of the warehouses, while the upper floors should also have rail or tramways through each division of goods, with the necessary turn-tables at their intersection with each other. These railways should be worked either by steam power or horse traction, as may be most advisable. All the quays and warehouses should be supplied with a sufficient number of cranes, of the requisite strength to lift and stow away the heaviest goods. These cranes should be worked either by hydraulic, steam, vacuum, manual, or animal power, as may be most advisable; in fact, they should be so designed that they may be worked either by the one or the other, as may be required.

Fresh water should be laid round all the quays and warehouses, through iron or glazed earthen pipes, and there should always be an ample supply, either for vessels frequenting the docks, or for extinguishing fires; and for this purpose capacious tanks or reservoirs should be established at the most convenient places; and if these reservoirs cannot be made at a sufficient height so as to command the highest warehouses, then the water should be forced through the hose attached to the supply-pipes by steam or other power, as shall be found most advisable. Gas, also, in properly fitted pipes, should be distributed over the quays and warehouses, and the movable lights should be as few as possible; those that are used should be properly guarded, so that all risk of fire from them may be avoided. No lucifer-matches should be permitted in any part of the establishment, nor should smoking be allowed. By these means the probabilities of fire will be reduced; and if, notwithstanding these precautions, a fire should break out, there will be the most ample provision for extinguishing it in the shortest possible time, and with the least damage to the property.

With regard to architecture, that strictly belonging to the office of the civil engineer is of the most simple character. The buildings should be laid out in the best manner, and the most convenient for their respective purposes, and thoroughly substantial. At the same time, their exterior appearance should possess a certain degree of symmetry and dignity, so as to impress upon the spectator the idea that they are thoroughly adapted for their purpose. The materials should be chiefly iron, stone, and brick; and timber should only be used when absolutely necessary. At the same time, although it is not altogether necessary, the civil engineer should have a thorough knowledge of the five orders of architecture, and the mode of applying them; the principles of constructing and equilibrating arches of all kinds must be thoroughly understood; and if he intends to combine the practice of domestic and public architecture with that which is only strictly necessary for civil engineering, then he must enter more largely into the subject, and study the different ancient and modern styles of building.

Surveying and levelling will also form an important part of his duties. In order to understand them it is necessary that he should know thoroughly plane and spherical trigonometry, and the calculations necessarily connected with them. He should also have a certain knowledge of astronomy, to enable him to calculate the tides and other phenomena connected therewith, and to be able to lay down correct charts of any harbour or sea coast, with the soundings, currents, and winds prevailing there.

Geology will form another important department of study, without which he cannot understand the nature of the materials that he will have to deal with, such as stone, lime, cements, earths, &c.; the angles at which they will stand in making deep cuttings and embankments; the best and most durable kind to be employed in any particular work, the proper mode of working it, and how to place it in the best position so as to resist the effects of the atmosphere or running water, the concussion of waves, &c., in the most effectual manner. The study of geology will further enable him to account for the formation of shoals and any given line of coast, together with the operation of the currents upon them, and the best mode of remedying their disastrous effects; also the best plan for designing and constructing harbours on each particular coast or situation.

Again, by having a thorough knowledge of the strata and formation of any given district of country, he will be enabled to ascertain where water may be found, and in what quantity; and if he practises mining, he will be able to predict with tolerable certainty where different kinds of minerals may be obtained, such as coal, iron, lead, copper, tin, gold, silver, &c., and the mode of working them to the greatest advantage.

In fact, geology combined with mineralogy he will find to be of most essential service in almost every department of civil engineering.

Embankments.—This is another department of engineering which requires a good deal of skill and judgment, particularly along an exposed open coast, where lowlands are to be protected against the encroachments of the sea. The first point is to select the line of embankment in such a manner that there shall always be in front of it a good foreshore, so that the force of the sea may be broken before it reaches the embankment; that is to say, where practicable, to have a certain extent of green or outlying marsh in front of it, so that the embankment when completed will seldom have a head of water to contend with at high tide of above six or seven feet. And even with this moderate depth at high water, when exposed to the action of a heavy gale of wind, there will for three or four hours be a considerable broken sea, calculated to do a great deal of damage, if the embankment be not properly constructed. Now, if the embankment have a good green foreshore in front, with sea slope of about 5 or 6 to 1, well sodded up, a facing of clay about 18 inches thick, 6 feet above the highest level of spring tides, the top being 6 feet wide, with back slopes of 2 to 1, with a back ditch 10 feet from the foot of the inner slope, the interior of the embankment being composed of sound earth well rammed or pressed together, so as to make it solid—an embankment of this kind will be able to resist such a pressure as we may ordinarily expect it to be exposed to.

There may be extraordinary cases where this will not be sufficient. When these occur it will be necessary to pave the surface with stone, about 9 inches thick, or with fagots. The former is, however, decidedly the best plan, as it will be permanent, whereas fagots are constantly rotting, and require renewal.

If the sea shows a tendency to carry away the foreshore, it must be prevented, by means of jetties so disposed as to collect the alluvial matter held by the sea water in suspension. These, if properly designed and constructed, will generally have the desired effect.

In cases where the water outside is deep and the sea face of the embankment may be exposed to a head of water of 12 feet and upwards, much greater precautions must be taken to guard against accident. The sea slopes of the embankment must be increased to 7 or 9 to 1, well faced with clay and paved with stone, having the foreshore in front well protected with jetties. In fact, no two cases will be alike: each must be treated separately according to the particular local circumstances, and therefore it is impossible to design a proper plan for any embankment without knowing all the local circumstances. The general principle is that the sea face of the embankment should never be less than from 4 or 5 to 1. In some particular cases a less slope will do, say 3 to 1. This, however, certainly depends upon local circumstances. The base of the outer slope should be particularly watched, and if any crack appears to be forming, it should be immediately stopped by jetties carried out as far as necessary. In forming embankments it is usual, when it can be done, to take the earth from the outside of the sea slope, but this should never be done within less than 10 yards from the base of the slope, and these *"floor pits,"* as they are termed, should generally not exceed 12 to 18 inches in depth, and be increased in width in proportion to the quantity of earth required for the bank; at every 10 or 15 yards, in the longitudinal direction, the earth should not be removed, but left to form small cross banks between the floor pits, so as to prevent any current being formed in them; thus these floor pits will soon be filled up by the alluvial matter brought in by the tide, when the outside slopes of the bank are neither exposed to the heavy lash of the waves nor to strong currents. Then if they are covered with good grass sods properly laid on and beaten into the face of the bank it may suffice, but not otherwise. If this should not answer the slope must be increased and, if necessary, paved with stone as above mentioned. When good clay cannot be obtained to face the bank, then the best of the earth that can be got must be employed, mixed with straw, well puddled with water, and laid upon the surface of the bank in a moist state about 18 inches thick, and then faced with stone about 9 inches thick, well rammed edgeways into it. In cases where it is necessary to protect any line of coast against the ravages of the ocean, the measures to be adopted will depend upon the form and geological character of the coast to be so protected, whether low flats and alluvial, or cliffs composed of rocks more or less hard, and easily acted upon by the waves, rain, and atmosphere. In the former case it will generally be found that the coast is surrounded by extensive flat sands, and that the water holds a large quantity of alluvial matter in suspension. The great object, therefore, should be to cause this alluvial matter to be deposited in such form and in such places as are best adapted to our purpose. Now this may generally be effected in an inexpensive manner, considering the object to be attained, by a series of jetties, either composed of stakes wattled together with fagots, or lines of loose stones disposed in such a manner that they shall break the rising and falling waters, and make them stagnant between the jetties, so that they may deposit their alluvial matter. In the first instance these jetties need not be raised more than two feet above the level of the sand, and when the sand or alluvial soil has accumulated up to the top, they may be again raised to a similar height, and so on until the soil in front of the coast has been converted into a green marsh; thus there will not only be formed a protection to the coast invaded by the sea, but fresh land may be gained in front of it and embanked from the sea. It is impossible to explain the precise disposition and direction of these jetties and works without a thorough knowledge of the locality, and such circumstances as its exposure to winds, tides, and currents. The principle however is to check the currents gradually, and in such a way as to prevent any strong current from being formed; for if a new and strong current should be created, not only will the alluvial matter not be deposited, but the works themselves will be carried away, and all the labour and expense will be wasted. It is generally advisable that such works should be commenced near the shore, and worked downwards towards the sea; thus, if they are properly managed, no deep pools or strong currents will be formed behind them; and the required process of filling or silting up will proceed regularly seaward, always increasing the protection required, and obtaining additional land as they proceed.

In some cases, where the sea is heavy, it may be necessary to have stronger jetties or works to relieve and protect the minor ones above described; but these should only be resorted to in places where the others are insufficient, or in greatly exposed situations; wherever the minor works will suffice, as they will in most cases if properly applied and constructed, the less heavy works are resorted to the better, as the great object is to lead not drive Nature; that is, to work with her instead of against her. By this means a few bricks and stakes will do a great deal more than far greater and more expensive works. So far as regards low alluvial coasts, these, if properly managed, will be found comparatively easy to deal with.

When we come to rocky coasts that are wearing away by the combined action of the sea below and the rains and atmosphere above, and where there is little or no alluvial matter held in suspension by the waters, that might be collected so as to form a protecting deposit at their base, then we must adopt a different system, but not altogether ignoring the other when it can be made useful. In this latter case we must secure the bases of the cliffs at least up to high-water mark by means of retaining walls, where the rock itself is not hard enough to resist the action of the sea. These walls need not be carried higher than absolutely necessary. In some cases a mere footing will do; in others, the wall may be carried up to half tide; and in others up to the full high-water mark; and although the rock may be naturally soft, yet if its surface be protected by harder stone, even of a very moderate thickness, it will be quite sufficient to resist further encroachment by the sea. As these retaining walls will be founded upon a base of solid rock, there is very little fear of their being undermined; therefore, when I said before that it would be necessary to carry these retaining walls up to high-water mark, it must be understood as applying only to those rocks that are easily abraded by the sea.

There is another point to be attended to. The base being secured, we must look to the cliff above. Here, from the effect of rains, the water frequently cannot get away, accumulates behind at the top, and sinks through the fissures, when partly by hydraulic pressure and partly by the effects of frost, large masses are detached and fall below; and as this is continually occurring, the progress of decay goes on increasing. Having secured the base, the next thing, where practicable, is to slope off the upper surface of the cliff, so as to prevent it from overhanging, and then to make a drain at the back to carry off any water that may lodge there. By these means, if properly carried into effect, the base of the cliff being protected against the sea from below and rainwater from above, there is every probability that it will be preserved, in all ordinary cases. In extraordinary cases additional measures must be taken to meet them upon the same principles. With regard to retaining walls of brickwork or masonry, these should be always in excess of strength beyond the pressure, whether vertical or lateral, that they may have to resist. When the pressure is simply lateral, then the mean thickness of wall built of masonry and brickwork—the mean thickness, generally speaking, of the main body of the wall—should be about one-fourth of the height, besides counterforts at the back at certain distances from each other, regulated according to the particular circumstances. These, upon the average, including the thickness of the main wall, will make the total mass to be equal to nearly one-third of the total height. My father frequently made these walls curved in the front as well as at the back, the front being struck from a radius whose centre was level with the top of the wall, and of such a length that the face of the wall should batter one-fifth of the total height; the back of the wall should be struck from a centre at the same level as the other, but a little longer, so that at the base the wall might be about 2 feet thicker than at the top, in addition to two or three footings of 6 inches each; and the base of the wall was made to incline backwards, according to the radius from whence it was struck.

These walls, when they are to rest upon alluvial soil, must be founded upon a platform composed of piles of a sufficient length and thickness, driven at right angles to the line of the foundation, until with the blow of a ram weighing 15 cwt. and falling 20 feet they will not move one-eighth of an inch. These piles should be driven in regular rows, longitudinally and transversely, about 3 feet apart, and hooped and shod with wrought-iron hoops and shoes. At the front, immediately under the tie of the wall, there should be a row of grooved and tongued sheeting piles driven close together, and to the same depth as the others, about 6 inches thick, having a waleing or longitudinal brace 6 inches thick and 12 inches wide, well bolted in each side of the top of the sheeting piles. The loose earth should be taken out to about a foot in depth, and the space filled in with stone or brickwork to the level of the pile heads, which should be carefully trimmed, then covered with sills about 12 inches square, well spiked down to them. The spaces between the sills should be well faced with brickwork, and the whole surface should then be covered with 6-inch plank, properly spiked down to the sills below. Upon this platform the masonry and brickwork of the wall should be built. The wall should be carefully backed up as it proceeds with sound earth or clay, or clay mixed with one-sixth of gravel or concrete, as shall be deemed most advisable. These curved walls, if properly constructed, are stronger and more economical than the ordinary walls.

In some cases, as in that of Sheerness, for example, the foundation is so bad that a totally different plan must be adopted. At Sheerness it was necessary that the base of the walls should be increased, distributing the weight over a wider area, so that each superficial foot of the superincumbent mass should have a larger bearing, thus greatly relieving the pressure over every part.

The foundation upon which the walls were built was as bad as possible, being composed of nothing but loose running silt and sand. Upon such a foundation walls of the ordinary kind would not have stood; my father therefore saw the necessity of designing some new construction, upon the principles above mentioned. He had previously adopted something similar for the docks at Great Grimsby, in Lincolnshire, in 1786, which design was carried into effect with great success. The walls at Sheerness and at Great Grimsby were built both upon the same principle, modified according to local circumstances. Sheerness docks were finished altogether in the year 1826, and they have stood ever since.

I believe that I have now enumerated all the chief points to which the education of a civil engineer should be directed. Whilst he continues in an engineer's office, whatever business is brought before him, he should always endeavour to thoroughly understand the reasons for which such and such a work is proposed to be made, and the principles upon which it is to be constructed; and if he finds, according to his previous education, difficulties either in the principle or construction, he should modestly state his doubts to his superior; if no explanation is given, he has simply to do as he is ordered, making notes of his doubts, and when the work is carried into effect he will then be able to ascertain how far he was right or wrong. If the work turns out to be a failure, his previous calculations will show him that he was right; but if the work succeeds, his calculations were wrong, and he should carefully go over them again to ascertain his error. He should follow the same process when he has to design and carry into effect any work upon his own responsibility, and if he is in doubt as to any point, let him consult some one of his professional brethren in whom he has confidence. When he is consulted on similar occasions by another engineer, let him give his advice and opinion to the best of his power; by this means he will gain the respect of his colleagues, and every one will be ready to help him when required.

Let him be particularly careful about his estimates; and after he has estimated *fully* the probable cost of a work, let him add an allowance of quite 15 per cent. for contingencies, which in all engineering works are so numerous and varied that it is almost impossible to foresee them.

We should always recollect that the great object of all engineering works is to produce a fair return for the capital expended upon them, or, in other words, that they should pay. If, after due calculation, it is found there is no chance of that, they should not be undertaken; for although it may be very gratifying to the professional reputation of an engineer to have executed a great work, it is but a poor consolation to his subscribers to find that their money has been comparatively thrown away without any adequate return.

Upon these grounds, therefore, I think it is better that the engineer should confine himself strictly to his business, that is, of designing and estimating any proposed work in the best possible manner to ensure the object intended. Let those who are most competent ascertain whether there is a sufficient prospect of traffic to pay a good return for the required capital; and so long as the engineer executes the work for his estimate, he cannot be blamed if the work does not pay a sufficient return. In fact, the whole commercial value of a work depends upon its cost, and therefore it is so important that the estimate should be adhered to as closely as possible, for if this be much exceeded the commercial calculation falls to the ground, and then the subscribers have just reason to complain. Against this I have heard it argued that if correct estimates were always made, and the ultimate cost of many works was known beforehand, they would never have been carried out, although notwithstanding the increased cost they have finally proved to be very valuable. This is certainly to some extent true; many inventions and discoveries have ruined the original promoters, yet have ultimately conferred the greatest benefits upon mankind; and many enterprises that have ruined the original undertakers have greatly enriched their successors. Still there can be no excuse for an engineer knowingly underestimating the cost of a work; he is undoubtedly bound to make a fair, honest estimate of every work committed to his charge, so far as his judgment goes; having done that his duty is discharged; nothing further can be expected of him than to see that the work entrusted to his care is strictly carried into effect according to that estimate.

Since the summer of 1866 I have done scarcely anything. The great crisis and subsequent panic that occurred at that time paralysed the commercial world. I considered my advancing years (I was then seventy-two), and the great hazard and uncertainty of carrying on business, and thought it most prudent to retire. After the harassing and anxious life that I had led for so many years, I felt my health so shaken as to require complete repose. But I hope, if God spares me, to be still useful to the profession and my country, by completing a work on the drainage of the fens and lowlands of Great Britain, and hydraulics generally. I also design to write a history of engineering, enlarged from my Address to the Institution of Civil Engineers, and a life of my revered father. All these I have already sketched out, and I hope to complete them, if it please God to spare my life a few years longer.

My apology for the present work is this: I think it is the duty of everyone who has led an active professional life faithfully to record the various works in which he has been engaged, the failures as well as the successes, detailing the causes of both; for we frequently learn more from the former than from the latter. I believe I have in this book faithfully done this. From unavoidable circumstances I have been obliged to trust entirely to memory while writing these pages, having been totally precluded from consulting notes or memoranda of any kind; I hope, therefore, that any inaccuracies that may be detected by the reader will be pardoned, though I believe that in the main my statements will be found correct.

Like others, I have had to contend with professional jealousy; but I believe I have on all occasions done justice to my rivals, and I have never wilfully attempted to injure anyone. Naturally of a very sanguine temperament, I am but too apt to view things in a favourable light, and to judge well of those with whom I come in contact; as a consequence of this I have often been deceived by those in whom I have placed the greatest confidence. This sanguine disposition has been the cause of many disappointments; but it has also enabled me to bear up successfully against failure, and still to look forward with hope to the future. Whenever a misfortune has occurred I have endeavoured to forget it as soon as possible; I always called to mind the words of the great Duke of Wellington, who said, *There is no use in looking back and brooding over the past; forget it, and apply your energies to the future, and do better next time.* This many people either cannot or will not do; hence they succumb. Doubtless everyone has his trials, and some are much better able to get through them than others; nevertheless, a very little reflection will show that what is past cannot be helped, and that by brooding over misfortune we do no good, but only waste our energies and invite failure in everything else.

The motto of life should be, Forward! We must expect to be checked, thwarted, and baffled in our endeavours to attain success; but these obstacles, instead of totally arresting our progress, should serve only to increase our energy. Like a river, impeded in its course, in silence waits till its accumulated strength sweeps the obstruction from its path, and it flows on majestically as before—so should we make every difficulty we encounter add to our strength, instead of increasing our weakness. Nevertheless, since "'tis not in mortals to command success," we may sometimes struggle in vain; and fortune ever against us, we may be overcome at the last; but even then we have this satisfaction—we have fought a good fight; we have done the best we could; we have done our duty to the best of our ability, and that is all that can be required of us. To do my duty has been my endeavour through life; and probably if I had adhered to it more strictly I might have done a great deal better. Nevertheless, little as I have done, I should not have accomplished half so much had I not kept that one object in view, as far as my physical and mental powers would permit; and this is no small consolation. The old motto, "*Nil desperandum,*" should be constantly on our lips, and should act like the spur on a jaded steed. Affairs are never so bad but they might have been worse, and they may generally be mended by energy and perseverance, and a determination to make the best of everything. We may not be able to accomplish all we aspire to achieve; nevertheless by refusing to yield to misfortune we shall escape the reproach of cowardice and faintheartedness. When we suffer a defeat, let us calmly consider the cause of it, and nine times out of ten we shall find that it is through our own fault; these lessons of experience should be carefully laid to heart, and serve for our future guidance.

I have never deemed wealth desirable for mere personal gratification, but only in so far as it would have enabled me to help others, to promote the advancement of science and the well-being of my fellow creatures; this would have conferred the greatest happiness upon me, but it has been denied by the Almighty Disposer of events, and most probably with justice, that it might be done better by other hands. I therefore humbly bow to the Almighty's decision; and if I have done the best I could in His sight, I am amply rewarded. I, however, most deeply regret that I have not done more. I return my most fervent thanks to the Almighty that He, out of His great mercy, has allowed me to do the little I have done; and I most devoutly hope that He through His Son Jesus Christ will pardon my shortcomings; and I say with all reverence, Bless the Lord for all His mercies!

DAWLISH, *December 9, 1867*.

THE END.

INDEX

C

Lector House believes that a society develops through a two-fold approach of continuous learning and adaptation, which is derived from the study of classic literary works spread across the historic timeline of literature records. Therefore, we aim at reviving, repairing and redeveloping all those inaccessible or damaged but historically as well as culturally important literature across subjects so that the future generations may have an opportunity to study and learn from past works to embark upon a journey of creating a better future.

This book is a result of an effort made by Lector House towards making a contribution to the preservation and repair of original ancient works which might hold historical significance to the approach of continuous learning across subjects.

HAPPY READING & LEARNING!

LECTOR HOUSE LLP
E-MAIL: lectorpublishing@gmail.com

www.ingramcontent.com/pod-product-compliance
Lightning Source LLC
LaVergne TN
LVHW091146180726
843490LV00004B/1326